# LIQUID CRYSTALS

# LIQUID CRYSTALS
## Physical Properties and Nonlinear Optical Phenomena

**IAM-CHOON KHOO**

A Wiley-Interscience Publication

**JOHN WILEY & SONS, INC.**

New York / Chichester / Brisbane / Toronto / Singapore

This text is printed on acid-free paper.



***Library of Congress Cataloging in Publication Data:***

Khoo, Iam-Choon.
Liquid crystals : physical properties and nonlinear optical phenomena / I. C. Khoo.
p. cm.

Includes index.
ISBN 0-471-30362-3
1. Liquid crystals. I. Title.
QD923.K49 1995
530.4′29—dc20 94-10371
CIP

Printed in the United States of America

10 9 8 7 6 5 4 3 2 1

To Chor-San, Richard, and Paul

# CONTENTS

# PREFACE

Liquid crystals possess very interesting physical and optical properties; they are also highly nonlinear optical materials. This book is written for seniors and beginning graduate students and researchers who are newcomers to the field of liquid crystals and/or nonlinear optics. The book is devoted to a detailed treatment of the basic principles underlying these wonderful properties of liquid crystals. A self-contained chapter on nonlinear optical principles and phenomena is also included for completeness.

The contents of this book are carefully chosen and ordered in such a way that it could be offered as either a first general course in liquid crystals or a specialized study of liquid crystal nonlinear optics. The book is organized into 10 chapters. Chapters 1 through 5 deal with the physical properties of thermotropic liquid crystals. Following a discussion of their molecular and chemical structures, theories for the isotropic and liquid crystalline phases are presented. In particular, some of the physical parameters, such as the order parameter, elastic constant, free energy, viscosity and flow, refractive index, and birefringence, are discussed. "New" materials such as polymeric liquid crystals, polymer-dispersed liquid crystals, dye-doped liquid crystals, and ferroelectric liquid crystals are also included in our discussion. Light scatterings, including Raman, Brillouin, Rayleigh, and Rayleigh wing scatterings, are treated in Chapter 5.

The second part of this book is devoted to a more specialized field of study that has emerged in the last decade or so centering on the nonlinear optical responses of liquid crystals to light or laser fields. Chapters 6 through 8 treat all the known mechanisms for optical nonlinearities in the principal, mesophases of liquid crystals. The basic principles, phenomena, and terminology in nonlinear optics are presented in Chapter 9. Chapter 10 summarizes the major nonlinear optical phenomena observed in liquid crystals to date.

Studies of liquid crystals are highly interdisciplinary, encompassing physics, chemistry, optics, and engineering. In order to present these wide-ranging materials in a more readily understandable form, I have adhered to discussing only the fundamentals. I have discussed electrooptical and magnetooptical effects, but left out their applications; these materials have been adequately treated in many books and review volumes cited in the References. Wherever possible and without loss of the physics, I have replaced rigorous theoretical formalisms with their simplified versions, for the sake of clarity.

During the course of writing this book, as in my other work on liquid crystals over the years, I have enjoyed valuable encouragement and support from a wide spectrum of people. First and foremost is my family; their patience, understanding, and unqualified support are my source of strength and motivation. In particular, I would like to thank my son Richard for helping with some word processing work and technical illustrations. Debby Pruger at the Pennsylvania State University typed the entire manuscript. I must also express very special and personal thanks to my colleagues and all my students—past and present—for their tireless efforts in making all those experiments and theoretical computations successful and enjoyable. Support from the National Science Foundation, Air Force Office of Scientific Research, Air Force Phillips Laboratory, Army Research Office, Navy Joint Services Program, and the Defense Advanced Research Projects Agency over the years is also gratefully acknowledged.

IAM-CHOON KHOO

*University Park, Pennsylvania*

# CHAPTER 1

# INTRODUCTION TO LIQUID CRYSTALS

## 1.1 MOLECULAR STRUCTURES AND CHEMICAL COMPOSITIONS

Liquid crystals are wonderful materials (1). In addition to the solid crystalline and liquid phases, liquid crystals exhibit intermediate phases where they flow like liquids, yet possess some physical properties characteristic of crystals. Materials that exhibit such unusual phases are often called mesogens (i.e., they are mesogenic), and the various phases in which they could exist are termed mesophases (2). The well-known and widely studied ones are thermotropics, polymerics (3), and lyotropics. As a function of temperature, or depending on the constituents, concentration, substituents, and so on, these liquid crystals exist in many so-called mesophases—nematic, cholesteric, smectic, and ferroelectric. To understand the physical and optical properties of these materials, we will begin by looking into their constituent molecules (4).

### 1.1.1 Chemical Structures

Figure 1.1 shows the basic structures of the most commonly occurring liquid crystal molecules. They are aromatic, and, if they contain benzene rings, they are often referred to as benzene derivatives.

In general, aromatic liquid crystal molecules such as those shown in Figure 1.1 comprise a side chain R, two or more aromatic rings A and A′, connected by linkage groups X and Y, and at the other end connected to a terminal group R′.

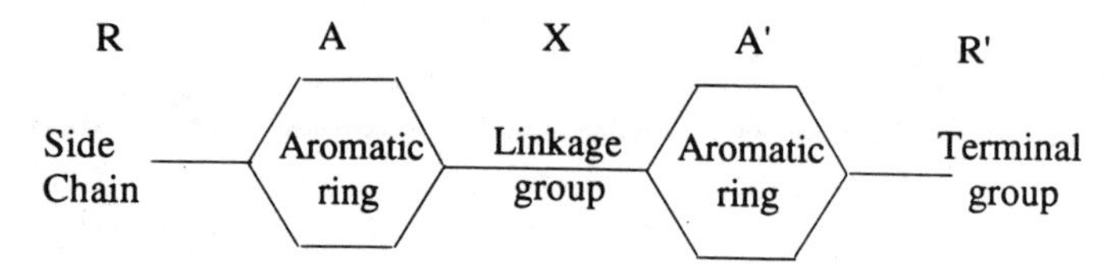


**Figure 1.1** Molecular structure of a typical liquid crystal.

Examples of side-chain and terminal groups are alkyl ($C_nH_{2n+1}$), alkoxy ($C_nH_{2n+1}O$), and others such as acyloxyl, alkylcarbonate, alkoxycarbonyl, and the nitro and cyano groups.

The linkage groups X and Y are simple bonds or groups such as stilbene (—CH=CH—), ester (—C(=O)—O—), tolane (—C≡C—), azoxy (—N=N—), Schiff base (—CH=N—), acetylene (—C≡C—), and diacetylene (—C≡C—C≡C—). The names of liquid crystals are often fashioned after the linkage group (e.g., Schiff-base liquid crystal).

There are quite a number of aromatic rings. These include saturated cyclohexane or unsaturated phenyl, biphenyl, and terphenyl in various combinations.

The majority of liquid crystals are benzene derivatives mentioned previously. The rest include heterocyclics, organometallics, sterols, and some organic salts or fatty acids. Their typical structures are shown in Figures 1.2 to 1.4.

Heterocyclic liquid crystals are similar in structure to benzene derivatives, with one or more of the benzene rings replaced by a pyridine, pyrimidine, or other similar group. Cholesterol derivatives are the most common chemical

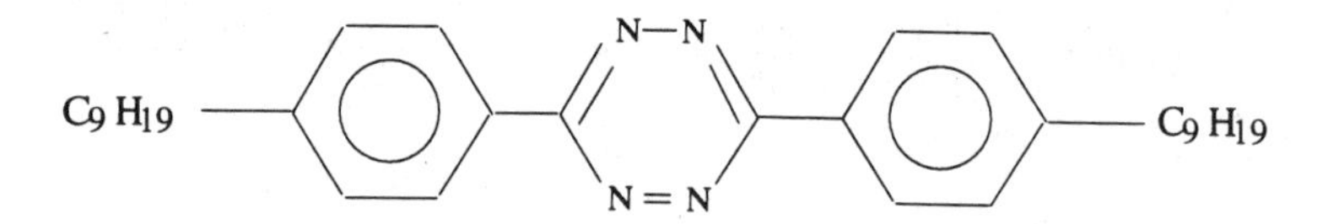


**Figure 1.2** Molecular structure of a heterocyclic liquid crystal.

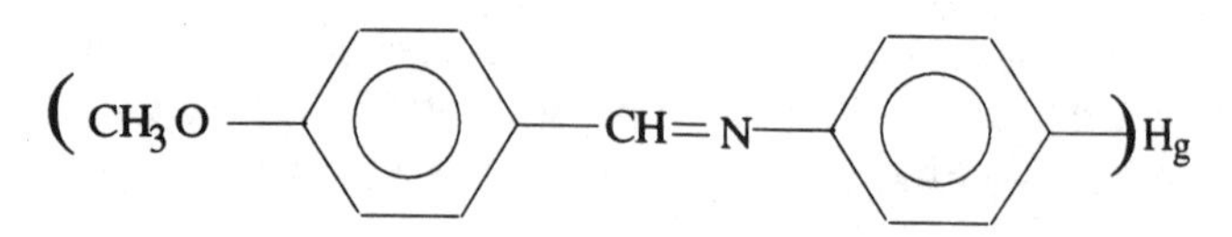


**Figure 1.3** Molecular structure of an organometallic liquid crystal.

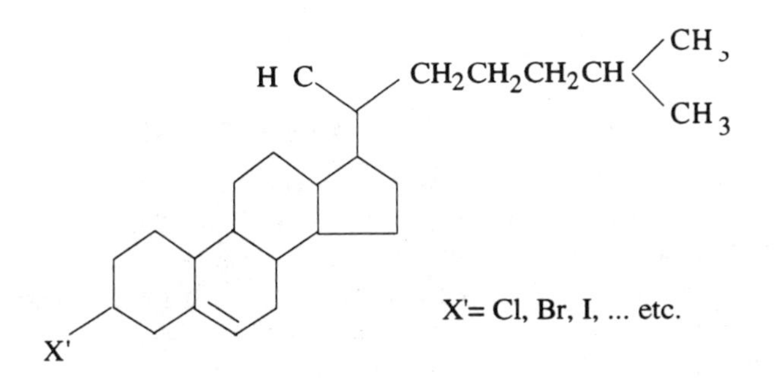


**Figure 1.4** Molecular structure of a sterol.

compounds that exhibit the cholesteric (or chiral nematic) phase of liquid crystals. Organometallic compounds are special in that they contain metallic atoms and possess interesting dynamical and magnetooptical properties (5).

All the physical and optical properties of liquid crystals are governed by the properties of these constituent groups and how they are chemically synthesized together. Dielectric constants, elastic constants, viscosities, absorption spectra, transition temperatures, existence of mesophases, anisotropies, and optical nonlinearities are all a consequence of how these molecules are engineered. Since these molecules are quite large and anisotropic, and therefore very complex, it is practically impossible to treat all the possible variations in the molecular architecture and the resulting changes in the physical properties. Nevertheless, there are some generally applicable observations on the dependence of the physical properties on the molecular constituents. These will be highlighted in the appropriate sections.

The chemical stability of liquid crystals depends very much on the central linkage group. Schiff-base liquid crystals are usually quite unstable. Ester, azo, and azoxy compounds are more stable, but are also quite susceptible to moisture, temperature change, and ultraviolet radiation (UV). Compounds without a central linkage group are among the most stable liquid crystals ever synthesized. The most widely studied one is 5CB (pentylcyanobiphenyl), whose structure is shown in Figure 1.5. Other compounds such as pyrimide and phenylcyclohexane are also quite stable.

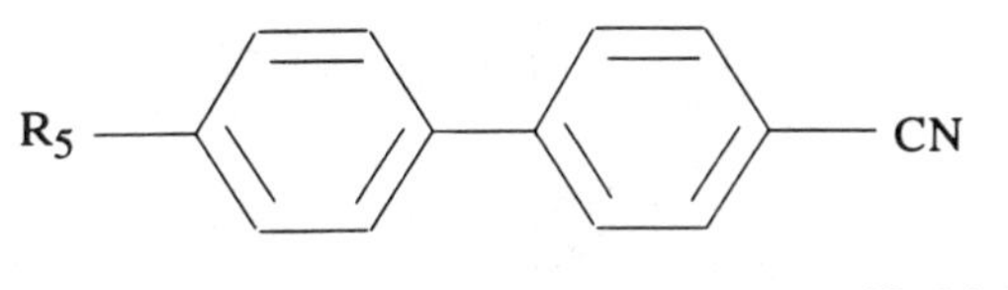


K 24 N 35.3 I

**Figure 1.5** Molecular structure of 5CB (pentylcyanobiphenyl).

## 1.2 ELECTRONIC PROPERTIES

### 1.2.1 Electronic Transitions and Ultraviolet Absorption

The electronic properties and processes occurring in liquid crystals are decided largely by the electronic properties of the constituent molecules. Since liquid crystal constituent molecules are quite large, their energy level structures are rather complex. As a matter of fact, just the process of writing down the Hamiltonian for an isolated molecule itself can be a very tedious undertaking. To also take into account interactions among the molecular groups and to account for the difference between individual molecules' electronic properties and the actual liquid crystals' responses will be a monumental task. It is fair to say that existing theories are still not sufficiently precise in relating the molecular structures and the liquid crystal responses. We shall limit ourselves here to stating some of the well-established results, mainly from molecular theory and experimental observations.

In essence, the basic framework of molecular theory is similar to the one described in Chapter 8, except that many more energy levels, or bands, are involved.

Generally, the energy levels are referred to as orbitals. There are $\pi$, $n$, and $\sigma$ orbitals, with their excited counterparts labeled as $\pi^*$, $n^*$, and $\sigma^*$. The energy differences between these electronic states which are connected by dipole transitions give the so-called resonant frequencies (or, if the levels are so large that bands are formed, give rise to absorption bands) of the molecule; the dependence of the molecular susceptibility on the frequency of the probing light gives the dispersion of the optical dielectric constant (6).

Since most liquid crystals are aromatic compounds, containing one or more aromatic rings, the energy levels or orbitals of aromatic rings play a major role. In particular, the $\pi \rightarrow \pi^*$ transitions in a benzene molecule have been extensively studied. Figure 1.6 shows three possible $\pi \rightarrow \pi^*$ transitions in a benzene molecule:

$$^1A_{1g} \rightarrow {}^1B_{1u} \qquad {}^1A_{1g} \rightarrow {}^1B_{2u} \qquad {}^1A_{1g} \rightarrow {}^1E_{1u}$$

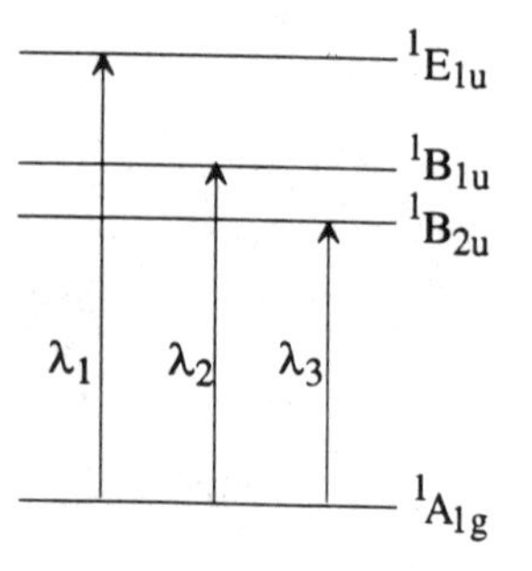


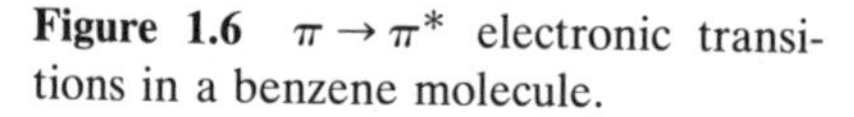

**Figure 1.6** $\pi \rightarrow \pi^*$ electronic transitions in a benzene molecule.

In general, these transitions correspond to the absorption of light in the near-UV spectral region ($\leq 200$ nm). These results for a benzene molecule can also be used for interpreting absorption of liquid crystals containing phenyl rings.

On the other hand, in a saturated cyclohexane ring or band, usually only $\sigma$ electrons are involved. The $\sigma \rightarrow \sigma^*$ transitions correspond to absorption of light of shorter wavelength ($\leq 180$ nm) in comparison to the $\pi \rightarrow \pi^*$ transition mentioned previously.

These electronic properties are often also viewed in terms of the presence or absence of conjugation (i.e., alternations of single and double bonds, as in the case of a benzene ring). In such conjugated molecules the $\pi$ electron's wave function is delocalized along the conjugation length, resulting in absorption of light in a longer wavelength region compared to, for example, that associated with the $\sigma$ electron in compounds that do not possess conjugation. Absorption data and spectral dependence for a variety of molecular constituents, including phenyl rings, biphenyls, terphenyls, tolanes, and diphenyl-diacetylenes, may be found in Khoo and Wu (6).

### 1.2.2 Visible and Infrared Absorption

From the preceding discussion, one can see that, in general, liquid crystals are quite absorptive in the UV region, as are most organic molecules. In the visible and near-infrared regime (i.e., from 0.4 nm to 5 $\mu$m), there are relatively fewer absorption bands, and thus liquid crystals are quite transparent in this regime.

As the wavelength is increased toward the infrared (e.g., $\geq 9$ $\mu$m), rovibrational transitions begin to dominate. Since rovibrational energy levels are omnipresent in all large molecules, in general, liquid crystals are quite absorptive in the infrared regime.

The spectral transmission dependence of two typical liquid crystals is shown in Figures 1.7*a* and *b*. The absorption coefficient $\alpha$ in the ultraviolet ($\sim 0.2$ $\mu$m) regime is on the order of $10^3$ cm$^{-1}$; in the visible ($\sim 0.5$ $\mu$m), regime, $\alpha \approx 10^0$ cm$^{-1}$; in the near-infrared ($\sim 1$ $\mu$m) regime, $\alpha \approx 10^{-1}$ cm$^{-1}$; and in the infrared ($\sim 10$ $\mu$m) regime, $\alpha \lesssim 10^2$ cm$^{-1}$. There are, of course, large variations among the thousands of liquid crystals "discovered" or engineered so far.

## 1.3 LYOTROPIC, POLYMERIC, AND THERMOTROPIC LIQUID CRYSTALS

One can classify liquid crystals in accordance with the physical parameters controlling the existence of the liquid crystalline phases. There are three distinct types of liquid crystals: lyotropic, polymeric, and thermotropic. These

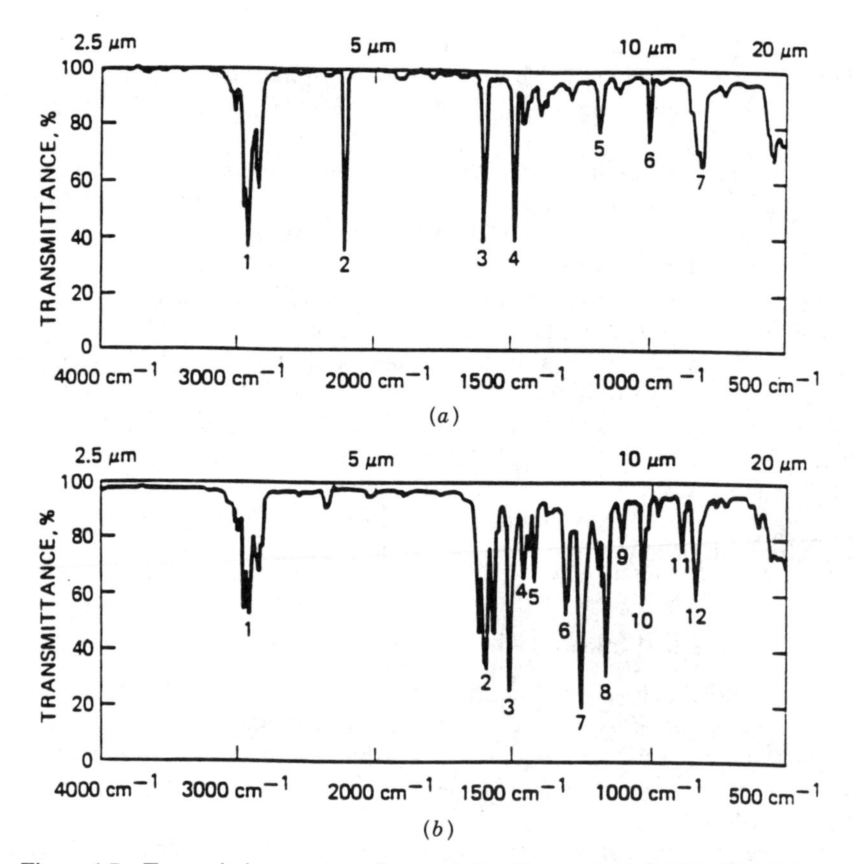


**Figure 1.7** Transmission spectra of nematic liquid crystals: (*a*) 5CB; (*b*) MBBA.

materials exhibit liquid crystalline properties as a function of different parameters.

### 1.3.1 Lyotropic Liquid Crystals

Lyotropic liquid crystals are obtained when an appropriate concentration of a material is dissolved in some solvent. The most common systems are those formed by water and amphiphilic molecules (molecules that possess a hydrophilic part that interacts strongly with water and a hydrophobic part that is water insoluble) such as soaps, detergents, and lipids. Here the most important variable controlling the existence of the liquid crystalline phase is the amount of solvent (or concentration). There are quite a number of phases observed in such water–amphiphilic systems, as the composition and temperature are varied; some appear as spherical micelles, and others possess ordered structures with one-, two-, or three-dimensional positional order.

Lyotropic liquid crystals are mainly of interest in biological studies; a good review may be found in Mariani et al. (2).

### 1.3.2 Polymeric Liquid Crystals

Polymeric liquid crystals are basically the polymer versions of the monomers discussed in Section 1.1. There are three common types of polymers, as shown in Figures 1.8*a*–*c*, which are characterized by the degree of flexibility. The vinyl type (Figure 1.8*a*) is the most flexible; the Dupont Kevlar polymer (Figure 1.8*b*) is semirigid; and the polypeptide chain (Figure 1.8*c*) is the most rigid. Mesogenic (or liquid crystalline) polymers are classified in accordance with the molecular architectural arrangement of the mesogenic monomer. Main-chain polymers are built up by joining together the rigid mesogenic groups in a manner depicted schematically in Figure 1.9*a*; the link may be a direct bond or some flexible spacer. Liquid crystal side-chain polymers are formed by pendant side attachment of mesogenic monomers to a conventional polymeric chain, as depicted in Figure 1.9*b*. A good account of polymeric liquid crystals may be found in Ciferri et al. (3). In general,

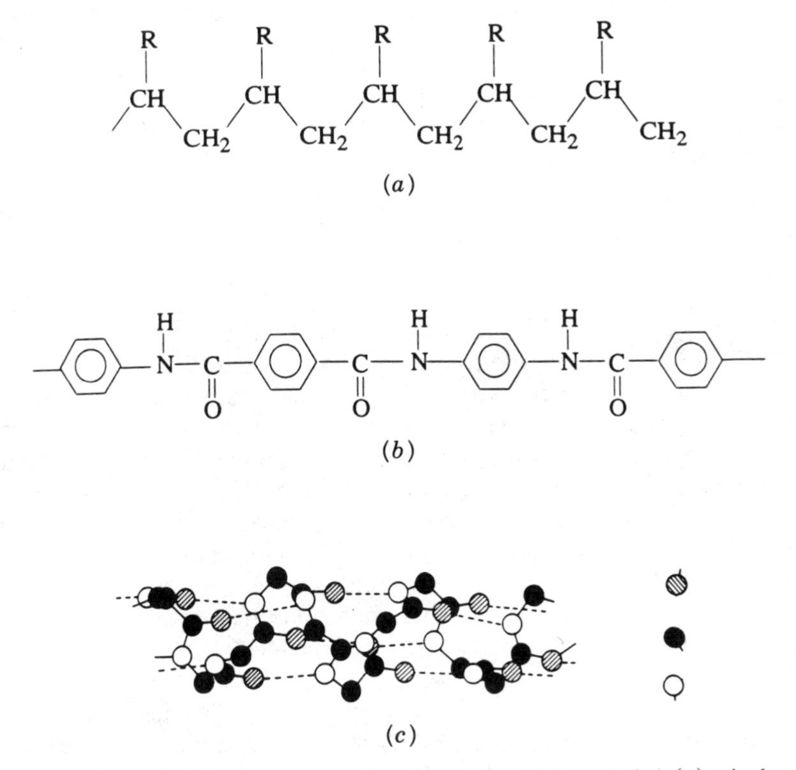


**Figure 1.8** Three different types of polymeric liquid crystals: (*a*) vinyl type; (*b*) Kevlar polymer; (*c*) polypeptide chain.

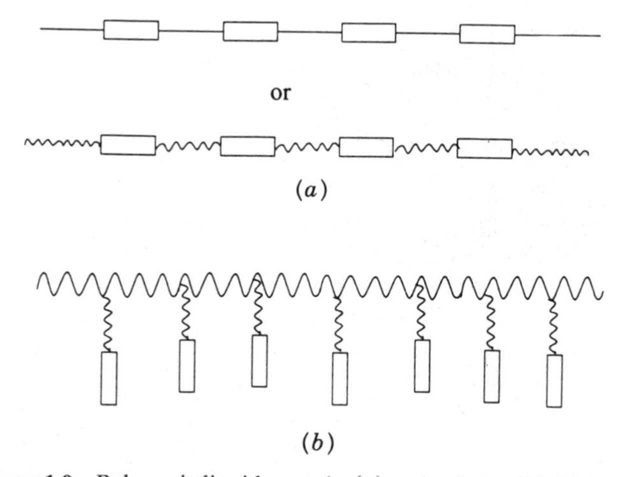


**Figure 1.9** Polymeric liquid crystals: (*a*) main chain; (*b*) side chain.

polymeric liquid crystals are characterized by much higher viscosity than monomers, and they appear to be useful for optical storage applications.

### 1.3.3 Thermotropic Liquid Crystals: Nematics, Cholesterics, and Smectics

The most widely used, and extensively studied for their linear as well as nonlinear optical properties, are thermotropic liquid crystals. They exhibit various liquid crystalline phases as a function of temperature. Although their molecular structures, as discussed in Section 1.1, are, in general, quite complicated, they are often represented as "rigid rods." These rigid rods interact with one another and form distinctive ordered structures. There are three main classes of thermotropic liquid crystals: nematic, cholesteric, and smectic. In smectic liquid crystals there are several subclassifications in accordance with the positional and directional arrangement of the molecules.

These mesophases are defined and characterized by many physical parameters such as long- and short-range order, orientational distribution functions, and so on. They are explained in greater detail in the following chapters. Here we continue to use the rigid-rod model and pictorially describe these phases in terms of their molecular arrangement.

Figure 1.10*a* depicts schematically the collective arrangement of the rodlike liquid crystals molecules in the nematic phase. The molecules are positionally random, very much like liquids; X-ray diffraction from nematics does not exhibit any diffraction peak. These molecules are, however, directionally correlated; they are aligned in a general direction defined by a unit vector $\hat{n}$, the so-called "director" axis.

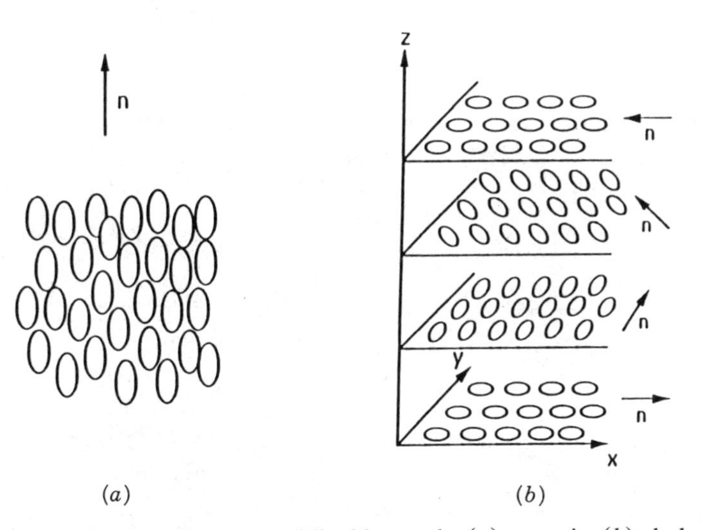


**Figure 1.10** Molecular alignments of liquid crystals: (*a*) nematic; (*b*) cholesteric or chiral nematic.

Generally, nematic molecules are centrosymmetric; their physical properties are the same in the $+\hat{n}$ and the optically uniaxial $-\hat{n}$ directions. In other words, if the individual molecules carry a permanent electric dipole (such a polar nature is typically the case), they will assemble in such a way that the bulk dipole moment is vanishing.

Cholesterics, now often called chiral nematic liquid crystals, resemble nematic liquid crystals in all physical properties except that the molecules tend to align in a helical manner as depicted in Figure 1.10*b*. This property results from the synthesis of cholesteric liquid crystals; they are obtained by adding a chiral molecule to a nematic liquid crystal. Some materials, such as cholesterol esters, are naturally chiral.

Smectic liquid crystals, unlike nematics, possess positional order; that is, the position of the molecules is correlated in some ordered pattern. Several subphases of smectics have been "discovered," in accordance with the arrangement or ordering of the molecules and their structural symmetry properties (2). We discuss here three representative ones: smectic-A, smectic-C, and smectic-C* (ferroelectrics).

Figure 1.11*a* depicts the layered structure of a smectic-A liquid crystal. In each laser the molecules are positionally random, but directionally ordered with their long axis normal to the plane of the layer. Similar to nematics, smectic-A liquid crystals are optically uniaxial; that is, there is a rotational symmetry around the director axis.

The smectic-C phase is different from the smectic-A phase in that the material is optically biaxial, and the molecular arrangement is such that the long axis, is tilted away from the layer normal $\hat{z}$ (cf. Figure 1.11*b*).

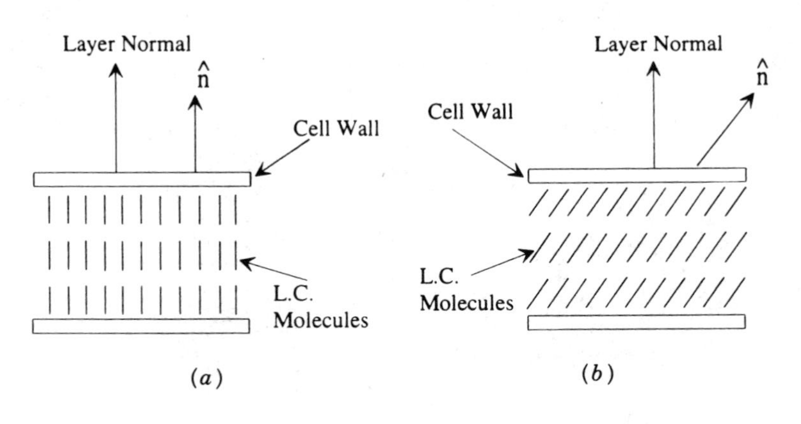


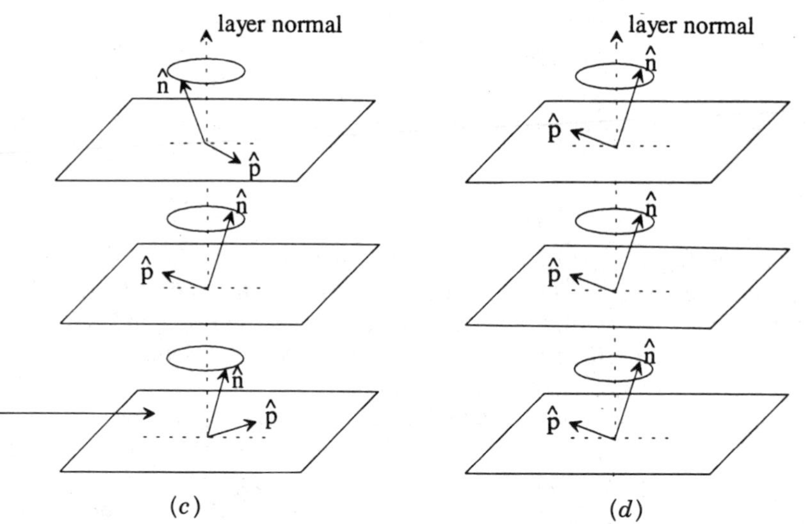


**Figure 1.11** Molecular arrangements of liquid crystals: (*a*) smectic-A; (*b*) smectic-C; (*c*) smectic-C* or ferroelectric ; (*d*) unwound smectic-C*.

In smectic-C* liquid crystals, as depicted in Figure 1.11*c*, the director axis $\hat{n}$ is tilted away from the layer normal $\hat{z}$ and "precesses" around the $\hat{z}$ axis in successive layers. This is analogous to cholesterics and is due to the introduction of optical-active or chiral molecules to the smectic-C liquid crystals.

Smectic-C* liquid crystals are interesting in one important respect, namely, they comprise a system that permits, by the symmetry principle, the existence of a spontaneous electric polarization. This can be explained simply in the following way.

The spontaneous electric polarization $\hat{p}$ is a vector and represents a breakdown of symmetry; that is, there is a directional preference. If the liquid crystal properties are independent of the director axis $\hat{n}$ direction (i.e., $+\hat{n}$ is the same as $-\hat{n}$), $\hat{p}$, if it exists, must be locally perpendicular to $\hat{n}$. In the case of smectic-A, which possesses rotational symmetry around $\hat{n}$, $\hat{p}$ must therefore be vanishing. In the case of smectic-C, there is a reflection symmetry (mirror symmetry) about the plane defined by the $\hat{n}$ and $\hat{z}$ axes, so $\hat{p}$ is also vanishing.

This reflection symmetry is broken if a chiral center is introduced to the molecule, resulting in a smectic-C* system. By convention, $\hat{p}$ is defined as positive if it is along the direction of $\hat{z} \times \hat{n}$, and negative otherwise. Figure 1.11*c* shows that since $\hat{n}$ precesses around $\hat{z}$, $\hat{p}$ also precesses around $\hat{z}$. If, by some external field, the helical structure is unwound and $\hat{n}$ points in a fixed direction, as in Figure 1.11*d*, then $\hat{p}$ will point in one direction. Clearly, this and other director axis reorientation processes are accompanied by considerable change in the optical refractive index and other properties of the system, and they can be utilized in practical electro- and optooptical modulation devices. A detailed discussion of smectic liquid crystals is given in Chapter 4.

## 1.4 MIXTURES AND COMPOSITES

In general, temperature ranges for the various mesophases of pure liquid crystals are quite limited. This and other physical limitations impose severe shortcomings on practical usage of these materials. Accordingly, while much fundamental research is still performed with pure liquid crystals, industrial applications employ mostly mixtures, composites, or specially doped liquid crystals with tailor-made physical and optical properties. Current progress and large-scale application of liquid crystals in optical technology are largely the result of tremendous advances in such new-material development efforts.

There are many ways and means of modifying a liquid crystal's physical properties. At the most fundamental level, various chemical groups such as bonds or atoms can be substituted into a particular class of liquid crystals. A good example is the cyanobiphenyl homologous series $n$CB ($n = 1, 2, 3, \ldots$). As $n$ is increased through synthesis, the viscosities, anisotropies, molecular sizes, and many other parameters are greatly modified. Some of these physical properties can also be modified by substitution. For example, the hydrogen in the 2, 3, and 4 positions of the phenyl ring may be substituted by some fluoro (F) or chloro (Cl) group (7).

Besides these molecular synthesis techniques, there are other physical processes that can be employed to dramatically improve the so-called performance characteristics of liquid crystals. In the following sections we describe three well-developed ones, focusing our discussion on nematic liquid crystals.

### 1.4.1 Mixtures

A large majority of liquid crystals in current device usage are eutectic mixtures of two or more mesogenic substances. A good example is E7 (from EM Chemicals), which is a mixture of four liquid crystals (cf. Figure 1.12).

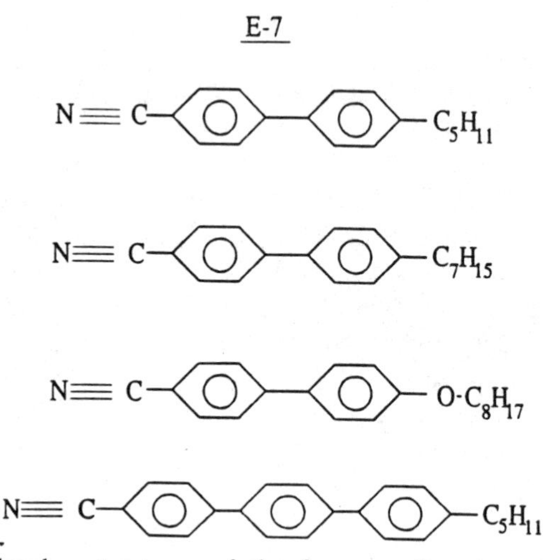


**Figure 1.12** Molecular structures of the four constituents making up the liquid crystal E7 (from EM Chemicals).

The optical properties, dielectric anisotropies, viscosities of E7 are very different from those of the individual mixture constituents. Creating mixtures is an art, guided, of course, by some scientific principles (8).

One of the guiding principles for making the right mixture can be illustrated by the exemplary phase diagram of two materials with different melting (i.e., crystal → nematic) and clearing (i.e., nematic → isotropic) points, as shown in Figure 1.13. Both substances have small nematic ranges ($T_i - T_n$

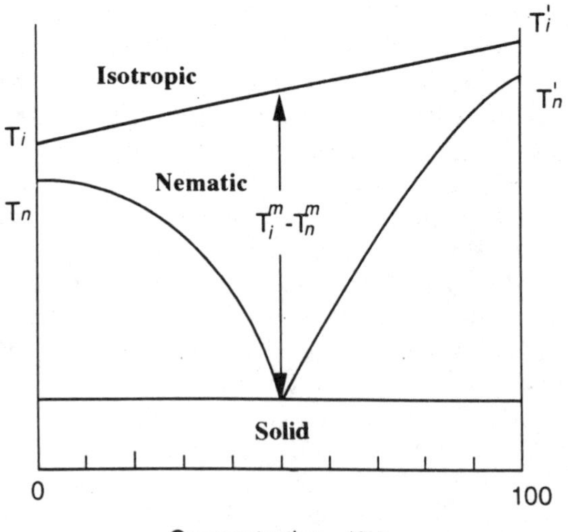

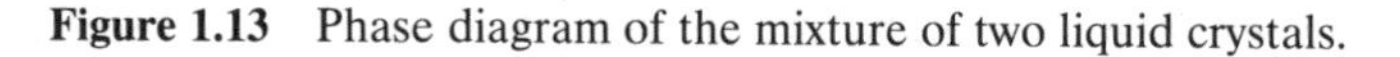

**Figure 1.13** Phase diagram of the mixture of two liquid crystals.

and $T_i' - T_n'$). When mixed at the right concentration (4), however, the nematic range $(T_i^m - T_n^m)$ of the mixture can be several magnitudes larger.

If the mixture components do not react chemically with one another, clearly their bulk physical properties, such as dielectric constant, viscosity, and anisotropy, are some weighted sum of the individual responses; that is, the physical parameter $\alpha_m$ of the mixture is related to the individual responses $\alpha_i$'s by $\alpha_m = \Sigma c_i \alpha_i$, where $c_i$ is the corresponding molar fraction. However, because of molecular correlation effects and the critical dependence of the constituents on their widely varying transition temperatures and other collective effects, the simple linear additive representation of the mixture's response is at best a rough approximation. Generally, one would expect that optical and other parameters (e.g., absorption lines or bands), which depend largely on the electronic responses of individual molecules, will follow the simple additive rule more closely than physical parameters (e.g., viscosities), which are highly dependent on intermolecular forces.

In accordance with the foregoing discussion, liquid crystal mixtures formed by different concentrations of the same set of constituents should be regarded as physically and optically different materials.

### 1.4.2 Dye-Doped Liquids Crystals

From the standpoint of optical properties, the doping of liquid crystals by appropriately dissolved concentrations and types of dyes clearly deserves special attention. The most important effect of dye molecules on liquid crystals is the modification of their well-known linear, and more recently observed nonlinear, optical properties (cf. Chapter 6).

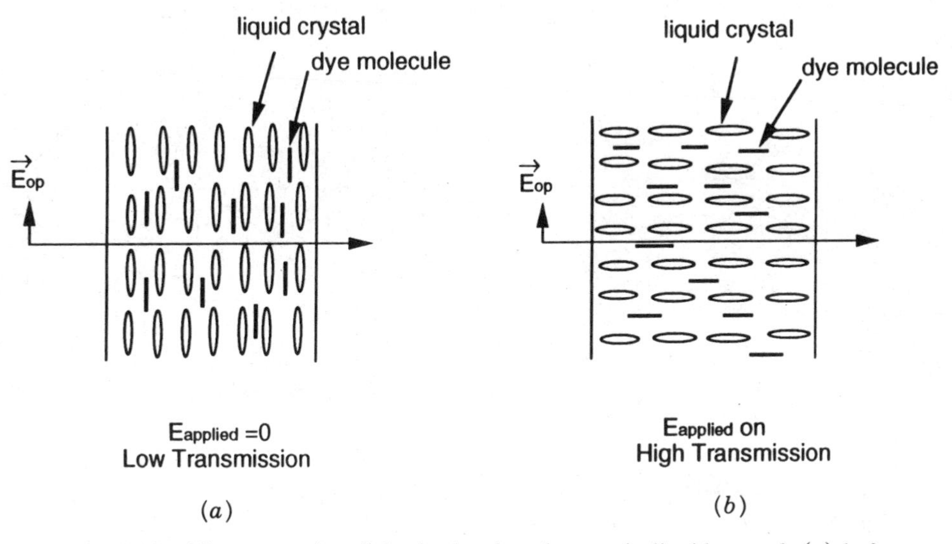


**Figure 1.14** Alignment of a dichroic dye-doped nematic liquid crystal: (*a*) before application of switching electric field; (*b*) switching field on.

An obvious effect of dissolved dye is to increase the absorption of a particular liquid crystal at some specified wavelength region. If the dye molecules undergo some physical or orientational changes following photon absorption, they could also affect the orientation of the host liquid crystal, giving rise to nonlinear or storage-type optical effects. These effects are discussed in greater detail in Chapter 6.

In linear optical and electrooptical applications, another frequently employed effect is the so-called guest–host effect. This utilizes the fact that the coefficients of the dissolved dichroic dyes are different for optical fields polarized parallel or perpendicular to the dye molecule's long (optical) axis. In general, a dichroic dye molecule absorbs much more for optical field polarization parallel to its long axis than for optical field polarization perpendicular to its long axis. These molecules are generally elongated in shape and can be oriented and reoriented by the host nematic liquid crystals. Accordingly, the transmission of the cell can be switched with the application of an external field (cf. Figure 1.14).

### 1.4.3 Polymer-Dispersed Liquid Crystals

Just as the presence of dye molecules modifies the absorption characteristics of liquid crystals, the presence of material interdispersed in the liquid crystals of a different refractive index modifies the scattering properties of the resulting "mixed" system. Polymer-dispersed liquid crystals are formed by introducing liquid crystals as micron-sized droplets into a polymer matrix. The optical indices of these randomly oriented liquid crystal droplets, in the absence of an external alignment field, depend on the liquid crystal–polymer interaction at the boundary, and therefore assume a random distribution (cf. Figure 1.15*a*). This causes large scattering. Upon the application of an

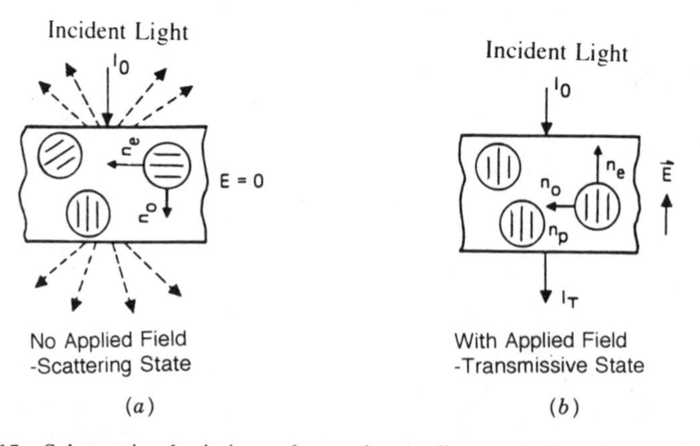


**Figure 1.15** Schematic depiction of a polymer-dispersed liquid crystal material: (*a*) in the absence of an external alignment field, highly scattered state; (*b*) when an external alignment field is on, transparent state.

external field, the droplets will be aligned (Figure 1.15*b*), and the system will become clear as the refractive index of the liquid crystal droplets matches the isotropic polymer backgrounds.

Polymer-dispersed liquid crystals were introduced a few years ago (9). There are now many techniques for preparing such composite liquid crystalline materials, including the phase separation and the encapsulation methods (10).

## 1.5 LIQUID CRYSTAL CELLS AND SAMPLE PREPARATION

Liquid crystals, particularly nematics which are commonly employed in many electrooptical devices, behave physically very much like liquids. Milk is often a good analogy to liquid crystals in such bulk, "unaligned" states. Their crystalline properties become apparent when such milky liquids are contained in (usually) flat thin cells.

The alignment of the liquid crystal axis in such cells is essentially controlled by the cell walls, whose surfaces are treated in a variety of ways to achieve various director axis alignments.

### 1.5.1 Bulk Thin Film

For nematics, two commonly used alignments are the so-called homeotropic and homogeneous (or planar) alignments, as shown in Figures 1.16*a* and *b*, respectively. To create homeotropic alignment, the cell walls are treated with a surfactant such as HTAB (hexadecyl-trimethyl-ammonium-bromide). These surfactants are basically soaps, whose molecules tend to align themselves perpendicular to the wall and thus impart the homeotropic alignment on the liquid crystal.

In the laboratory, a quick and effective way to make a homeotropic nematic liquid crystal sample is as follows: Dissolve 1 part of HTAB in 50 parts of distilled deionized water by volume. Clean two glass slides (or other optical flats appropriate for the spectral region of interest). Dip the slides in the HTAB solution and slowly withdraw them. This effectively introduces a coating of HTAB molecules on the glass slides. The glass slides should then be dried in an oven or by other means. To prepare the nematic liquid crystal sample, prepare a spacer (Mylar or some nonreactive plastic) of desirable dimension and thinness and place the spacer on one of the slides. Fill the inner spacer with the nematic liquid crystal under study (it helps to first warm it to the isotropic phase). Place the second slide on top of this and clamp the two slides together. Once assembled, the sample should be left alone, and it will slowly (in a few minutes) settle into a clear homeotropically aligned state.

Planar alignment can be achieved in many ways. A simple but effective method is to first coat the cell wall with some polymer such as PVA (polyvinyl alcohol) and then rub it unidirectionally with a lens tissue. This process

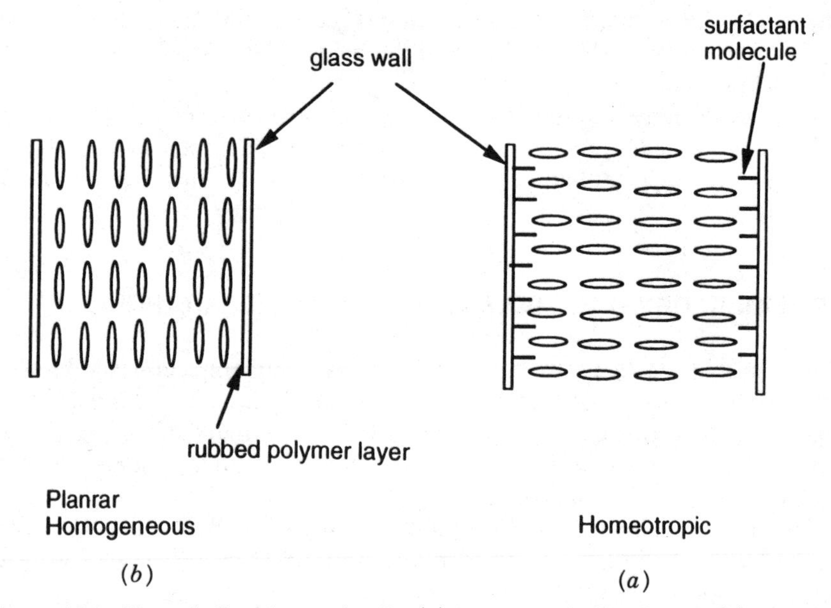


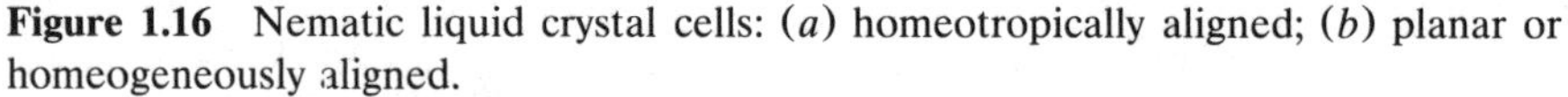

**Figure 1.16** Nematic liquid crystal cells: (*a*) homeotropically aligned; (*b*) planar or homeogeneously aligned.

creates elongated stress/strain on the polymer and facilitates the alignment of the long axis of the liquid crystal molecules along the rubbed direction (i.e., on the plane of the cell wall). Another method is to deposit silicon oxide obliquely onto the cell wall.

For preparing a PVA-coated planar sample in the laboratory, the following technique has proven to be quite reliable. Dissolve chemically pure polyvinyl alcohol (which is solid at room temperature) in distilled deionized water at an elevated temperature (near the boiling point) at a concentration of about 0.2%. Dip the cleaned glass slide into the PVA solution at room temperature and slowly withdraw it, thus leaving a film of the solution on the slide. (Alternatively, one could place a small amount of the PVA solution on the slide and spread it into a thin coating.) The coated slide is then dried in an oven, followed by unidirectional rubbing of its surfaces with a lens tissue. The rest of the procedure for cell assembly is the same as that for homeotropic alignment.

Ideally, of course, these cell preparation processes should be performed in a clean room and preferably in an enclosure free of humidity or other chemicals (e.g., a nitrogen-filled enclosure) in order to prolong the lifetime of the sample. Nevertheless, the liquid crystal cells prepared with the techniques outlined previously have been shown to last several months and can withstand many temperature cyclings through the nematic–isotropic phase transition point, provided the liquid crystals used are chemically stable.

Generally, nematics such as 5CB and E7 are quite stable, whereas MBAA (*p*-methoxybenzylidene-*p*′-*n*-butylaniline) tends to degrade in a few days.

Besides these two standard cell alignments, there are many other variations such as hybrid, twisted, supertwisted, and fingerprint (6). Industrial processing of these nematic cells, as well as the transparent conductive coating of the cell windows for electrooptical device applications is understandably more elaborate.

For chiral nematic liquid crystals, the method outlined previously for a planar nematic cell has been shown to be quite effective. For smectic-A the preparation method is similar to that for a homeotropic nematic cell. In this case, however, it helps to have an externally applied field to help maintain the homeotropic alignment as the sample (slowly) cools down from the nematic to the smectic phase. The cell preparation method for a ferroelectric liquid crystal, smectic-C*, is similar to the method for a planar nematic cell (11).

### 1.5.2 Liquid Crystal Optical Slab Waveguide and Fiber

Besides the bulk thin film structures discussed in the preceding section, liquid crystals could also be fabricated into optical waveguide forms. Both slab (13) and cylindrical (fiber) (14, 15) waveguide structures have been investigated.

A typical liquid crystal slab waveguide (12, 13) is shown in Figure 1.17. A thin film (approximately 1 $\mu$m) of liquid crystal is sandwiched between two glass slides (of lower refractive index than the liquid crystal), one of which has been deposited with an organic film into which an input laser is introduced via the coupling prism. The laser excites the transverse electric (TE) and/or transverse magnetic (TM) modes in the organic film, which are then guided into the nematic liquid crystal region. Using such optical waveguides, Whinnery et al. (12) and Giallorenzi et al. (13) have measured the scattering losses in nematic and smectic liquid crystals and introduced electrooptical

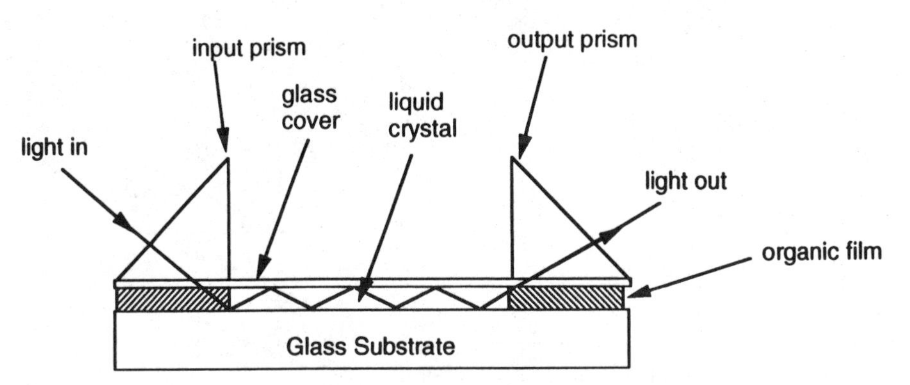


**Figure 1.17** Schematic depiction of a liquid crystal slab waveguide structure.

and integrated optical switching devices. However, the large losses in nematics (about 20 dB/cm) and their relatively slow responses impose serious limitations in practical integrated electrooptical applications. The scattering losses in smectic waveguides are generally much lower, and they may be useful in nonlinear optical applications (cf. Chapter 10).

Liquid crystals "fibers" are usually made by filling hollow fibers (microcapillaries) made of material of lower indices of refraction (14, 15). The microcapillaries are usually made of Pyrex or silica glass whose refractive indices are 1.47 and 1.45, respectively. It was observed by Green and Madden (14) that the scattering losses of the nematic liquid crystal fiber core are considerably reduced for a core diameter smaller than 10 $\mu$m; typically, the loss is about 3 dB/cm (compared to 20 dB/cm for a slab waveguide or bulk thin film). Also, the director axis alignment within the core is highly dependent on the liquid crystals–capillary interface interaction (i.e., the capillary material). In silica or Pyrex capillaries the nematic director tends to align along the axis of the fiber (Figure 1.18*a*), whereas in borosilicate capillaries the nematic director tends to align in a radial direction, occasionally mixed in with a thread of axially aligned material running down the axis of the fiber (Figure 1.18*b*).

Fabrications of such liquid crystal fibers with isotropic phase liquid crystals are much easier. Because of the fluid property and much lower scattering loss, liquid crystal fibers of much longer dimension have been fabricated and

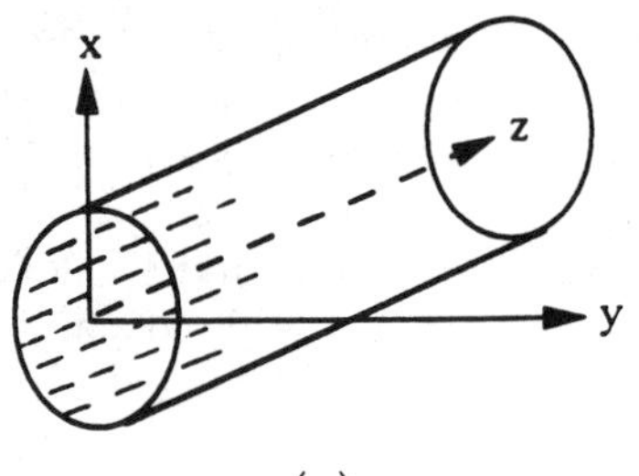


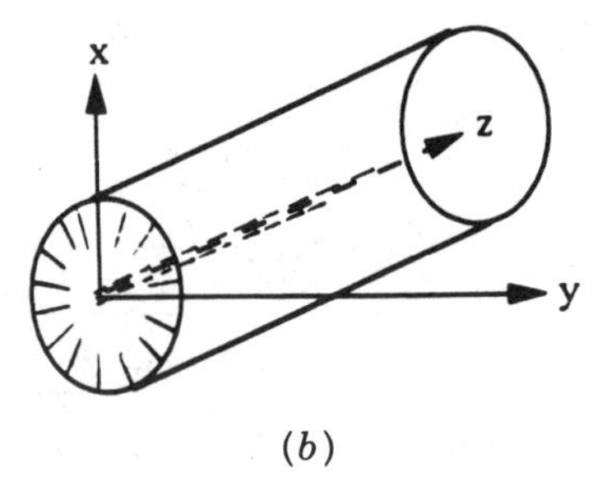


**Figure 1.18** (*a*) Axial alignment of a nematic liquid crystal cored fiber. (*b*) Mixed radial and axial alignment of a nematic liquid crystal cored fiber.

shown to exhibit interesting nonlinear optical properties (15) in conjunction with nanosecond and laser pulse propagation and backscattering. Some of these details are given in Chapter 10.

## REFERENCES


1. P. G. deGennes, "The Physics of Liquid Crystals," Clarendon Press, Oxford, 1974.
2. P. Mariani, F. Rustichelli, and G. Torquati, *in* "Physics of Liquid Crystalline Materials" (I. C. Khoo and F. Simoni, eds.). Gordon & Breach, Philadelphia, 1991.
3. A. Ciferri, W. R. Krigbaum, and R. B. Meyer eds., "Polymer Liquid Crystals," Academic Press, New York.
4. See, for example, L. M. Blinov, "Electro-Optical and Magneto-Optical Properties of Liquid Crystals." Wiley (Interscience), New York, 1983.
5. M. Ghedini, S. Armentano, R. Bartolino, F. Rustichelli, G. Torquati, M. Kirov, and M. Petrov, New liquid crystalline compounds containing transition metals. *Mol. Cryst. Liq. Cryst.* **151**, 75 (1987).
6. I. C. Khoo and S. T. Wu, "Optics and Nonlinear Optics of Liquid Crystals." World Scientific, Singapore, 1992.
7. G. W. Gray, M. Hird, and K. J. Toyne, The synthesis of several lateral difluoro-substituted 4,4″-dialkyl- and 4,4″-alkoxyalkyl-terphenyls. *Mol. Cryst. Liq. Cryst.* **204**, 43 (1991); see also S. T. Wu, D. Coates, and E. Bartmann, Physical properties of chlorinated liquid crystals. *Liq. Cryst.* **10**, 635 (1991).
8. See, for example, D. Demus, C. Fietkau, R. Schubert, and H. Kehlen, Calculation and experimental verification of eutectic systems with nematic phases. *Mol. Cryst. Liq. Cryst.* **25**, 215 (1974).
9. J. W. Doane, N. A. Vaz, B. G. Wu, and S. Zumer, Field controlled light scattering from nematic microdroplets. *Appl. Phys. Lett.* **48**, 269 (1986).
10. J. L. West, Phase separation of liquid crystals in polymers. *Mol. Cryst. Liq. Cryst.* **157**, 428 (1988); P. Drzaic, Polymer dispersed nematic liquid crystal for large area displays and light valves. *J. Appl. Phys.* **60**, 2142 (1986).
11. R. Macdonald, J. Schwartz, and H. J. Eichler, Laser-induced optical switching of a ferroelectric liquid crystal. *Int. J. Nonlinear Opt. Phys.* **1**, 103 (1992).
12. J. R. Whinnery, C. Hu, and Y. S. Kwon, Liquid crystal waveguides for integrated optics. *IEEE J. Quantum Electron. QE13*, 262 (1977).
13. G. Giallorenzi, J. A. Weiss, and J. P. Sheridan, Light scattering from smectic liquid-crystal waveguides. *J. Appl. Phys.* **47**, 1820 (1976).
14. M. Geren and S. J. Madden, Low loss nematic liquid crystal cored fiber waveguide. *Appl. Opt.* **28**, 5202 (1989).
15. I. C. Khoo, H. Li, P. G. LoPresti, and Yu Liang, Observation of optical limiting and backscattering of nanosecond laser pulses in liquid crystal fibers. *Opt. Lett.* **19**, 530 (1994); see also I. C. Khoo, Sukho Lee, P. G. LoPresti, R. G. Lindquist, and H. Li, Isotropic liquid crystalline film and fiber structures for optical limiting application. *Int. J. Nonlinear Opt. Phys.* **2**, No. 4 p. 559 (1993).

# CHAPTER 2

# ORDER PARAMETER, PHASE TRANSITION, AND FREE ENERGIES

## 2.1 BASIC CONCEPTS

### 2.1.1 Introduction

Generally speaking, we can divide liquid crystalline phases into two distinctly different types: the ordered and the disordered. For the ordered phase, the theoretical framework invoked for describing the physical properties of liquid crystals is closer in form to that pertaining to solids; it is often called elastic or continuum theory. In this case various terms and definitions typical of solid materials (e.g., elastic constant, distortion energy, torque, etc.) are commonly used. Nevertheless, the interesting fact about liquid crystals is that in such an ordered phase they still possess many properties typical of liquids. In particular, they flow like liquids and thus require hydrodynamical theories for their complete description. These are explained in further detail in the next chapter.

Liquid crystals in the disordered or isotropic phase behave very much like ordinary fluids of anisotropic molecules. They can thus be described by theories pertaining to anisotropic fluids. There is, however, one important difference.

Near the isotropic → nematic phase transition temperature, liquid crystals exhibit some highly correlated pretransitional effects. In general, the molecules become highly susceptible to external fields, and their responses tend to slow down considerably.

In the next few sections we introduce some basic concepts and definitions, such as order parameter, short- and long-range order, phase transition, and so on, which form the basis for describing the ordered and disordered phases of liquid crystals.

### 2.1.2 Scalar and Tensor Order Parameter

Perhaps the most prominent physical characteristic of liquid crystals is their directoral correlation, which is manifested in the two directionally ordered phases discussed briefly in the preceding section. The physics of liquid crystals is best described in terms of the so-called order parameters (1,2).

If we use the long axis of the molecule as a reference and denote it as $\hat{k}$, the microscopic scalar order parameter $S$ is defined (1,2) as follows:

$$\begin{aligned} S &= \tfrac{1}{2}\langle(\hat{k}\cdot\hat{n})(\hat{k}\cdot\hat{n})-1\rangle \\ &= \tfrac{1}{2}\langle 3\cos^2\theta - 1\rangle \end{aligned} \tag{2.1}$$

With reference to Figure 2.1, $\theta$ is the angle made by the molecular axis with the director axis. The average $\langle\ \rangle$ is taken over the whole ensemble; this kind of order is usually termed long-range order. It is called microscopic because it describes the average response of a molecule. The scalar order parameter

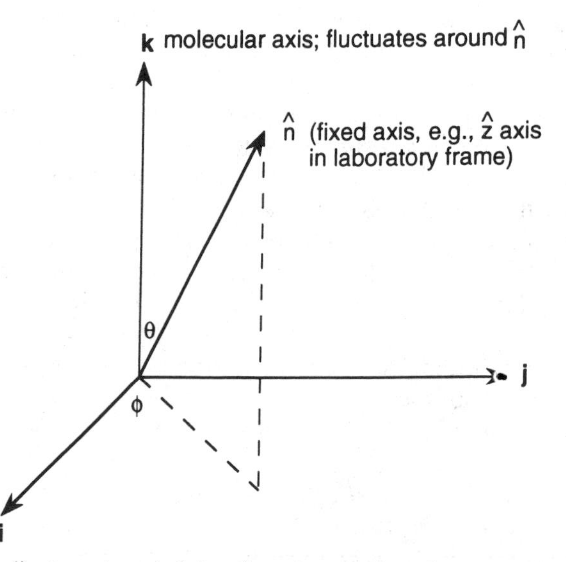


**Figure 2.1** Coordinate system defining the microscopic order parameter of a nematic liquid crystal molecule. $\hat{\imath}$, $\hat{\jmath}$, and $\hat{k}$ are the molecular axes, whereas $\hat{n}$ is the laboratory axis that denotes the average direction of liquid crystal alignment.

defined previously is sufficient for describing liquid crystalline systems comprised of molecules that possess cylindrical or rotational symmetry around the long axis $\hat{k}$.

On the other hand, for molecules lacking such symmetry, or in cases where such rotational symmetry is "destroyed" by the presence of asymmetric dopants or intramolecular material interactions, a more general tensor order parameter $S_{ij}$ is needed.

$S_{ij}$ is defined as

$$S_{ij} = \tfrac{1}{2}\langle 3(\hat{n}\cdot\hat{\imath})(\hat{n}\cdot\hat{\jmath}) - 1\rangle \tag{2.2}$$

where $\hat{\imath}$, $\hat{\jmath}$, and $\hat{k}$ are unit vectors along the molecular axes. With reference to Figure 2.1, the three diagonal components $S_{ii}$, $S_{jj}$, and $S_{kk}$ are given by

$$S_{ii} = \tfrac{1}{2}\langle 3\sin^2\theta\cos^2\phi - 1\rangle \tag{2.3a}$$

$$S_{jj} = \tfrac{1}{2}\langle 3\sin^2\theta\sin^2\phi - 1\rangle \tag{2.3b}$$

$$S_{kk} = \tfrac{1}{2}\langle 3\cos^2\theta - 1\rangle \tag{2.3c}$$

Note that $S_{ii} + S_{jj} + S_{kk} = 0$. Put another way, $S$ is a traceless tensor because its diagonal elements sum to zero.

For a more complete description of the statistical properties of the liquid crystal orientation, functions involving higher powers of $\cos^2\theta$ are needed. The most natural functions to use are the Legendre polynomials $P_l(\cos\theta)$ $(l = 0, 1, 2, \ldots)$, in terms of which we can write (2.1) as $S = \langle P_2\rangle$, which measures the average of $\cos^2\theta$. The next nonvanishing term is $\langle P_4\rangle$, which provides a measure of the dispersion of $\langle\cos^2\theta\rangle$.

The order parameters defined previously in terms of the directional averages can be translated into expressions in terms of the anisotropies in the physical parameters such as magnetic, electric, and optical susceptibilities. For example, in terms of the optical dielectric anisotropies $\Delta\varepsilon = \varepsilon_{\parallel} - \varepsilon_{\perp}$, one can define a so-called macroscopic order parameter which characterizes the bulk response

$$Q_{\alpha\beta} \equiv \varepsilon_{\alpha\beta} - \frac{1}{3}\delta_{\alpha\beta}\sum_{\gamma}\varepsilon_{\gamma\gamma} \equiv \delta\varepsilon_{\alpha\beta} \tag{2.4}$$

It is called macroscopic because it describes the bulk property of the material. To be more explicit, consider a uniaxial material such that $\varepsilon_{\alpha\beta}$ in the molecular axis system is of the form

$$\varepsilon_{\alpha\beta} = \begin{pmatrix} \varepsilon_{\perp} & 0 & 0 \\ 0 & \varepsilon_{\perp} & 0 \\ 0 & 0 & \varepsilon_{\parallel} \end{pmatrix} \tag{2.5}$$

Writing $Q_{\alpha\beta}$ explicitly in terms of their diagonal components, we thus have

$$Q_{xx} = Q_{yy} = -\Delta\varepsilon \tag{2.6}$$

and

$$Q_{zz} = 2\,\Delta\varepsilon \tag{2.7}$$

It is useful to note here that, in tensor form, $\varepsilon_{\alpha\beta}$ can be expressed as

$$\varepsilon_{\alpha\beta} \equiv \varepsilon_{\perp}\delta_{\alpha\beta} + \Delta\varepsilon\, n_\alpha n_\beta \tag{2.8}$$

Note that this form shows that $\varepsilon = \varepsilon_{\parallel}$ for an optical field parallel to $\hat{n}$, and $\varepsilon = \varepsilon_{\perp}$ for an optical field perpendicular to $\hat{n}$.

Similarly, other parameters such as the magnetic ($\chi^m$) and electric ($\chi$) susceptibilities may be expressed as

$$\chi^m_{\alpha\beta} = \chi^m_{\perp}\delta_{\alpha\beta} + \Delta\chi^m n_\alpha n_\beta \tag{2.9a}$$

and

$$\chi_{\alpha\beta} = \chi_{\perp}\delta_{\alpha\beta} + \Delta\chi\, n_\alpha n_\beta \tag{2.9b}$$

respectively, in terms of their respective anisotropies $\Delta\chi^m$ and $\Delta\chi$.

Generally, however, optical dielectric anisotropy and its dc or low-frequency counterpart (the dielectric anisotropy) provide a less reliable measure of the order parameter because they involve electric fields. This is because of the so-called local field effect: The effective electric field acting on a molecule is a superposition of the electric field from the externally applied source and the field created by the induced dipoles surrounding the molecules. For systems where the molecules are not correlated, the effective field can be fairly accurately approximated by some local field correction factor (3); in liquid crystalline systems these correction factors are much less accurate. For a more reliable determination of the order parameter, one usually employs nonelectric field-related parameters, such as the magnetic susceptibility anisotropy:

$$Q_{\alpha\beta} = \chi^m_{\alpha\beta} - \frac{1}{3}\delta_{\alpha\beta}\sum_{\gamma}\chi^m_{\gamma\gamma} \tag{2.10}$$

### 2.1.3 Long- and Short-Range Order

The order parameter, defined by (2.2) and its variants such as (2.4) and (2.8), is an average over the whole system and therefore provides a measure of the long-range orientational order. The smaller the fluctuation of the molecular axis from the director axis orientation direction, the closer is the magnitude of $S$ to unity. In a perfectly aligned liquid crystal, as in other crystalline material, $\langle \cos^2 \theta \rangle = 1$ and $S = 1$; on the other hand, in a perfectly random system, such as ordinary liquids or the isotropic phase of liquid crystals, $\langle \cos^2 \theta \rangle = \frac{1}{3}$ and $S = 0$.

An important distinction between liquid crystals and ordinary anisotropic or isotropic liquids is that, in the isotropic phase, there could exist a so-called short-range order (1,2); that is, molecules within a short distance of one another are correlated by intermolecular interactions (4). These molecular interactions may be viewed as remnants of those existing in the nematic phase. Clearly, the closer the isotropic liquid crystal is to the phase transition temperature, the more pronounced will be the short-range order and its manifestations in many physical parameters. Short-range order in the isotropic phase gives rise to interesting critical behavior in the response of the liquid crystals to externally applied fields (electric, magnetic, optical) (cf. Section 3.2).

As pointed out at the beginning of this chapter, the physical and optical properties of liquid crystals may be roughly classified into two types: one pertaining to the ordered phase, characterized by long-range order and crystalline-like physical properties; the other pertaining to the so-called disordered phase, where a short-range order exists. All these order parameters show critical dependences as the temperature approaches the phase transition temperature $T_c$ from the respective directions.

## 2.2 MOLECULAR INTERACTIONS AND PHASE TRANSITIONS

In principle, if the electronic structure of a liquid crystal molecule is known, one can deduce the various thermodynamical properties. This is a monumental task in quantum statistical chemistry that has seldom, if ever, been attempted in a quantitative or conclusive way. There are some fairly reliable guidelines, usually obtained empirically, that relate molecular structures with the existence of the liquid crystal mesophases and, less reliably, the corresponding transition temperatures.

One simple observation is that to generate liquid crystals, one should use elongated molecules. This is best illustrated by the $n$cB homolog (5) ($n = 1,2,3,\ldots$). For $n \leq 4$, the material does not exhibit a nematic phase. For $n = 5,6,7$, the material possesses a nematic range. For $n > 8$, smectic phases begin to appear.

Another reliable observation is that the nematic → isotropic phase transition temperature $T_c$ is a good indicator of the thermal stability of the nematic phase (6); the higher the $T_c$, the greater is the thermal stability of the nematic phase. In this respect, the types of chemical groups used as substituents in the terminal groups or side chain play a significant role—an increase in the polarizability of the substituent tends to be accompanied by an increase in $T_c$.

Such molecular-structure-based approaches are clearly extremely complex and often tend to yield contradictory predictions, because of the wide variation in the molecular electronic structures and intermolecular interactions present. In order to explain the phase transition and the behavior of the order parameter in the vicinity of the phase transition temperature, some simpler physical models have been employed (6). For the nematic phase, a simple but quite successful approach was introduced by Maier and Saupe (7). The liquid crystal molecules are treated as rigid rods, which are correlated (described by a long-range order parameter) with one another by Coulomb interactions. For the isotropic phase, deGennes introduced a Landau type of phase transition theory (1–3) which is based on a short-range order parameter.

The theoretical formalism for describing the nematic → isotropic phase transition and some of the results and consequences are given in the next section. This is followed by a summary of some of the basic concepts introduced for the isotropic phase.

## 2.3 MOLECULAR THEORIES AND RESULTS FOR THE LIQUID CRYSTALLINE PHASE

Among the various theories developed for describing the order parameter and phase transitions in the liquid crystalline phase, the most popular and successful one is that first advanced by Maier and Saupe and corroborated in studies by others (8). In this formalism Coulombic intermolecular dipole–dipole interactions are assumed. The interaction energy of a molecule with its surroundings is then shown to be of the form (6):

$$W_{\text{int}} = -\frac{A}{V^2} S\left(\frac{3}{2}\cos^2\theta - 1\right) \tag{2.11}$$

where $V$ is the molar volume ($V = M/\rho$), $S$ is the order parameter, and $A$ is a constant determined by the transition moments of the molecules. Both $V$ and $S$ are functions of temperature. Comparing (2.11) and (2.1) for the definition of $S$, we note that $W_{\text{int}} \sim S^2$, so this mean field theory by Maier and Saupe is often referred to as the $S^2$ interaction theory (1). This interaction energy is included in the free enthalpy per molecule (chemical potential) and is used in conjunction with an angular distribution function $f(\theta, \phi)$ for statistical mechanics calculations.

### 2.3.1 Maier–Saupe Theory: Order Parameter Near $T_c$

Following the formalism of deGennes, the interaction energy may be written as

$$G_1 = -\tfrac{1}{2}U(p,T)S\left(\tfrac{3}{2}\cos^2\theta - 1\right) \tag{2.12}$$

The total free enthalpy per molecule is therefore

$$G(p,T) = G_i(p,T) + K_B T \int f(\theta,\phi) \log 4\pi f(\theta,\phi)\, d\Omega + G_1(p,T,S) \tag{2.13}$$

where $G_i$ is the free enthalpy of the isotropic phase. Minimizing $G(p,T)$ with respect to the distribution function $f$, one gets

$$f(\theta) = \frac{\exp(m\cos^2\theta)}{4\pi z} \tag{2.14}$$

where

$$m = \frac{3}{2}\frac{US}{K_B T} \tag{2.15}$$

and the partition function $z$ is given by

$$z = \int_0^1 e^{mx^2}\, dx \tag{2.16}$$

From the definition of $S = -\frac{1}{2} + \frac{3}{2}\langle\cos^2\theta\rangle$, we have

$$\begin{aligned} S &= -\frac{1}{2} + \frac{3}{2z}\int_0^1 x^2 e^{mx^2}\, dx \\ &= -\frac{1}{2} + \frac{3}{2}\frac{\partial z}{z\,\partial m} \end{aligned} \tag{2.17}$$

The coupled equations (2.15) and (2.17) for $m$ and $S$ may be solved graphically for various values of $U/K_B T$, the relative magnitude of the intermolecular interaction to the thermal energies. Figure 2.2 depicts the case for $T$ below a temperature $T_c$ defined by

$$\frac{k_B T_c}{U(T_c)} = 4.55 \tag{2.18}$$

Figure 2.2 shows that curves 1 and 2 for $S$ intersect at the origin $O$ and two points $N$ and $M$. Both points $O$ and $N$ correspond to minima of $G$, whereas $M$ corresponds to a local maximum of $G$. For $T < T_c$, the value of $G$ is lower at point $N$ than at point $O$; that is, $S$ is nonzero and corresponds to the nematic phase. For temperatures above $T_c$, the stable (minimum energy) state corresponds to $O$; that is, $S = O$ and corresponds to the isotropic phase.

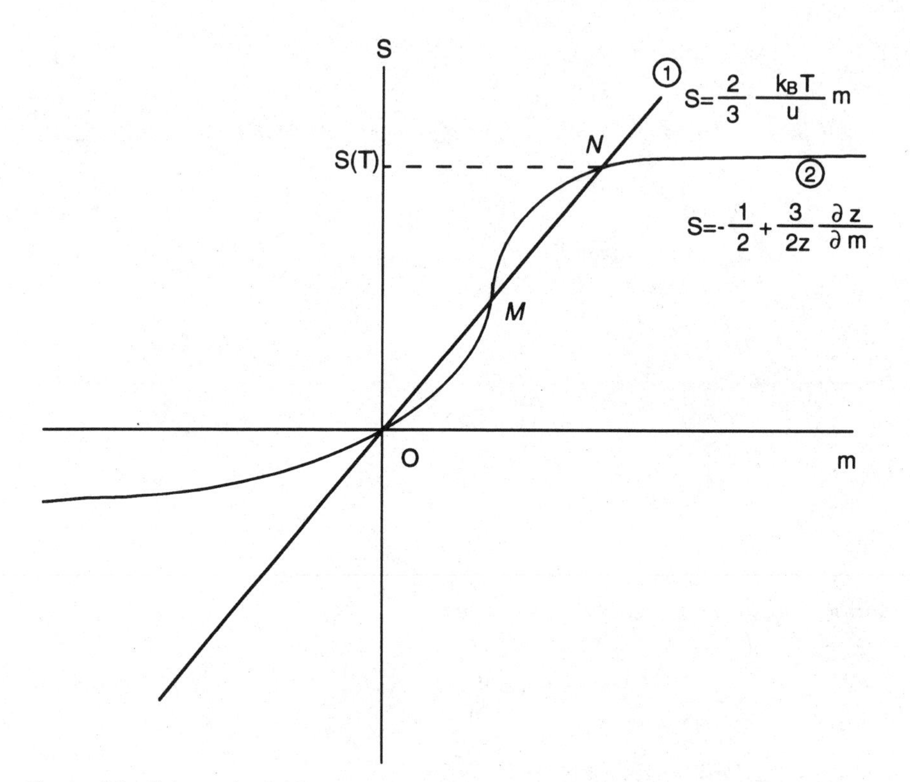


**Figure 2.2** Schematic depiction of the numerical solution of the two transcendental equations for the order parameter for $T < T_c$; there is only one intersection point (at the origin).

The transition at $T = T_c$ is a first-order one. The order parameter just below $T_c$ is

$$S_c \equiv S(T_c) = 0.44 \tag{2.19}$$

It has also been demonstrated that the temperature dependence of the order parameter of most nematics is well approximated by the expression (9):

$$S = \left(1 - \frac{0.98TV^2}{T_c V_c^2}\right)^{0.22} \tag{2.20}$$

where $V$ and $V_c$ are the molar volumes at $T$ and $T_c$, respectively.

In spite of some of these predictions which are in good agreement with the experimental results, the Maier–Saupe theory is not without its shortcomings. For example, the universal temperature dependence of $S$ on $T/T_c$ is really not valid (10); agreement with experimental results requires an improved theory that accounts for the noncylindrical shape of the molecules (8). The temperature variation given in (2.20) also cannot account for the critical dependence of the refractive indices. Nevertheless, the Maier–Saupe theory remains an effective, clear, and simple theoretical framework and starting point for understanding nematic liquid crystal complexities.

### 2.3.2 Nonequilibrium and Dynamical Dependence of the Order Parameter

While the equilibrium statistical mechanics of the nematic liquid crystal order parameter and related physical properties near $T_c$ are now quite well understood, the dynamical responses of the order parameter remain relatively unexplored. This is probably due to the fact that most studies of the order parameter near the phase transition point, such as the critical exponent, changes in molar volume, and other physical parameters, are directed at understanding the phase transition processes themselves and are usually performed with temperature changes occurring at very slow rates.

There have been several studies in recent years, however, in which the temperature of the nematics are abruptly raised by very short laser pulses (5, 11, 12). The pulse duration of the laser is in the nanosecond or picosecond time scale, which, as we shall see presently, is much shorter than the response time of the order parameter. As a result the nematic film under study exhibits delayed signals.

Figures 2.3*a* and *b* show the observed diffraction from a nematic film in a dynamic grating experiment (11). In such an experiment, explained in more detail in Chapter 7, the diffracted signal is a measure of the dynamical change in the refractive index, $\Delta n(t)$, following an instantaneous (delta function like) pump pulse. For a nematic liquid crystal the principal change in the refractive index associated with a rise in temperature is through the density and order parameter (12), that is,

$$\Delta n = \frac{dn}{d\rho} d\rho + \frac{dn}{dS} dS \tag{2.21}$$

Unlike the change in order parameter, which is a collective molecular effect, the change in density $d\rho$ arises from the individual responses of the molecules and responds relatively quickly to the temperature change.

These results are reflected in Figures 2.3*a* and *b*. The diffracted signal contains an initial "spike," which rises and decays away in the time scale on the order of the laser pulse. On the other hand, the order parameter contribution to the signal builds up rather slowly. In Figure 2.3*a* the buildup time is about 30 $\mu$s for nanosecond and visible laser pulse excitation, while in Figure 2.3*b* the buildup time is as long as 175 $\mu$s for infrared laser pulse

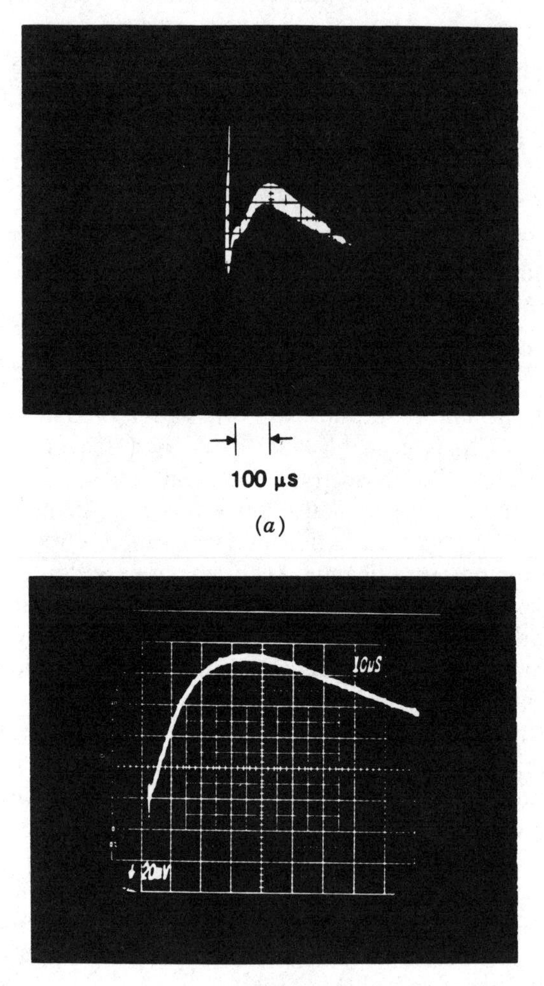


**Figure 2.3** (*a*) Observed oscilloscope trace of the diffracted signal in a dynamical scattering experiment involving microsecond infrared ($CO_2$ at 10.6 $\mu$m) laser pump pulses. Sample used is a planar aligned nematic (E7) film. (*b*) Observed oscilloscope trace of the diffracted signal from a nematic film under nanosecond visible (Nd:Yag at 0.53 $\mu$m) laser pump pulse excitation. Sample used is a planar aligned nematic (E7) film.

excitation. Figure 2.4 shows the observed "slowing down" in the response of the order parameter as the temperature approaches $T_c$.

One can also see from the relative heights of the density and order parameter components in Figures 2.3*a* and *b* that the overall response of the nematic film is different for the two forms of excitation. Absorption of infrared photons ($\lambda = 10.6$ $\mu$m) corresponds to the excitation of the ground

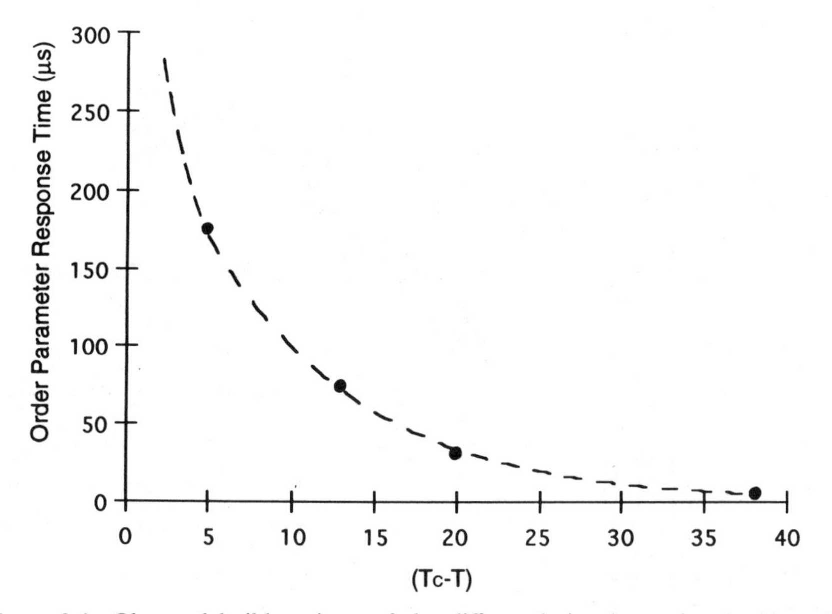


**Figure 2.4** Observed buildup times of the diffracted signal associated with order parameter change as a function of the temperature vicinity $(T_c - T)$; excitation by infrared microsecond laser pulses on E7 nematic film.

(electronic) state's rovibrational manifold, whereas the visible photoabsorption ($\lambda = 0.53$ $\mu$m) corresponds to excitation of the molecules to the electronically excited states (cf. Figure 2.5). The electronic molecular structures of these two excited states are different and may therefore account for the different dynamical response behavior of the order parameter, which is dependent on the intermolecular Coulombic dipole–dipole interaction. From this observation one may conclude that the dynamical grating technique would be an interesting technique for probing the different dynamical behaviors of the order parameters for molecules in various states of excitation, as well as the critical "slowing down" of the order parameter near $T_c$. This information will be important in the operation of practical devices based on the laser-induced order parameter changes in liquid crystals (cf. Chapter 10).

## 2.4 ISOTROPIC PHASE OF A LIQUID CRYSTAL

Above $T_c$ liquid crystals lose their directional order and behave in many respects like liquids. All bulk physical parameters also assume an isotropic form although the molecules are anisotropic.

The isotropic phase is, nevertheless, a very interesting and important phase for both fundamental and applied studies. It is fundamentally interest-

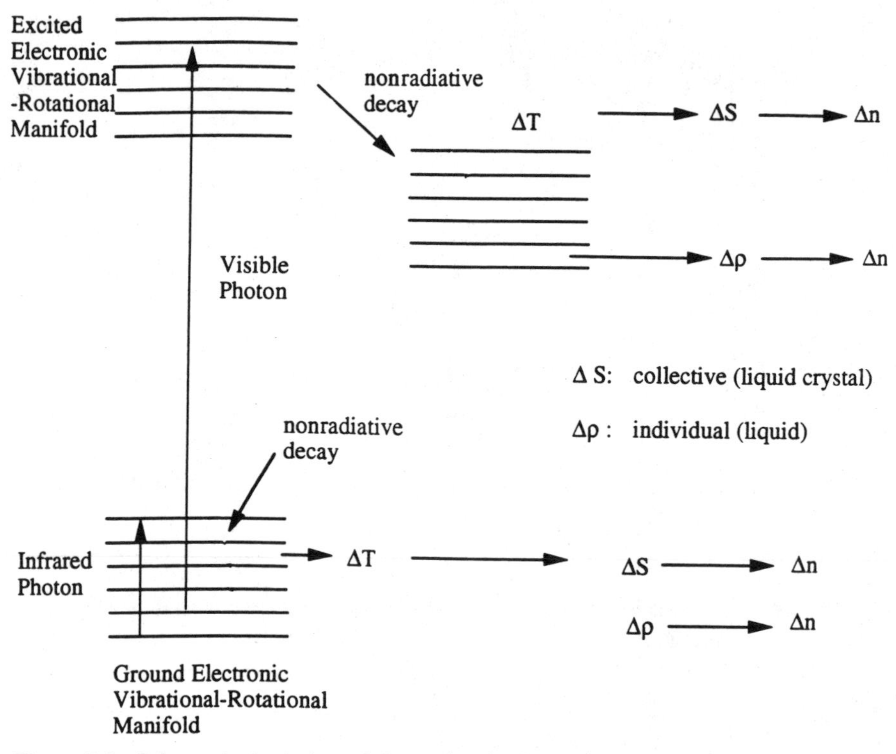


**Figure 2.5** Schematic depiction of the molecular levels involved in ground electronic state rovibrational excitations by infrared photoabsorptions and excited electronic state excitation by visible photoabsorptions.

ing because of the existence of short-range order, which gives rise to the critical temperature dependence of various physical parameters just above the phase transition temperature. These critical behaviors provide a good testing ground for the liquid crystal physics.

On the other hand, recent studies have also shown that isotropic liquid crystals may be superior in many ways for constructing practical nonlinear optical devices, in comparison to the other liquid crystalline phases (cf. Chapter 6). In general, the scattering loss is less and thus allowing longer interaction lengths, and relaxation times are on a much faster scale. These properties easily make up for the smaller optical nonlinearity for practical applications.

### 2.4.1 Free Energy and Phase Transition

We begin our discussion of the isotropic phase of liquid crystals with the free energy of the system, following deGennes' pioneering theoretical development (1, 2). The starting point is the order parameter, which we denote by $Q$.

In the absence of an external field, the isotropic phase is characterized by $Q = 0$; the minimum of the free energy also corresponds to $Q = 0$. This means that, in the Landau expansion of the free energy in terms of the order parameter $Q$, there is no linear term in $Q$, that is,

$$F = F_0 + \frac{1}{2}A(T) \sum_{\rho,\alpha} Q_{\alpha\beta}Q_{\beta\alpha} + \frac{1}{3}B(T) \sum_{\alpha,\beta,\gamma} Q_{\alpha\beta}Q_{\beta\gamma}Q_{\gamma\alpha} + O(Q4) \quad (2.22)$$

where $F_0$ is a constant and $A(T)$ and $B(T)$ are temperature-dependent expansion coefficients:

$$A(T) = a(T - T^*) \quad (2.23)$$

where $T^*$ is very close to, but lower than, $T_c$. Typically, $T_c - T_c^* = 1$ K.

Note that $F$ contains a nonzero term of order $Q^3$. This odd function of $Q$ ensures that states with some nonvanishing value of $Q$ (e.g., due to some alignment of molecules) will have different free-energy values depending on the direction of the alignment. For example, the free energy for a state with an order parameter $Q$ of the form

$$Q_1 = \begin{pmatrix} -\xi & 0 & 0 \\ 0 & -\xi & 0 \\ 0 & 0 & 2\xi \end{pmatrix} \quad (2.24a)$$

(i.e., with some alignment of the molecule in the $z$ direction) is not the same as the state with a negative $Q$ parameter

$$Q_2 = \begin{pmatrix} \xi & 0 & 0 \\ 0 & \xi & 0 \\ 0 & 0 & -2\xi \end{pmatrix} = -Q_1 \quad (2.24b)$$

(which signifies some alignment of the molecules in the $x-y$ plane).

The cubic term in $Q$ is also important in that it dictates the phase transition at $T = T_c$ is of first order (i.e., the first-order derivative of $F$, $\partial F/\partial\theta$, is vanishing at $T = T_c$, as shown in Figure 2.6). The system has two stable minima, corresponding to $Q = 0$ or $Q \neq 0$ (i.e., the coexistence of the isotropic and nematic phases). On the other hand, for $T = T_c^*$ ($< T_c$), there is only one stable minimum at $Q \neq 0$; this translates into the existence of a single liquid crystalline phase (e.g., nematic or smectic).

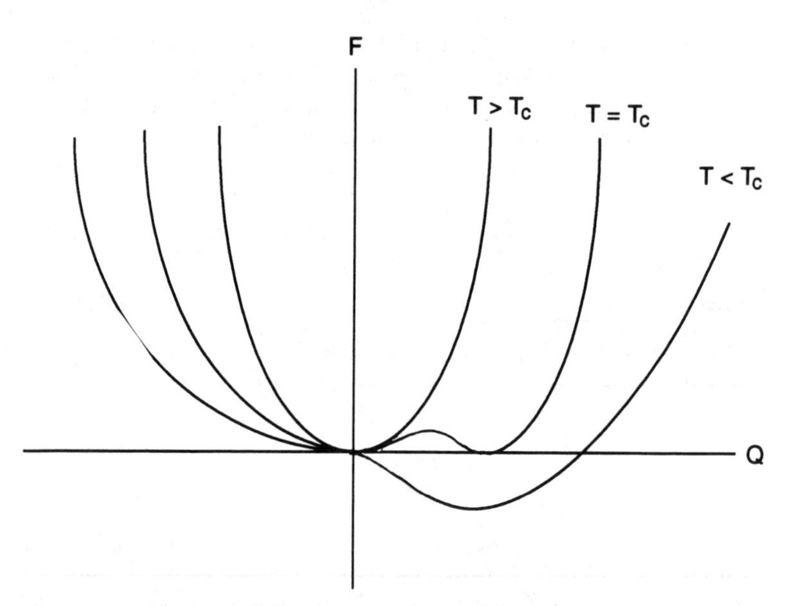


**Figure 2.6** Free energies $F(Q)$ for different temperatures $T$. At $T = T_c$, $\partial F/\partial Q = 0$ at two values of $Q$, where $F$ has two stable minima. On the other hand, at $T = T_c^*$ ($< T_c$), there is only one stable minimum where $\partial F/\partial Q = 0$.

### 2.4.2 Free Energy in the Presence of an Applied Field

In the presence of an externally applied field (e.g., dc or low-frequency electric, magnetic, or optical electric field), a corresponding interaction term should be added to the free energy.

For an applied magnetic field **H**, the energy associated with it is

$$F_{\text{int}} = -\int_0^H \mathbf{M} \cdot d\mathbf{H} \tag{2.25}$$

where **M** is the magnetization given by

$$M_\alpha = \sum_\beta \chi_{\alpha\beta} H_\beta \tag{2.26}$$

Thus

$$F_{\text{int}} = -\frac{1}{2} \sum_\alpha \chi_{\alpha\beta}^m H_\beta H_\alpha \tag{2.27}$$

Using (2.9), we can rewrite $F_{\text{int}}$ as

$$F_{\text{int}} = -\frac{1}{2}\chi_{\perp}^{m} \sum_{\alpha} H_{\alpha} H_{\alpha} - \frac{1}{2} \sum_{\alpha,\beta} \Delta\chi^{m} n_{\alpha} n_{\beta} H_{\alpha} H_{\beta} \tag{2.28}$$

The first term on the right-hand side of (2.28) is independent of the orientation of the (anisotropic) molecules, and it can therefore be included in the constant $F_0$.

On the other hand, the second term is dependent on the orientation of the molecules. Using (2.10) for the order parameter $Q_{\alpha\beta}$, we can write it as

$$F_{\text{int}}^{H} = -\frac{1}{2} \sum_{\alpha,\beta} Q_{\alpha\beta} H_{\alpha} H_{\beta} \tag{2.29}$$

Therefore, the total free energy of a liquid crystal in the isotropic phase, under the action of an externally applied magnetic field, is given by

$$F = F_0' + \frac{1}{2}A(T) \sum_{\alpha,\beta} Q_{\alpha\beta} Q_{\beta\alpha} + \frac{1}{3}B(T) \sum_{\alpha,\beta,\gamma} Q_{\alpha\beta} Q_{\beta\gamma} Q_{\gamma\alpha}$$
$$- \frac{1}{2} \sum_{\alpha,\beta} Q_{\alpha\beta} H_{\alpha} H_{\beta} \tag{2.30}$$

Without solving the problem explicitly, we can infer from the magnetic interaction term that a lower energy state corresponds to some alignment of the molecules in the direction of the magnetic field (for $\Delta\chi^{m} > 0$).

Using a similar approach, we can also deduce that the electric interaction contribution to the free energy is given by (in mks units)

$$F_{\text{int}}^{E} = -\frac{1}{2}\int_{0}^{E} \mathbf{D}\cdot\mathbf{E} = -\frac{1}{2}\varepsilon_{\perp} E^{2} - \frac{1}{2}\Delta\varepsilon(\hat{n}\cdot\mathbf{E})^{2} \tag{2.31}$$

The orientational dependent term is therefore

$$F^{E} = -\frac{1}{2}\Delta\varepsilon(\hat{n}\cdot\mathbf{E})^{2} = -\frac{1}{2}\sum_{\alpha\beta} Q_{\alpha\beta} E_{\alpha} E_{\beta} \tag{2.32}$$

where $Q_{\alpha\beta}$ is defined in (2.4).

In Chapter 6 we will present a detailed discussion of the isotropic phase molecular orientation effects by an applied optical field, from a short intense laser pulse. It is shown that both the response time and the induced order $Q$ depend on the temperature vicinity $(T - T_c)$ in a critical way; they both vary as $(T - T_c)^{-1}$, which becomes very large near $T_c$. This near-$T_c$ critical slowing down behavior of the order parameter $Q$ of the isotropic phase is similar to the slowing down behavior of the order parameter $S$ of the nematic phase discussed in the previous section. Besides the nematic $\leftrightarrow$ isotropic phase transition, which is the most prominent order $\leftrightarrow$ disorder transition exhibited by liquid crystals, there are other equally interesting phase transition processes among the various mesophases (13), such as smectic-A $\leftrightarrow$ smectic-C*, which is discussed in Chapter 4.

## REFERENCES

1. P. G. deGennes, "The Physics of Liquid Crystals." Clarendon Press, Oxford, 1974.
2. P. G. deGennes, *Mol. Cryst. Liq. Cryst.* **12**, 193 (1971).
3. L. D. Landau, *in* "Collected Papers" (D. Ter Haar, ed.). Gordon & Breach, New York, 1965.
4. J. D. Litster, *in* "Critical Phenomena" (R. E. Mills, ed.). McGraw-Hill, New York, 1971.
5. I. C. Khoo and S. T. Wu, "Optics and Nonlinear Optics of Liquid Crystals." World Scientific, Singapore.
6. See, for example, L. M. Blinov, "Electro-Optical and Magneto-Optical Properties of Liquid Crystals." Wiley, Chichester, 1983.
7. W. Maier and A. Saupe, *Z. Naturforsch.* **14A**, 882 (1959); for a concise account of the theory, see Khoo and Wu (5).
8. R. L. Humphries and G. R. Lukhurst, *Chem. Phys. Lett.* **17**, 514 (1972); G. R. Luckhurst, C. Zannoni, P. L. Nordio, and U. Segré, *Mol. Phys.* **30**, 1345 (1975); M. J. Freiser, *Mol. Cryst. Liq. Cryst.* **14**, 165 (1971).
9. L. M. Blinov, V. A. Kizel, V. G. Rumyantsev, and V. V. Titov, *J. Phys.* (*Paris*) **36**, Colloq. C1, C1–C69 (1975); see also Blinov (6).
10. W. H. DeJeu, "Physical Properties of Liquid Crystalline Materials." Gordon & Breach, 1980.
11. I. C. Khoo, R. G. Lindquist, R. R. Michael, R. J. Mansfield, and P. G. LoPresti, *J. Appl. Phys.* **69**, 3853 (1991).
12. I. C. Khoo and R. Normandin, *IEEE J. Quantum Electron.* **QE21**, 329 (1985).
13. See, for example, C. Giannessi, *in* "Physics of Liquid Crystalline Materials" (I. C. Khoo and F. Simoni, eds.). Gordon & Breach, Philadelphia, 1991; see also deGennes (1).

# CHAPTER 3

# NEMATIC LIQUID CRYSTALS

## 3.1 INTRODUCTION

Nematics are the most widely studied liquid crystals. They are also the most widely used. As a matter of fact, nematics best exemplify the dual nature of liquid crystals—fluidity and crystalline structure. To describe their liquid-like properties, one needs to involve hydrodynamics. On the other hand, their crystalline properties necessitate theoretical formalisms pertaining to solids or crystals. To study their optical properties, it is also necessary that we invoke their electronic structures and properties.

In this chapter we discuss all three aspects of nematogen theory: solid-state continuum theory, hydrodynamics, and electrooptical properties, in that order.

## 3.2 ELASTIC CONTINUUM THEORY

### 3.2.1 The Vector Field: Director Axis $\hat{n}(\mathbf{r})$

In elastic continuum theory, introduced and refined over the last several decades by several workers (1–3), nematics are basically viewed as crystalline in form. An aligned sample may thus be regarded as a single crystal, in which the molecules are, on the average, aligned along the direction defined by the director axis $\hat{n}$.

The crystal is uniaxial and is characterized by a tensorial order parameter

$$S_{\alpha\beta} = S(T)\left(n_\alpha n_\beta - \tfrac{1}{3}\delta_{\alpha\beta}\right) \tag{3.1}$$

As a result of externally applied fields, stresses/constraints from the boundary surfaces, the director axis will be "distorted," and, accordingly, the order parameter will also vary spatially. The characteristic length over which significant variation in the order parameter will occur, in most cases, is much larger than the molecular size. Typically, for distortions of the form shown in Figures 3.1*a*–*c*, the characteristic length is on the order of 1 $\mu$m, whereas the molecular dimension is on the order of at most a few tens of angstroms. Under this circumstance, as in other similar systems or media [e.g., ferromagnets ( )], the continuum theory is valid.

The first principle of continuum theory is therefore neglecting the details of the molecular structures. Liquid crystal molecules are viewed as rigid rods; their entire collective behavior may be described in terms of the director axis $\hat{n}(\mathbf{r})$, a vector field. In this picture the spatial variation of the order parameter

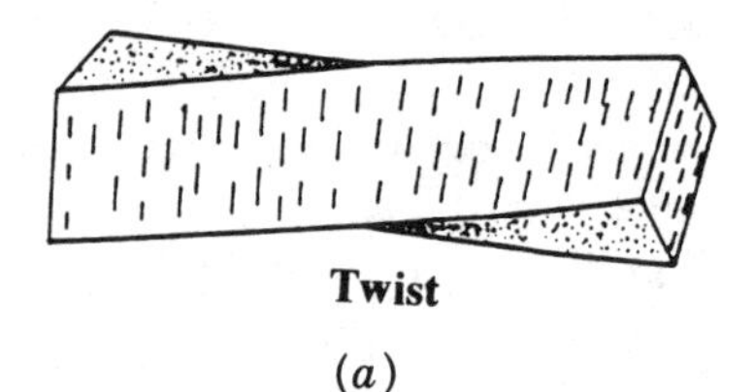


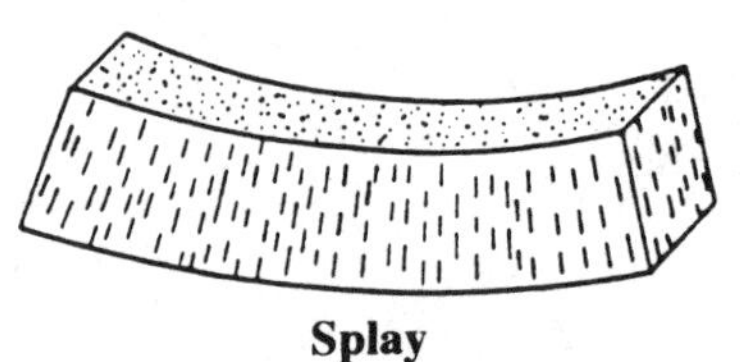


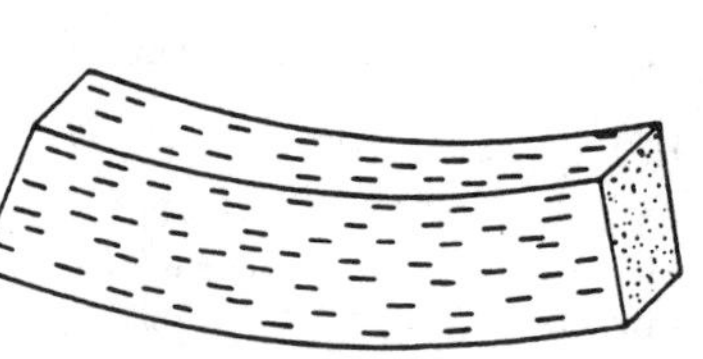


**Figure 3.1** (*a*) Twist deformation in a nematic liquid crystal; (*b*) splay deformation; (*c*) bend deformation.

is described by

$$S_{\alpha\beta}(\mathbf{r}) = S(T)\left[\left(n_\alpha(\mathbf{r})n_\beta(\mathbf{r}) - \tfrac{1}{3}\delta_{\alpha\beta}\right)\right] \tag{3.2}$$

In other words, in a spatially "distorted" nematic crystal, the local optical properties are still those pertaining to a uniaxial crystal and remain unchanged; it is only the orientation (direction) of $\hat{n}$ that varies spatially.

For nematics, the states corresponding to $\hat{n}$ and $-\hat{n}$ are indistinguishable. In other words, even if the individual molecules possess permanent dipoles (actually most liquid crystal molecules do), the molecules are collectively arranged in such a way that the net dipole moment is vanishingly small; that is, there are just as many dipoles up as there are dipoles down in the collection of molecules represented by $\hat{n}$.

### 3.2.2 Elastic Constants, Free Energies, and Molecular Fields

Upon application of an external perturbation field, a nematic liquid crystal will undergo deformation just as any solid. There is, however, an important difference. A good example is shown in Figure 3.1*a*, which depicts a "solid" subjected to torsion, with one end fixed. In ordinary solids this would create a very large stress, arising from the fact that the molecules are translationally displaced by the torsional stress. On the other hand, such twist deformations in liquid crystals, owing to the fluidity of the molecules, simply involve a rotation of the molecules in the direction of the torque; there is no translational displacement of the center of gravity of the molecules, and thus the elastic energy involved is quite small.

Similarly, other types of deformations such as splay and bend deformations, as shown in Figures 3.1*b* and *c*, respectively, involving mainly changes in the director axis $\hat{n}(\mathbf{r})$, will incur much less elastic energy change than the corresponding ones in ordinary solids. It is evident from Figures 3.1*a*–*c* that the splay and bend deformations necessarily involve flow of the liquid crystal, whereas the twist deformation does not. In Section 3.5 we will return to these couplings between flow and director axis deformation.

Twist, splay, and bend are the three principal distinct director axis deformations in nematic liquid crystals. Since the correspond to spatial changes in $\hat{n}(\mathbf{r})$, the basic parameters involved in the deformation energies are various *spatial* derivatives [i.e., curvatures of $n(\mathbf{r})$, such as $\nabla \times \hat{n}(\mathbf{r})$, $\nabla \cdot \hat{n}(\mathbf{r})$, etc.]. Following the theoretical formalism first developed by Frank (1), the free-energy densities (in units of energy per volume) associated with these deformations are given by

$$\text{Splay:} \quad F_1 = \tfrac{1}{2}K_1(\nabla \cdot \hat{n})^2 \tag{3.3}$$

$$\text{Twist:} \quad F_2 = \tfrac{1}{2}K_2(\hat{n} \cdot \nabla \times \hat{n})^2 \tag{3.4}$$

$$\text{Bend:} \quad F_3 = \tfrac{1}{2}K_3(\hat{n} \times (\nabla \times \hat{n}))^2 \tag{3.5}$$

where $K_1$, $K_2$, and $K_3$ are the respective Frank elastic constants.

In general, the three elastic constants are different in magnitude. Typically, they are on the order of $10^{-6}$ dyne in cgs units (or $10^{-11}$ N in mks units). For MBBA (*p*-methoxybenzylidene-*p*′-butylaniline), $K_1$, $K_2$, and $K_3$ are, respectively, $5.8 \times 10^{-7}$ dyne, $3.4 \times 10^{-7}$ dyne, and $7 \times 10^{-7}$ dyne. For almost all nematics $K_3$ is the largest, as a result of the rigid-rod shape of the molecules.

Generally, more than one form of deformation will be induced by an applied external field. If all three forms of deformation are created, the total distortion free-energy density is given by

$$F_d = \tfrac{1}{2}K_1(\nabla\cdot\hat{n})^2 + \tfrac{1}{2}K_2(\hat{n}\cdot\nabla\times\hat{n})^2 + \tfrac{1}{2}K_3(\hat{n}\times\nabla\times\hat{n})^2 \tag{3.6}$$

This expression, and the resulting equations of motion and analysis, can be greatly simplified if one makes a frequently used assumption, namely, the one-constant approximation ($K_1 = K_2 = K_3 = K$). In this case (3.6) becomes

$$F_d = \tfrac{1}{2}K\left[(\nabla\cdot n)^2 + (\nabla\times\hat{n})^2\right] \tag{3.7}$$

Equation (3.6), or its simplified version, describes the deformation of the director axis vector field $\hat{n}(\mathbf{r})$ in the bulk of the nematic liquid crystal. A complete description should include the surface interaction energy at the nematic liquid crystal cell boundaries. Accounting for this, the total energy density of the system should be

$$F'_d = F_d + F_{\text{surface}} \tag{3.8}$$

where the surface energy term is dependent on the surface treatment.

In other words, the equilibrium configuration of the nematic liquid crystal is obtained by a minimization of the total free energy of the system, $F_{\text{total}} = \int F'_d\, dV$. If external fields (electric, magnetic, or optical) are applied, the corresponding free-energy terms (see the following sections) will be added to the total free-energy expression.

Under the so-called "hard-boundary" condition, in which the liquid crystal molecules are strongly anchored to the boundary and do not respond to the applied perturbation fields (cf. Figure 3.2), the surface energy may thus be regarded as a constant; the surface interactions therefore do not enter into the dynamical equations describing the field-induced effects in nematic liquid crystals.

On the other hand, if the molecules are not strongly anchored to the boundary, that is, the so-called "soft-boundary" condition (Figure 3.3), an applied field will perturb the orientation of the molecules at the cell boundaries. In this case a quantitative description of the dynamics of the field-induced effects must account for these surface energy terms. A good account

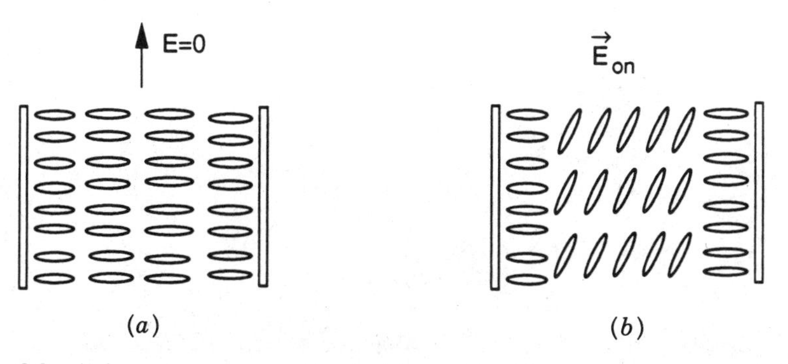


**Figure 3.2** A homeotropic nematic liquid crystal with strong surface anchoring: (*a*) external field off; (*b*) External field on—only the bulk director axis is deformed.

of surface energy interactions may be found in Barbero et al. (4) which treat the case of optical field-induced effects in a hybrid aligned nematic liquid crystal cell.

From (3.6) for the free energy, one can obtain the corresponding so-called molecular fields **f** using the Lagrange equation (3). In spatial coordinate component form, we have

$$f_\alpha = \frac{-\partial F}{\partial n_\beta} + \sum_\beta \frac{\partial}{\partial x_\beta} \frac{\partial F}{\partial g_{\beta\alpha}} \tag{3.9}$$

where

$$g_{\beta\alpha} = \frac{\partial n_\beta}{\partial x_\alpha} \qquad \alpha, \beta = x, y, z \tag{3.10}$$

More explicitly, (3.9) gives, for the total molecular field association with splay, twist, and bend deformations,

$$\mathbf{f} = \mathbf{f}_1 + \mathbf{f}_2 + \mathbf{f}_3 \tag{3.11}$$

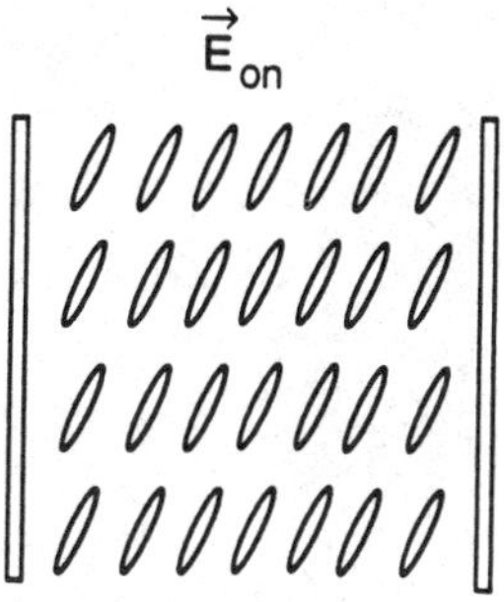

$\vec{E}_{on}$

**Figure 3.3** Soft-boundary condition. Under the action of an applied field, both the surface and bulk director axes are distorted.

where

$$\text{Splay:} \quad \mathbf{f}_1 = K_1 \, \nabla(\nabla \cdot \hat{n}) \tag{3.12}$$

$$\text{Twist:} \quad \mathbf{f}_2 = -K_2[A \nabla \times \hat{n} + \nabla \times (A\hat{n})] \tag{3.13}$$

$$\text{Bend:} \quad \mathbf{f}_3 = K_3[\mathbf{B} \times (\nabla \times \hat{n}) + \nabla \times (\hat{n} \times \mathbf{B})] \tag{3.14}$$

with $A = \hat{n} \cdot (\nabla \times \hat{n})$ and $\mathbf{B} = \hat{n} \times (\nabla \times \hat{n})$.

## 3.3 DIELECTRIC CONSTANTS AND REFRACTIVE INDICES

Dielectric constants and refractive indices, as well as electrical conductivities of liquid crystals, are physical parameters characterizing the electronic responses of liquid crystals to externally applied fields (electric, magnetic, or optical). Because of the molecular and energy level structures of nematic molecules, these responses are highly dependent on the direction and the frequencies of the field. Accordingly, we shall classify our studies of dielectric permittivities and other electrooptical parameters into two distinctive frequency regimes: (1) dc and low frequency and (2) optical frequency. Where the transition from regime (1) to (2) occurs, of course, is governed by dielectric relaxation processes and the dynamical time constant; typically, the Debye relaxation frequency in nematics is on the order of $10^{10}$ Hz.

### 3.3.1 dc and Low Frequency Dielectric Permittivities, Conductivities, and Magnetic Susceptibility

The dielectric constant $\varepsilon$ is defined by the Maxwell equation (5):

$$\mathbf{D} = \bar{\bar{\varepsilon}} : \mathbf{E} \tag{3.15a}$$

where $\mathbf{D}$ is the displacement current, $\mathbf{E}$ is the electric field, and $\bar{\bar{\varepsilon}}$ is a tensor. For a uniaxial nematic liquid crystal, we have

$$\bar{\bar{\varepsilon}} = \begin{bmatrix} \varepsilon_\perp & 0 & 0 \\ 0 & \varepsilon_\perp & 0 \\ 0 & 0 & \varepsilon_\parallel \end{bmatrix} \tag{3.15b}$$

Equations (3.15$a$) and (3.15$b$) yield, for the two principal axes,

$$D_\parallel = \varepsilon_\parallel E_\parallel \tag{3.16}$$

and

$$D_\perp = \varepsilon_\perp E_\perp \tag{3.17}$$

Typical values of $\varepsilon_{\parallel}$ and $\varepsilon_{\perp}$ are on the order of $5\varepsilon_0$, where $\varepsilon_0$ is the permittivity of free space.

Similarly, the electric conductivities $\sigma_{\parallel}$ and $\sigma_{\perp}$ of nematics are defined by

$$J_{\parallel} = \sigma_{\parallel} E_{\parallel} \tag{3.18}$$

and

$$J_{\perp} = \sigma_{\perp} E_{\perp} \tag{3.19}$$

where $J_{\parallel}$ and $J_{\perp}$ are the currents flowing along, and perpendicularly to, the director axis, respectively. In conjunction with an applied dc electric field, the conductivity anisotropy could give rise to space charge accumulation and space charge fields (6). These space charge field and applied dc fields could create strong director axis reorientation in a nematic film, giving rise to an orientational photorefractive effect as observed recently by Khoo et al. [cf. Section 10.6]

Most nematics (e.g., E7, 5CB, etc., $\varepsilon_{\parallel} > \varepsilon_{\perp}$) are said to possess positive (dielectric) anisotropy. On the other hand, some nematics, such as MBBA, possess negative anisotropy (i.e., $\varepsilon_{\perp} < \varepsilon_{\parallel}$). The controlling factors are the molecular constituents and structures.

In general, $\varepsilon_{\parallel}$ and $\varepsilon_{\perp}$ have different dispersion regions, as shown in Figure 3.4 for 4-methoxy-4′-*n*-butylazoxy-benzene (7), which possesses negative dielectric anisotropy ($\Delta\varepsilon < 0$). Also plotted in Figure 3.4 is the dispersion of $\varepsilon_{\text{iso}}$, the dielectric constant for the isotropic case. Notice that for frequencies of $10^9$ Hz or less, $\varepsilon_{\perp} > \varepsilon_{\parallel}$. At higher frequencies and in the optical regime, $\varepsilon_{\parallel} > \varepsilon_{\perp}$ (i.e., the dielectric anisotropy changes sign).

For some nematic liquid crystals this changeover in the sign of $\Delta\varepsilon = \varepsilon_{\parallel} - \varepsilon_{\perp}$ occurs at a much lower frequency [cf. Figure 3.5 for phenylbenzoates (8). This changeover frequency $f_{\text{co}}$ is lower because of the long three-ringed molecular structure, which is highly resistant to rotation of the molecules around the short axes.

For electrooptical applications, the dielectric relaxation behavior of $\varepsilon_{\parallel}$ and $\varepsilon_{\perp}$ for the different classes of nematic liquid crystals, and the relationships between the molecular structures and the dielectric constant, are obviously very important. This topic, however, is beyond the scope of this

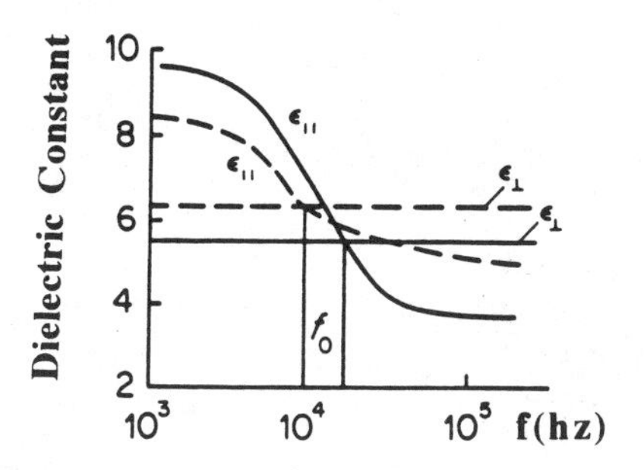


**Figure 3.4** Dispersion data of the dielectric constants $\varepsilon_{\parallel}$ and $\varepsilon_{\perp}$ for the nematic and isotropic phase of the liquid crystal 4-methoxy-4′-*n*-butylazoxy-benzene.

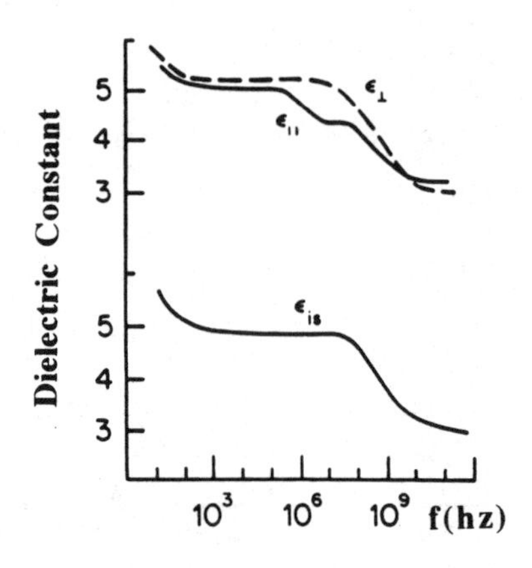


**Figure 3.5** Dispersion data of the dielectric constants $\varepsilon_{\parallel}$ and $\varepsilon_{\perp}$ for the nematic and isotropic phases of liquid crystal phenylbenzoates.

chapter, and the reader is referred to Blinov (8) and Khoo and Wu (9) and the references quoted therein for more detailed information.

Pure organic liquids are dielectric [i.e., nonconducting ($\sigma = 0$)]. The electric conductivities of liquid crystals are due to some impurities or ions. Generally, $\sigma_{\parallel}$ is larger than $\sigma_{\perp}$. Electrical conduction plays an important role in electrooptical applications of liquid crystals in terms of stability and instability, chemical degradation and lifetime of the device, (9). Typically, $\sigma_{\parallel}$ and $\sigma_{\perp}$ are on the order of $10^{-10}$ $S^{-1}$-$cm^{-1}$ for pure nematics. As in almost all materials, these conductivities could be varied by several orders of magnitude with the use of appropriate dopants.

The magnetic susceptibility of a material is defined in terms of the magnetization **M**, the magnetic induction **B**, and the magnetic field strength **H** by

$$\mathbf{M} = \frac{\mathbf{B}}{\mu_0} - \mathbf{H} = \bar{\bar{\chi}}_m : \mathbf{H} \tag{3.20}$$

and

$$\mathbf{B} = \mu_0 \left(1 + \bar{\bar{\chi}}_m\right) : \mathbf{H} \tag{3.21}$$

The magnetic susceptibility tensor $\bar{\bar{\chi}}_m$ is anisotropic. For a uniaxial material such as a nematic, the magnetic susceptibility takes the form

$$\bar{\bar{\chi}}_m = \begin{bmatrix} \chi_{\perp}^m & 0 & 0 \\ 0 & \chi_{\perp}^m & 0 \\ 0 & 0 & \chi_{\parallel}^m \end{bmatrix} \tag{3.22}$$

Note that this is similar to the dielectric constants $\bar{\bar{\varepsilon}}$.

Nematic liquid crystals, in fact, liquid crystals in general, are diamagnetic. Therefore, $\chi_{\parallel}^m$ and $\chi_{\perp}^m$ are small and negative, on the order of $10^{-5}$ (in SI units). As a result of the smallness of these magnetic susceptibilities, the magnetic interactions among the molecules comprising the liquid crystal are small (in comparison with their interactions with the external applied field). Consequently, the local field acting on the molecules differs very little from the external field, and, in general, magnetic measurements are the preferred way to studying liquid crystal order parameters and other physical processes.

### 3.3.2 Free Energy and Torques by Electric and Magnetic Fields

In this section we consider the interactions of nematic liquid crystals with applied fields (electric or magnetic); we will limit our discussion to only dielectric and diamagnetic interactions.

For a general applied (dc, low-frequency, or optical) electric field **E**, the displacement **D** may be written in the form

$$\mathbf{D} = \varepsilon_{\perp} \mathbf{E} + (\varepsilon_{\parallel} - \varepsilon_{\perp})(\hat{n} \cdot \mathbf{E})\hat{n} \tag{3.23}$$

The electric interaction energy density is therefore

$$u_E = -\int_0^E \mathbf{D} \cdot \mathbf{E} = -\frac{1}{2}\varepsilon_{\perp}(\mathbf{E} \cdot \mathbf{E}) - \frac{\Delta\varepsilon}{2}(\hat{n} \cdot \mathbf{E})^2 \tag{3.24}$$

Note that the first term on the right-hand side of (3.24) is independent of the orientation of the director axis. It can therefore be neglected in the director axis deformation energy. Accordingly, the free-energy density term associated with the application of an electric field is given by

$$F_E = -\frac{\Delta\varepsilon}{2}(\hat{n} \cdot \mathbf{E})^2 \tag{3.25}$$

in SI units [in cgs units, $F_E = -\Delta\varepsilon/8\pi(n \cdot E)^2$]. In analogy to the molecular fields considered in the preceding section, one could associate a molecular torque with the electric field given by

$$\begin{aligned} \mathbf{\Gamma}_E &= \mathbf{D} \times \mathbf{E} \\ &= \Delta\varepsilon(\hat{n} \cdot \mathbf{E})(\hat{n} \times \mathbf{E}) \end{aligned} \tag{3.26}$$

Similar considerations for the magnetic field yield a magnetic energy density term $U_m$ given by

$$\begin{aligned} U_m &= -\int_0^M \mathbf{B} \cdot dM \\ &= \frac{1}{2\mu_0}\chi_{\perp}^m B^2 - \frac{1}{2\mu_0}\Delta\chi^m(\mathbf{B} \cdot \hat{n})^2 \end{aligned} \tag{3.27}$$

a magnetic free-energy density (associated with director axis reorientation)

$F_m$ given by

$$F_m = \frac{1}{2\mu_0} \Delta\chi^m (\mathbf{B}\cdot\hat{n})^2 \tag{3.28}$$

and a magnetic torque density

$$\begin{aligned}\mathbf{\Gamma}_H &= \mathbf{M}\times\mathbf{H} \\ &= \Delta\chi^m(\hat{n}\cdot\mathbf{H})(\hat{n}\times\mathbf{H})\end{aligned} \tag{3.29}$$

These electric and magnetic torques play a central role in various field-induced effects in liquid crystals.

## 3.4 OPTICAL DIELECTRIC CONSTANTS AND REFRACTIVE INDICES

### 3.4.1 Linear Susceptibility and Local Field Effect

In the optical regime, $\varepsilon_{\parallel} > \varepsilon_{\perp}$. Typically, $\varepsilon_{\parallel}$ is on the other order of $2.89\varepsilon_0$ and $\varepsilon_{\perp}$ is $2.25\varepsilon_0$. This corresponds to refractive indices $n_{\parallel} = 1.7$ and $n_{\perp} =$ 1.5. An interesting property of nematic liquid crystals is that such a large birefringence ($\Delta n = n_{\parallel} - n_{\perp} \approx 0.2$) is manifested throughout the whole optical spectral regime [from near ultraviolet ($\approx 400$ nm), to visible ($\approx 500$ nm) and near infrared (1–3 $\mu$m), to the infrared regime (8–12 $\mu$m), i.e., from 400 nm to 12 $\mu$m]. Figure 3.6 shows the measured birefringence of three typical nematic liquid crystals from the UV to the far infrared ($\lambda = 16$ $\mu$m).

The optical dielectric constants originate from the linear polarization **P** generated by the incident optical field $E_{\text{op}}$ on the nematic liquid crystal:

$$\mathbf{P} = \varepsilon_0 \bar{\bar{\chi}}\cdot\mathbf{E} \tag{3.30a}$$

From the defining equation

$$\mathbf{D} = \varepsilon_0\mathbf{E} + \mathbf{P} = \bar{\bar{\varepsilon}} : \mathbf{E} \tag{3.30b}$$

we have

$$\bar{\bar{\varepsilon}} = \varepsilon_0(1 + \bar{\bar{\chi}}^{(1)}), \tag{3.30c}$$

Here $\bar{\bar{\chi}}^{(1)}$ is the linear (sometimes termed first order) susceptibility tensor of the nematics. $\bar{\bar{\chi}}^{(1)}$ is a macroscopic parameter, and is related to the microscopic (molecular) parameter, the molecular polarizabilities tensor $\alpha_{ij}$, in the

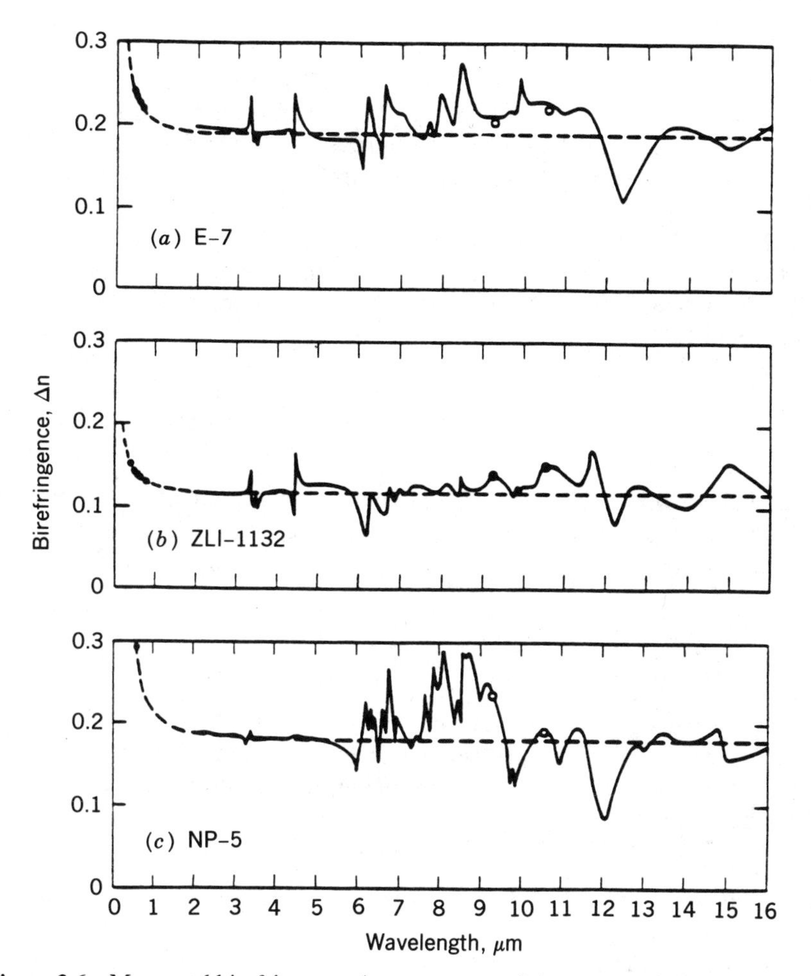


**Figure 3.6** Measured birefringence $\Delta\varepsilon = \varepsilon_{\parallel} - \varepsilon_{\perp}$ of three nematic liquid crystals.

following way:

$$d_i = \alpha_{ij} E_j^{\text{loc}}$$

$$\mathbf{d} = \bar{\bar{\alpha}} : \mathbf{E}^{\text{loc}} \qquad (3.31a)$$

$$\mathbf{P} = N\mathbf{d} \qquad (3.31b)$$

where $d_i$ is the $i$th component of the induced dipole $\mathbf{d}$ and $N$ is the number density. In Chapter 8 a rigorous quantum mechanical derivation of $\alpha$ in terms of the dipole matrix elements or oscillator strengths and the energy levels and level populations will be presented.

The connection between the microscopic parameter $\alpha_{ij}$ and the macroscopic parameter $\chi_{ij}$ is the local field correction factor (i.e., the difference between the externally applied field and the actual field as experienced by the molecules). Several theoretical formalisms have been developed to evaluate the field correction factor, ranging from simplified to complex and sophisticated ones.

Most of the approaches for obtaining the local field correction factor are based on the Lorentz results (5), which state that the internal field (i.e., the local field as experienced by a molecule $\mathbf{E}^{\text{loc}}$ in a solid) is related to the applied field $\mathbf{E}^{\text{app}}$ by

$$\mathbf{E}^{\text{loc}} = \frac{n^2+2}{3}\mathbf{E}^{\text{app}} \tag{3.32a}$$

In particular, Vuks (10) analyzed experimental data and proposed that the local field in an anisotropic crystal may be taken as isotropic and expressed in the form

$$E^{\text{loc}} = \frac{\langle n^2\rangle+2}{3}\langle E\rangle \tag{3.32b}$$

where $\langle n^2\rangle = \frac{1}{3}(n_x^2+n_y^2+n_z^2)$ and $n_x$, $n_y$, and $n_z$ are the principal refractive indices of the crystal. This approach has been employed in the study of liquid crystals (11). A more generalized expression for anisotropic crystals is given in Dunmur (12):

$$E_i^{\text{loc}} = \left[1+\bar{\bar{L}}_{ii}(n_i^2-1)\right]E_i^{\text{app}} \qquad i = x, y, z \tag{3.33}$$

More generally, one can write (3.33) as

$$\mathbf{E}^{\text{loc}} = \bar{\bar{K}}:\mathbf{E}^{\text{app}} \tag{3.34}$$

where $\bar{\bar{K}}$, the local field correction "factor" is a second-rank tensor, which states that the local field $\mathbf{E}^{\text{loc}}$ is linearly related to the macroscopic applied field $\mathbf{E}^{\text{app}}$. In general, experimental measurements show that Vuks and Dunmur's treatments are qualitatively in agreement.

### 3.4.2 Equilibrium Temperature and Order Parameter Dependence of Refractive Indices

In two principal refractive indices $n_\perp$ and $n_\parallel$ of a uniaxial liquid crystal and the anisotropy $n_\parallel - n_\perp$ have been the subject of intensive studies for their fundamental importance in the understanding of liquid crystal physics and for their vital roles in applied electrooptic devices. Since it is the dielectric constants ($\varepsilon_{\parallel}$ and $\varepsilon_{\perp}$) that enter directly and linearly into the constitutive equations (3.30$a$) to (3.30$c$), it is theoretically more convenient to discuss the fundamentals of these temperature dependences in terms of the dielectric constants.

From (3.34) for the local field $\mathbf{E}^{\text{loc}}$ and (3.31) for the induced dipole moments, we can express the polarization $\mathbf{P}$ ($\equiv N\mathbf{d}$) by

$$\mathbf{P} = N\bar{\bar{\alpha}}:(\bar{\bar{K}}:\mathbf{E}) \tag{3.35}$$

where $\bar{\bar{\alpha}}$ is the polarizability tensor of the molecule, $N$ is the number of molecules per unit volume, and the parentheses denote averaging over the orientations of all molecules.

The dielectric constant $\bar{\bar{\varepsilon}}$ (in units if $\varepsilon_0$) is therefore given by

$$\bar{\bar{\varepsilon}} = 1 + \left(\frac{N}{\varepsilon_0}\right)\bar{\bar{\alpha}}:\bar{\bar{K}} \tag{3.36}$$

and

$$\begin{aligned}\Delta\varepsilon &= \varepsilon_{\parallel} - \varepsilon_{\perp} \\ &= \frac{N}{\varepsilon_0}\left(\langle\bar{\bar{\alpha}}:\bar{\bar{K}}\rangle_{\parallel} - \langle\bar{\bar{\alpha}}:\bar{\bar{K}}\rangle_{\perp}\right)\end{aligned} \tag{3.37}$$

From these considerations and the observation by deJeu and Bordewijk (13) that

$$\Delta\varepsilon \propto \rho S \tag{3.38}$$

and

$$\langle\bar{\bar{\alpha}}\cdot\bar{\bar{K}}\rangle_{\parallel} - \langle\bar{\bar{\alpha}}\cdot\bar{\bar{K}}\rangle_{\perp} \sim S \tag{3.39}$$

we can write $\varepsilon_{\parallel} = \varepsilon_{\perp}$ as

$$\varepsilon_{\parallel} = n_{\parallel}^2 = 1 + \left(\frac{N}{3\varepsilon_0}\right)\left[\alpha_l k_l(2S+1) + \alpha_t k_t(2-2S)\right] \tag{3.40}$$

and

$$\varepsilon_{\perp} = n_{\perp}^2 = 1 + \left(\frac{N}{3\varepsilon_0}\right)\left[\alpha_l k_l(1-S) + \alpha_t k_t(2+S)\right] \tag{3.41}$$

respectively, where $k_l$ and $k_t$ are the values of $\bar{\bar{K}}$ along the principal axis and $S$ is the order parameter.

One can rewrite (3.40) and (3.41) as

$$\varepsilon_{\parallel} = \varepsilon_l + \tfrac{2}{3}\Delta\varepsilon \tag{3.42}$$

and

$$\varepsilon_{\perp} = \varepsilon_l - \tfrac{1}{3}\Delta\varepsilon \tag{3.43}$$

where

$$\begin{aligned}\Delta\varepsilon &= \left(\frac{N}{\varepsilon_0}\right)[\alpha_l k_l - \alpha_t k_t]S \\ &= \frac{N_A\rho}{\varepsilon_0 M}(\alpha_l K_l - \alpha_t K_t) \sim \rho S\end{aligned} \tag{3.44}$$

and

$$\varepsilon_l = 1 + \frac{N_A\rho}{3\varepsilon_0 M}(\alpha_l K_l + 2\alpha_t k_t) \tag{3.45}$$

Notice that we have replaced $N$ by $(N_A\rho)/M$, where $N_A$ is Avogardo's number, $\rho$ is the density, and $M$ is the mass number.

The final explicit forms of the $\varepsilon$'s depend on the determination of the internal field tensor. However, it is important to note that, in terms of the temperature dependence of the $\varepsilon$'s (14),

$$\varepsilon_l \sim 1 + \text{const}\,\rho = 1 + C_1\rho \tag{3.46}$$

and

$$\Delta\varepsilon \sim \text{const}\,\rho S = C_2\rho S \tag{3.47}$$

In other words, the temperature ($T$) dependence of $\varepsilon_{\parallel}$ and $\varepsilon_{\perp}$ (and the corresponding refractive indices $n_{\parallel}$ and $n_{\perp}$) are through the dependences of $\rho$ and $S$ on $T$.

One of the most striking features of the temperature dependence of the refractive indices of nematic liquid crystals is that the thermal index gradients ($dn_{\parallel}/dT$ and $dn_{\perp}/d_T$) become extraordinarily large near the phase transition temperature (Figures 3.7*a* and *b*). From (3.46), (3.47), (3.40) and (3.41), we can obtain $dn_{\parallel}/dT$ and $dn_{\perp}/dT$ as

$$\frac{dn_{\parallel}}{dT} = \frac{1}{n_{\parallel}}\left(C_1\frac{d\rho}{dT} + \frac{2}{3}C_2 S\frac{d\rho}{dT} + \frac{2}{3}C_2\rho\frac{dS}{dT}\right) \tag{3.48}$$

$$\frac{dn_{\parallel}}{dT} = \frac{1}{n_{\perp}} - \left(C_1\frac{d\rho}{dT} - \frac{1}{3}C_2 S\frac{d\rho}{dT} - \frac{1}{3}C_2\rho\frac{dS}{dT}\right) \tag{3.49}$$

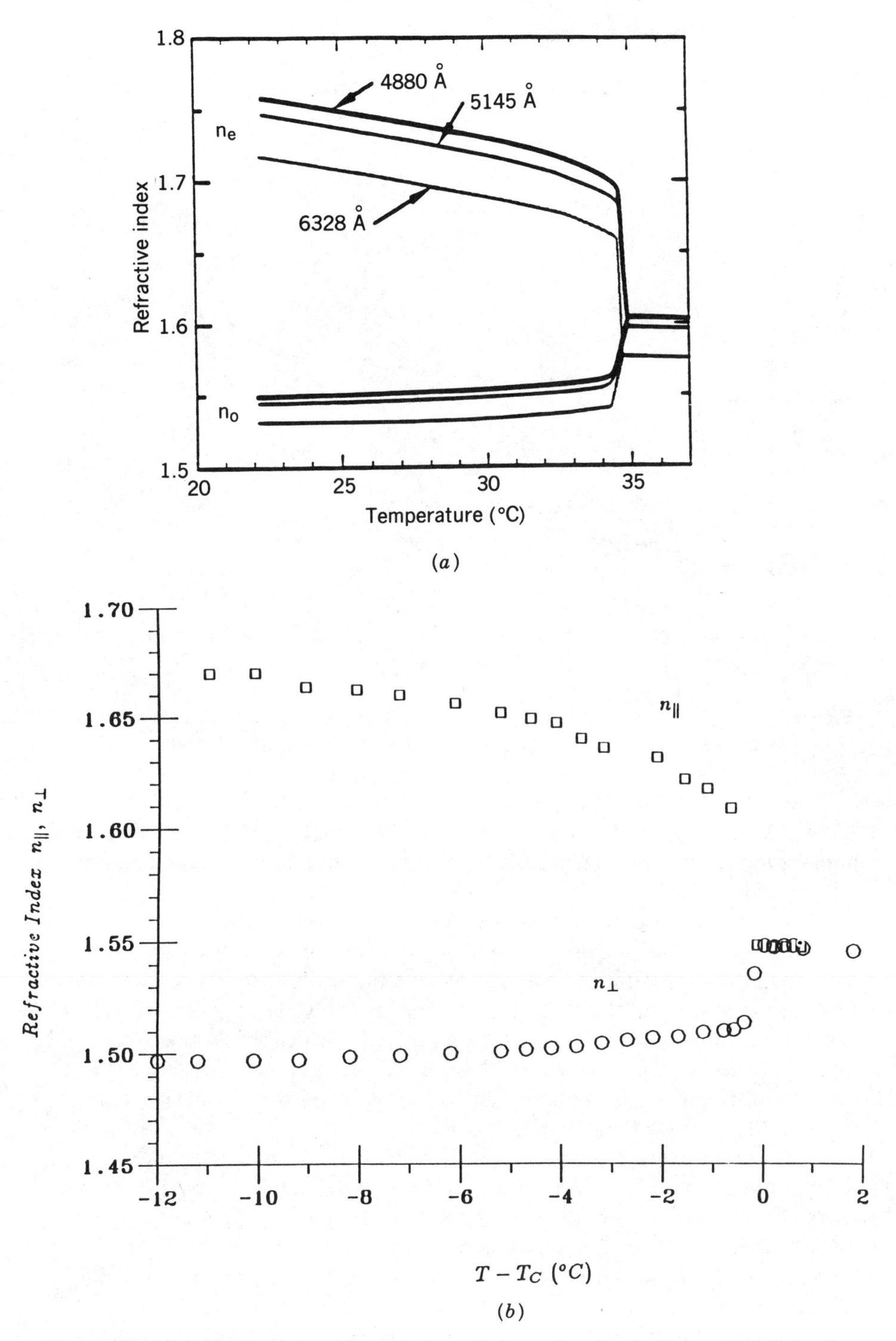


**Figure 3.7** (*a*) Temperature dependence of the refractive indices of 5CB in the visible region. (*b*) Temperature dependence of the refractive indices of 5CB in the infrared (10.6 $\mu$m) region.

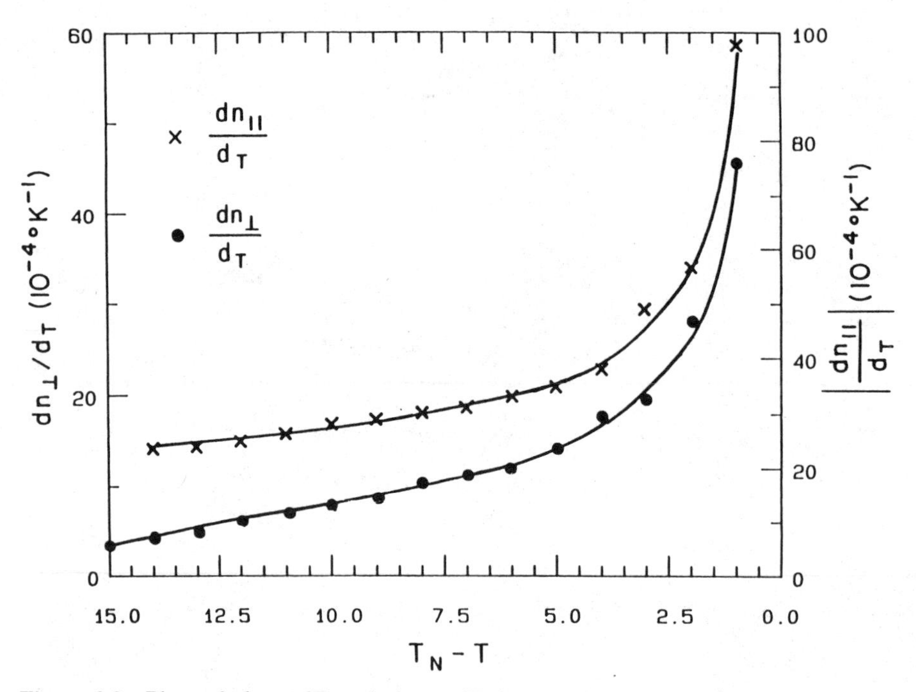


**Figure 3.8** Plots of $dn_{\parallel}/dT$ and $dn_{\perp}/dT$ for the liquid crystal [after Khoo and Normandin (14) and Horn (15)]. The solid curves are for visual aid only.

Figure 3.8 shows the plot of $dn_{\parallel}/dT$ and $dn_{\perp}/dT$ for 5CB as a function of temperature, with experimental data deduced from a more detailed measurement by Horn (15).

Studies of the optical refractive indices of liquid crystals, as presented previously, are traditionally confined to what one may term as the classical and steady-state regime. In this regime the molecules are assumed to be in the ground state, and the optical field intensity is a stationary one. Results or conclusions obtained from such an approach, which have been outlined previously and in the next section, have to be put in the proper context when these fundamental assumptions about the state of the molecules and the applied field are no longer true.

Detailed theories dealing with these quantum mechanical, nonlinear, or transient optical effects are given in Chapters 6 and 8. As an example consider the expression for the (linear) molecular polarizability given in (8.28). Note that the refractive indices of an excited molecule are completely different from those associated with a molecule in the ground state. These differences are due to the fact that a totally different set of dipole matrix elements $d_{ij}$ and frequency denominators $(\omega_i - \omega_j)$ are involved.

Second, if the intensities of impinging optical fields are fast oscillatory (e.g., picosecond laser pulses) and their time durations are comparable to the internal relaxation dynamics of the molecules, these optical fields will "see" the transient responses of the molecule. These transient responses in the internal motions (sometimes termed internal temperature) of the molecules are usually manifested in the form of spectral shifts; that is, the emission and absorption spectra of the molecules are momentarily shifted, usually in the picosecond time scale. Accordingly, the "effective" molecular polarizabilities, which translate into the refractive indices, will also experience a time-dependent change. These transient changes in the refractive index associated with the molecular excitations under ultrashort laser pulses should be clearly distinguished from the usual temperature effects associated with the stationary or equilibrium state. In the stationary case the fast molecular transients have relaxed, and the absorbed energy has been converted to an overall rise in the bulk temperature.

As remarked earlier in Chapter 1, even in the so-called stationary situation where the internal molecular excitations have relaxed, the order parameter $S$ may still not attain the equilibrium state. This usually happens in the nanosecond or microsecond time scale. Details of these considerations are given in Chapter 7 where we discuss pulsed laser-induced heating effects.

## 3.5 FLOWS AND HYDRODYNAMICS

One of the most striking properties of liquid crystals is their ability to flow freely while exhibiting various anisotropic and crystalline properties. It is this dual nature of liquid crystals that makes them very interesting materials to study; it also makes the theoretical formalism very complex.

The main feature that distinguishes liquid crystals in their ordered mesophases (e.g, the nematic phase) from ordinary fluids is that their

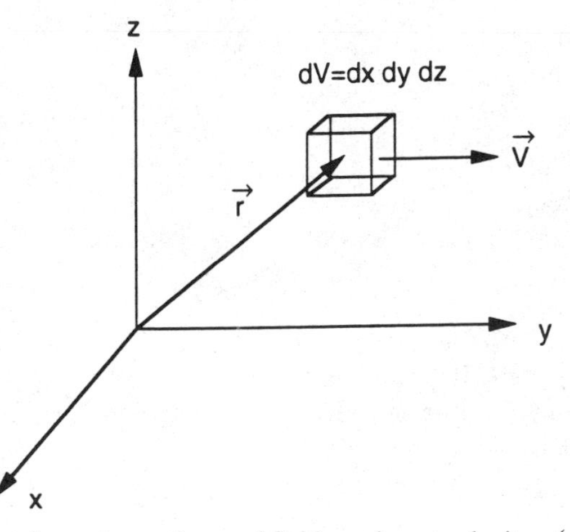


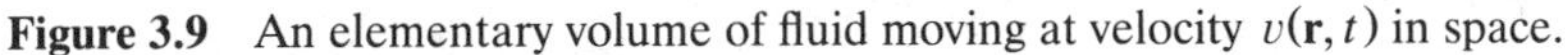
**Figure 3.9** An elementary volume of fluid moving at velocity $v(\mathbf{r}, t)$ in space.

physical properties are dependent on the orientation of the director axis $n(\mathbf{r})$. These orientation flow processes are necessarily coupled, except in very unusual cases (e.g., pure twist deformation). Therefore, studies of the hydrodynamics of liquid crystals will involve a great deal more (anisotropic) parameters, than in ordinary liquids.

We begin our discussion by reviewing first the hydrodynamics of an ordinary fluid. This is followed by a discussion of the general hydrodynamics of liquid crystals. Specific cases involving a variety of flow-orientational couplings are then treated.

### 3.5.1 Hydrodynamics of Ordinary Isotropic Fluids

Consider an elementary volume $dV = dx\,dy\,dz$ of a fluid moving in space as shown in Figure 3.9. The following parameters are needed to describe its dynamics:

Position vector: $\mathbf{r}$

Velocity: $\mathbf{v}(\mathbf{r}, t)$

Density: $\rho(\mathbf{r}, t)$

Pressure: $p(\mathbf{r}, t)$

Forces in general: $\mathbf{f}$

In later chapters where we study laser-induced acoustic (sound, density) waves in liquid crystals, or generally when one deals with acoustic waves, it is necessary to assume that the density $\rho(\mathbf{r}, t)$ is a spatially and temporally varying function. In this chapter, however, we "decouple" such density wave excitation from all the processes under consideration and basically limit our attention to the flow and orientational effects of an *incompressible* fluid. In this case we have

$$\rho(\mathbf{r}, t) = \text{const} \tag{3.50}$$

For all liquids, in fact, for all gas particles or charges in motion, the equation of continuity also holds

$$\nabla \cdot (\rho \mathbf{v}) = -\frac{\partial \rho}{\partial t} \tag{3.51}$$

This equation states that the total variation of $\rho \mathbf{v}$ over the surface of an enclosing volume is equal to the rate of decrease of the density. Since $\partial \rho / \partial t = 0$, we thus have, from (3.51),

$$\nabla \cdot \mathbf{v} = 0 \tag{3.52}$$

The equation of motion describing the acceleration **a** of the fluid elements is simply Newton's law:

$$\rho \frac{d\mathbf{v}}{dt} = \mathbf{f} \tag{3.53a}$$

Studies of the hydrodynamics of liquids may be said to be centered around this equation of motion, as we identify all the various origins and mechanisms of forces acting on the fluid elements and attempt to solve for their motion in time and space.

We shall start with the left-hand side of (3.53$a$). Since $\mathbf{v} = \mathbf{v}(\mathbf{r}, t)$,

$$\frac{d\mathbf{v}}{dt} = \frac{\partial \mathbf{v}}{\partial t} + (\nabla \cdot \mathbf{v})\mathbf{v} \tag{3.53b}$$

The force on the right-hand side of (3.53$a$) comes from a variety of sources, including the pressure gradient $-\nabla p$), viscous forces $\mathbf{f}_{\text{vis}}$, and external fields (electric, magnetic, optical, gravitational, etc.) $\mathbf{f}_{\text{ext}}$.

Equation (3.53$a$) thus becomes

$$\rho\left[\frac{\partial \mathbf{v}}{\partial t} + (\mathbf{v} \cdot \nabla)\mathbf{v}\right] = -\nabla p + \mathbf{f}_{\text{vis}} + \mathbf{f}_{\text{ext}} \tag{3.54}$$

Let us ignore the external field for the moment. The formulation of the equation of motion for a fluid element is complete once we identify the viscous forces. Note that, in analogy to the pressure gradient term, the viscous force $\mathbf{f}_{\text{vis}}$ is the space derivation of a quantity which has the unit of pressure (i.e., force per unit). Such a quantity is termed the stress tensor $\sigma$ (i.e., the force is caused by the gradient in the stress, cf. Figure 3.10). For example, the $\alpha$ component of **f** may be expressed as

$$f_\alpha = \frac{\partial}{\partial x_\alpha} \sigma_{\alpha\beta} \tag{3.55}$$

Accordingly, we may rewrite (3.54) as

$$\rho\left(\frac{\partial v_\alpha}{\partial t} + v_\beta \frac{\partial v_\alpha}{\partial x_\beta}\right) = -\frac{\partial p}{\partial x_\alpha} + \frac{\partial \sigma_{\alpha\beta}}{\partial x_\beta} \tag{3.56}$$

where summation over repeated indices is implicit.

By consideration of the fact that there are no forces acting when the fluid velocity is a constant, the stress tensor is taken to be linear in the gradients of

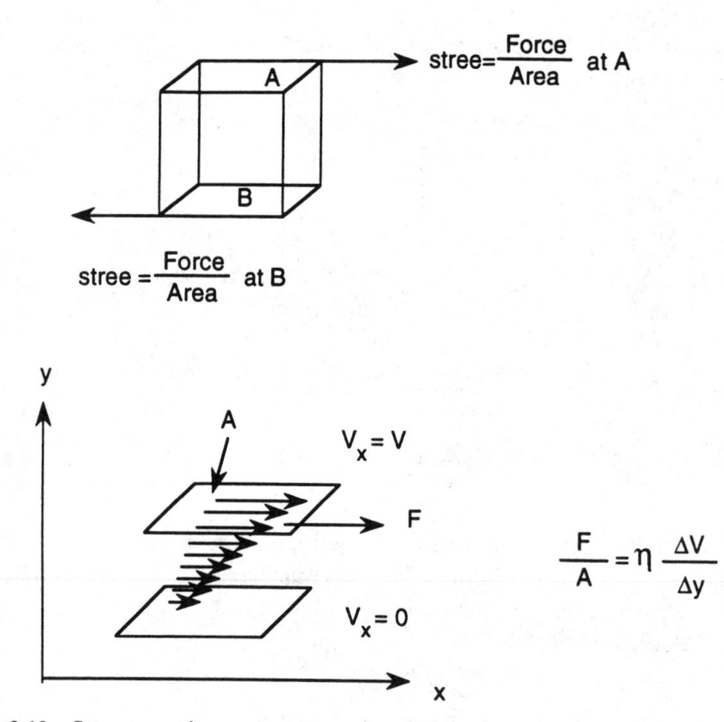

**Figure 3.10** Stresses acting on two opposite planes of an elementary volume of fluid.

the velocity (cf. Figure 3.10), that is,

$$\sigma_{\alpha\beta} = \eta\left(\frac{\partial v_\beta}{\partial x_\alpha} + \frac{\partial v_\alpha}{\partial x_\beta}\right) \tag{3.57}$$

The proportionality constant $\eta$ is the viscosity coefficient (in units of g-cm$^{-1}$-S$^{-1}$). Note that for a fluid under uniform rotation $\mathbf{w}_0$ (i.e., $\mathbf{v} = \mathbf{w}_0 \times \mathbf{r}$), we have $\partial v_\alpha / \partial x_\beta = -\partial v_\beta / \partial x_\alpha$, which means $\sigma_{\alpha\beta} = 0$.

Equations (3.56), together with (3.57), forms the basis for studying the hydrodynamics of an isotropic fluid. Note that since the viscous force $\mathbf{f}_{\text{vis}}$ is a spatial derivative of the stress tensor, which in turn is a spatial derivative of the velocity, $\mathbf{f}_{\text{vis}}$ is of the form $\eta\nabla^2\mathbf{v}$. Equation (3.54) therefore may be written as

$$\frac{\partial \mathbf{v}}{\partial t} + (\mathbf{v}\cdot\boldsymbol{\nabla})\mathbf{v} = -\frac{\boldsymbol{\nabla} p}{\rho} + \frac{\mathbf{f}_{\text{ext}}}{\rho} + \frac{\eta\nabla^2\mathbf{v}}{\rho} \tag{3.58}$$

which is usually referred to as the Navier–Stokes equation for an incompressible fluid.

### 3.5.2 General Stress Tensor for Nematic Liquid Crystals

The general theoretical framework for describing the hydrodynamics of liquid crystals has been developed principally by Leslie (16) and Ericksen (17). Their approaches account for the fact that the stress tensor depends not only on the velocity gradients, but also on the *orientation* and *rotation* of the director. Accordingly, the stress tensor is given by

$$
\begin{aligned}
\sigma_{\alpha\beta} = {} & \alpha_1 n_\gamma n_\delta A_{\gamma\delta} n_\alpha n_\beta \\
& + \alpha_2 n_\alpha n_\beta + \alpha_3 n_\beta n_\alpha \\
& + \alpha_4 A_{\alpha\beta} \\
& + \alpha_5 n_\gamma A_{\gamma\beta} + \alpha_6 n_\beta n_\gamma A_{\gamma\alpha}
\end{aligned} \tag{3.59}
$$

where the $A_{\alpha\beta}$'s are defined by

$$
A_{\alpha\beta} = \frac{1}{2}\left[\frac{\partial v_\beta}{\partial x_\alpha} + \frac{\partial v_\alpha}{\partial x_\beta}\right] \tag{3.60}
$$

Note that all the other terms on the right-hand side of (3.59) involve the director orientation, except the fourth term, $\alpha_4 A_{\alpha\beta}$. This is the same term as that for an isotropic fluid [cf. equation (3.57)], that is, $\alpha_4 = 2\eta$.

Therefore, in this formalism there are six so-called Leslie coefficients, $\alpha_1, \alpha_2, \ldots, \alpha_6$, which have the dimension of viscosity coefficients. It was shown by Parodi (18) that

$$
\alpha_2 + \alpha_3 = \alpha_6 - \alpha_5 \tag{3.61}
$$

and so there are really five independent coefficients.

In the next few sections we will study particular cases of director axis orientation and deformation and show how these Leslie coefficients are related to other commonly used viscosity coefficients.

### 3.5.3 Flows with Fixed Director Axis Orientation

Consider here the simplest case of flows in which the director axis orientation is held fixed. This may be achieved by a strong externally applied magnetic field (cf. Figure 3.11), where the magnetic field is along the direction $\hat{n}$. Consider the case of shear flow, where the velocity is in the $z$ direction and the velocity gradient is along the $x$ direction. This process could occur, for example, in liquid crystals confined by two parallel plates in the $y-z$ plane.

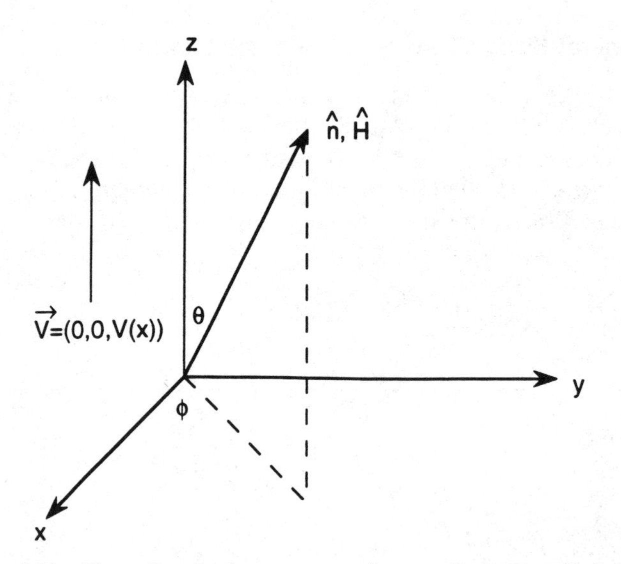


**Figure 3.11** Shear flow in the presence of an applied magnetic field **H**.

In terms of the orientation of the director axis, there are three distinct possibilities involving three corresponding viscosity coefficients:

1. $\eta_1$: $\hat{n}$ is parallel to the velocity gradient, that is, along the $x$ axis ($\theta = 90°$, $\phi = 0°$).
2. $\eta_2$: $\hat{n}$ is parallel to the flow velocity, that is, along the $z$ axis and lies in the shear plane $x - z$ ($\theta = 0°$, $\phi = 0°$).
3. $\eta_3$: $\hat{n}$ is perpendicular to the shear plane, that is, along the $y$ axis ($\theta = 90°$, $\phi = 90°$).

These three configurations have been investigated by Miesowicz (19), and the $\eta$'s are known as the Miesowicz coefficients. In the original paper, as well as the treatment by deGennes (3), the definitions of $\eta_1$ and $\eta_3$ are interchanged. In deGennes notation, in terms of $\eta_a$, $\eta_b$, and $\eta_c$, we have $\eta_a = \eta_3$, $\eta_b = \eta_2$, and $\eta_c = \eta_1$. The notation used here is attributed to Helfrich (20), which is now the conventional one.

To obtain the relationships between $\eta_{1,2,3}$ and the Leslie coefficients $\alpha_{1,2,\ldots,6}$, one could evaluate the stress tensor $\sigma_{\alpha\beta}$ and the shear rate $A_{\alpha\beta}$ for various director orientations and flow and velocity gradient directions. From these considerations, the following relationships are obtained (3):

$$\begin{aligned} \eta_1 &= \tfrac{1}{2}(\alpha_4 + \alpha_5 - \alpha_2) \\ \eta_2 &= \tfrac{1}{2}(\alpha_3 + \alpha_4 + \alpha_6) \\ \eta_3 &= \tfrac{1}{2}\alpha_4 \end{aligned} \tag{3.62}$$

In the shear plane $x-z$, the general effective viscosity coefficient is actually more correctly expressed in the form (21):

$$\eta_{\text{eff}} = \eta_1 + \eta_{12}\cos^2\theta + \eta_2 \tag{3.63}$$

in order to account for angular velocity gradients.

The coefficient $\eta_{12}$ is related to the Leslie coefficient $\alpha_1$ by

$$\eta_{12} = \alpha_1 \tag{3.64}$$

### 3.5.4 Flows with Director Axis Reorientation

The preceding section deals with the case where the director axis is fixed during fluid flow. In more general situations director axis reorientations often accompany fluid flows and vice versa. Taking into account the moment of inertia $I$ and the torque $\mathbf{\Gamma} = \hat{n} \times \mathbf{f}$, where $\mathbf{f}$ is the molecular internal elastic field defined in (3.11), $\mathbf{\Gamma}_{\text{ext}}$ is the torque associated with an externally applied field, and $\mathbf{\Gamma}_{\text{vis}}$ is the viscous torque associated with the viscous forces, the equation of motion describing the angular acceleration $d\Omega/dt$ of the director axis may be written as

$$I\frac{d\Omega}{dt} = (\hat{n} \times \mathbf{f} + \mathbf{\Gamma}_{\text{ext}}) - \mathbf{\Gamma}_{\text{vis}} \tag{3.65}$$

The viscous torque $\mathbf{\Gamma}_{\text{vis}}$ consists of two components (3): one arising from pure rotational effect (i.e., no coupling to the fluid flow) given by $\gamma_1 \hat{n} \times \mathbf{N}$ and another arising from coupling to the fluid motion given by $\gamma_2 \hat{n} \times \hat{A}\hat{n}$. Therefore, we have

$$\mathbf{\Gamma}_{\text{vis}} = \hat{n} \times \left[\gamma_1 \mathbf{N} + \gamma_2 \hat{A}\hat{n}\right] \tag{3.66}$$

Here $\mathbf{N}$ is the rate of change of the director with respect to the immobile background fluid, given by

$$\mathbf{N} = \frac{d\hat{n}}{dt} - \hat{\omega} \times \hat{n} \tag{3.67}$$

where $\boldsymbol{\omega}$ is the angular velocity of the liquid. In (3.66) $\hat{A}$ is the velocity gradient tensor defined by (3.60).

The viscosity coefficients $\gamma_1$ and $\gamma_2$ are related to the Leslie coefficient $\alpha$'s by (3):

$$\gamma_1 = \alpha_3 - \alpha_2 \tag{3.68$a$}$$

$$\gamma_2 = \alpha_2 + \alpha_3 = \alpha_6 - \alpha_5 \tag{3.68$b$}$$

Consider the flow configuration depicted in Figure 3.11. Without the magnetic field and setting $\phi = 0$, we have

$$\mathbf{v} = [0, 0, v(x)] \tag{3.69a}$$

$$\hat{n} = [\sin\theta, 0, \cos\theta] \tag{3.69b}$$

$$A_{xz} = \frac{1}{2}\frac{dv}{dx} \tag{3.69c}$$

$$N_z = \omega_y n_x = -A_{xz} n_x \tag{3.69d}$$

$$N_x = -\omega_y n_z = A_{xz} n_z \tag{3.69e}$$

From (3.66) the viscous torque along the $y$ direction is given by

$$\begin{aligned}\Gamma_{\text{vis}} &= -\gamma_1(n_z N_x - n_x N_z) - \gamma_2(n_z n_\mu A_{\mu x} - n_x n_\mu A_{\mu z}) \\ &= -\frac{1}{2}\frac{dv}{dx}\left[\gamma_1 + \gamma_2(\cos^2\theta - \sin^2\theta)\right] \\ &= -\frac{dv}{dx}\left[\alpha_3\cos^2\theta - \alpha_2\sin^2\theta\right]\end{aligned} \tag{3.70}$$

In the steady state, whence the shear torque vanishes, a stable director axis orientation is induced by the flow with an angle $\theta_{\text{flow}}$ given by

$$\cos 2\,\theta_{\text{flow}} = -\frac{\gamma_1}{\gamma_2} \tag{3.71}$$

For more complicated flow geometries, the director axis orientation will assume correspondingly complex profiles.

## 3.6 FIELD-INDUCED DIRECTOR AXIS REORIENTATION EFFECTS

We now consider the process of director axis reorientation by an external static or low-frequency field. Optical field effects are discussed in Chapter 6. The following examples will illustrate some of the important relationships among the various torques and dynamical effects discussed in the preceding sections. We will consider the magnetic field as it does not involve complicated local field effects and other electric phenomena (e.g., conduction). The electric field counterparts of the results obtained here for the magnetic field can be simply obtained by the replacement of $\Delta\chi^m H^2$ by $\Delta\varepsilon E^2$ [cf. equations (3.26) and (3.29)].

### 3.6.1 Field-Induced Reorientation without Flow Coupling

The following example demonstrates how the viscosity coefficient $\gamma_1$ comes into play in field-induced reorientational effects. Consider pure twist deformation caused by an externally applied field **H** on a planar sample as depicted in Figure 3.12. Let $\theta$ denote the angle of deformation.

The director axis $\hat{n}$ is thus given by $\hat{n} = (\cos\theta, \sin\theta, 0)$. From this and the preceding equations, the free energy (3.4) and elastic torque (3.13) are, respectively,

$$F_2 = \frac{K_2}{2}\left(\frac{\partial\theta}{\partial z}\right)^2 \tag{3.72a}$$

and

$$\mathbf{\Gamma} = K_2 \frac{\partial^2\theta}{\partial z^2}\hat{z} \tag{3.72b}$$

The viscous torque is given by

$$\mathbf{\Gamma}_{\text{vis}} = -\gamma_1 \frac{d\theta}{dt} \tag{3.73}$$

The torque exerted by the external field **H**, from (3.29), becomes

$$\mathbf{\Gamma}_{\text{ext}} = \Delta\chi^m H^2 \sin\theta\cos\theta \tag{3.74}$$

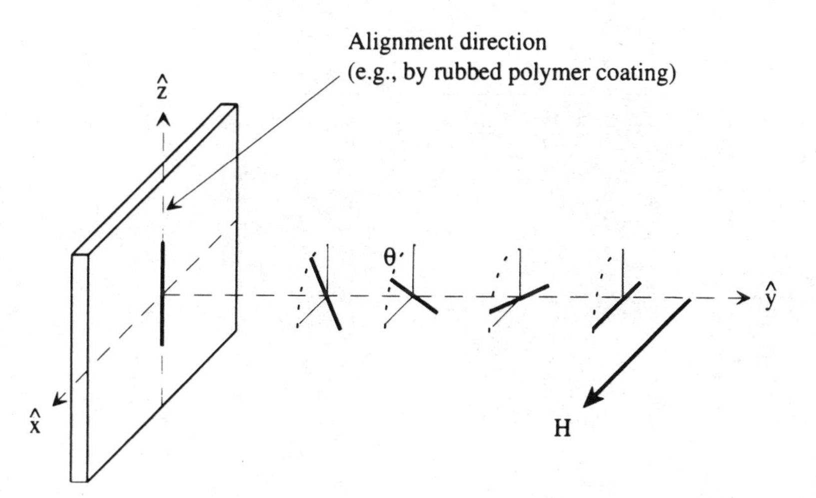


**Figure 3.12** Pure twist deformation induced by an external magnetic field **H** on a planar sample; there is no fluid motion.

Hence, the torque balance equation gives

$$\gamma_1 \frac{d\theta}{dt} = K_2 \frac{\partial^2 \theta}{\partial z^2} + \Delta\chi^m H^2 \sin\theta \cos\theta \tag{3.75a}$$

In the equilibrium situation, $\gamma_1(d\theta/dt) = 0$, and (3.75) becomes

$$K_2 \frac{\partial^2 \theta}{\partial z^2} + \Delta\chi^m H^2 \sin\theta \cos\theta = 0 \tag{3.75b}$$

An interesting result from this equation is the so-called Freedericksz transition (3). For an applied field strength less than a critical field $H_F$, $\theta = 0$. For $H > H_F$, reorientation occurs. The expression for $H_F$ is given by

$$H_F = \left(\frac{\pi}{d}\right)\left(\frac{K}{\Delta\chi^m}\right)^{1/2} \tag{3.76}$$

assuming that the reorientation obeys the hard-boundary (strong anchoring) condition (i.e., $\theta = 0$ at $z = 0$ and at $z = d$). For $H$ just above $H_F$, $\theta$ is given approximately by

$$\theta = \theta_0 \sin\left(\frac{\pi z}{d}\right) \tag{3.77a}$$

where

$$\theta_0 \sim 2\frac{(H - H_F)^{1/2}}{H_F} \tag{3.77b}$$

For the case where $H$ is abruptly reduced from its value above $H_F$ to 0, (3.75$a$) becomes

$$\gamma_1 \frac{d\theta}{dt} = K_2 \frac{\partial^2 \theta}{\partial z^2} \tag{3.78}$$

writing $\theta(z,t) = \theta_0(t)\sin(\pi z/d)$ gives

$$\dot{\theta} = \frac{-\pi^2 k_2}{d^2 \gamma_1}\theta \tag{3.79}$$

that is,

$$\theta_0(t) = \theta_0 e^{-t/\tau} \tag{3.80}$$

where the relaxation time constant $\tau$ is given by

$$\tau = \frac{\gamma_1 d^2}{\pi^2 k_2} \tag{3.81}$$

### 3.6.2 Reorientation with Flow Coupling

Field-induced director axis reorientation, accompanied by fluid flow, is clearly complicated as it involves many more physical parameters.

Consider the interaction geometry shown in Figure 3.13. A homeotropically aligned nematic liquid crystal film is acted on by an electric or a magnetic field in the $x$ direction. Let $\phi$ denote the director axis reorientation angle from the original alignment direction $z$. Assume hard-boundary conditions at the two cell walls at $z = 0$ and at $z = d$. The flow in the $x$ direction, with a $z$ dependence.

The following are the pertinent parameters involved:

$$\text{Director axis:} \quad \hat{n} = (\sin\phi, 0, \cos\phi) \tag{3.82a}$$

$$\text{Velocity field:} \quad \mathbf{v} = (v(z), 0, 0) \tag{3.82b}$$

$$\text{Free energies:} \quad F = \frac{1}{2}K_1(\nabla\cdot\hat{n})^2 + \frac{1}{2}K_3(\hat{n}\times\nabla\times\hat{n})^2$$
$$= \frac{1}{2}K_1 \sin^2\phi\left(\frac{d\phi}{dz}\right)^2 + \frac{1}{2}K_3\cos^2\phi\left(\frac{d\phi}{dz}\right)^2 \tag{3.82c}$$

$$\text{Elastic torques} = \left[K_1\sin^2\phi + K_3\cos^2\phi\right]\frac{d^2\phi}{dz^2} + \left[(K_1 - K_3)\sin\phi\cos\phi\right]\left(\frac{d\phi}{dz}\right)^2 \tag{3.82d}$$

$$\text{Field-induced torque} = \varepsilon_0\,\Delta\varepsilon\, E^2\sin\phi\cos\phi \tag{3.82e}$$

$$\text{Rotational viscous torque} = \gamma_1\frac{d\phi}{dt} \tag{3.82f}$$

$$\text{Flow-orientational viscous torque} = \frac{dv}{dz}\left[\alpha_2\sin^2\phi - \alpha_3\cos^2\phi\right] \tag{3.82g}$$

Using (3.65), the equation of motion taking into account these torques, as well as the moment of inertia $I$ of the molecules involved, is given by

$$I\frac{d^2\phi}{dt^2} + \gamma_1\frac{d\phi}{dt} = [K_1\sin^2\phi + K_3\cos^2\phi]\frac{d^2\phi}{dz^2} + [(K_1 - K_3)\sin\phi\cos\phi]\left(\frac{d\phi}{dz}\right)^2 + (\alpha_2\sin^2\phi - \alpha_3\cos^2\phi)\frac{dv}{dz} + \varepsilon_0\,\Delta\varepsilon\, E^2\sin\phi\cos\phi \tag{3.83}$$

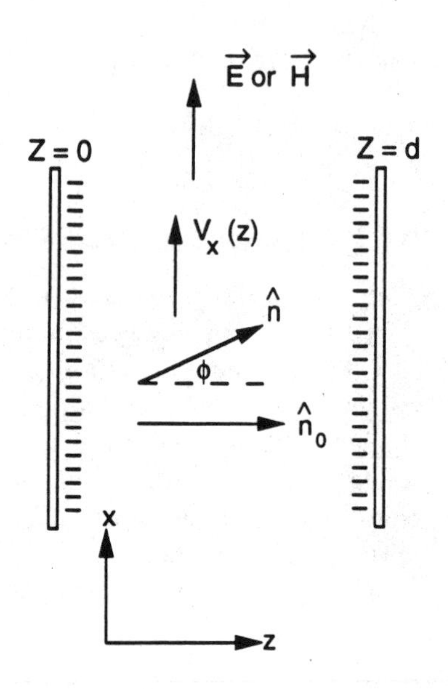


**Figure 3.13** Director axis reorientation causing flows.

This equation may be solved for various experimental conditions. Optically induced director axis reorientation and flow effects have been studied by two groups (22, 23) using picosecond laser pulses. A solution of the previous equation is also presented in Eichler and Macdonald (23).

## REFERENCES


1. F. C. Frank, *Discuss, Faraday Soc.* **25**, 19 (1958).
2. J. L. Ericksen, *in* "Liquid Crystals" (G. Brown, ed.). Gordon & Breach, New York, 1969.
3. P. G. deGennes, "Physics of Liquid Crystals." Clarendon Press, Oxford, 1974.
4. See, for example, G. Barbero and F. Simoni, *Appl. Phys. Lett.* **41**, 504 (1992); G. Barbero, F. Simoni, and P. Aiello, *J. Appl. Phys.*, **55**, 304 (1984), see also S. Faetti, *in* "Physics of Liquid Crystalline Materials" (I. C. Khoo and F. Simoni, eds.). Gordon & Breach, Philadelphia, 1991.
5. J. D. Jackson, "Classical Electrodynamics." Wiley, New York, 1975.
6. W. Helfrich, *J. Chem. Phys.* **51**, 4092 (1969). See also reference 8, chapter 5.
7. J. P. Parneix, A. Chapoton, and E. Constant, *J. Phys. (Paris)* **36**, 1143 (1975).
8. L. M. Blinov, "Electro-Optical and Magneto-Optical Properties of Liquid Crystals." Wiley (Interscience), Chichester, 1983.
9. I. C. Khoo and S. T. Su, "Optics and Nonlinear Optics of Liquid Crystals." World Scientific, Singapore, 1993.
10. M. F. Vuks, *Opt. Spektrosk.* **60**, 644 (1966).
11. S. Chandrasekhar and N. V. Madhusudana, *J. Phys. (Paris)* **30**, Colloq. C4, C4–C25 (1969).

12. D. A. Dunmar, *Chem. Phys. Lett.* **10**, 49 (1971).
13. W. H. deJeu and P. Bordewijk, *J. Chem. Phys.* **68**, 109 (1978).
14. I. C. Khoo and R. Normandin, *IEEE J. Quantum Electron.* **QE-21**, 329 (1985).
15. R. G. Horn, *J. Phys. (Paris)* **39**, 105 (1978).
16. F. M. Leslie, *Quantum J. Mech. Appl. Math.* **19**, 357 (1966).
17. J. L. Ericksen, *Phys. Fluids* **9**, 1205 (1966).
18. O. Parodi, *J. Phys. (Paris)* **31**, 581 (1970).
19. M. Miesowicz, *Nature (London)* **17**, 261 (1935); **158**, 27 (1946).
20. W. Helfrich, *J. Chem. Phys.* **51**, 4092 (1969).
21. See, for example, W. H. deJeu, "Physical Properties of Liquid Crystalline Materials." Gordon & Breech, New York, 1980.
22. I. C. Khoo, R. G. Lindquist, R. R. Michael, R. J. Mansfield, and P. G. LoPresti, *J. Appl. Phys.* **69**, 3853 (1991); I. C. Khoo and R. Normandin, *ibid.* **55**, 1416 (1984).
23. H. J. Eichler and R. Macdonald, *Phys. Rev. Lett.* **67**, 2666 (1991).

# CHAPTER 4

# CHOLESTERIC, SMECTIC, AND FERROELECTRIC LIQUID CRYSTALS

## 4.1 CHOLESTERIC LIQUID CRYSTALS

The physical properties of cholesteric liquid crystals are in almost all aspects similar to nematics, except that the director axis assumes a helical form [Figure 4.1*a*] with a finite pitch $p = 2\pi/q_0$. Figures 4.1*b* and *c* show two commonly occurring director axis alignments: (*b*) planar twisted and (*c*) fingerprint. From this point of view, we may regard nematics as a special case of cholesterics with $p \to \infty$.

Since the optical property of the nematic, a uniaxial material, is integrally related to the director axis, the helical arrangement of the latter in a cholesteric certainly introduces new optical properties, particularly in the propagation and reflection of light from cholesteric liquid crystal cells. In this section we summarize the main physical properties associated with the helical structure.

### 4.1.1 Free Energies

Since the equilibrium configuration of a cholesteric liquid crystal is a helical structure with a pitch wave vector $q_0$ ($q_0 = 2\pi/p_0$), its elastic free energy will necessarily reflect the presence of $q_0$. The evolution of a cholesteric to a nematic liquid crystal may be viewed as the "untwisting" of the helical

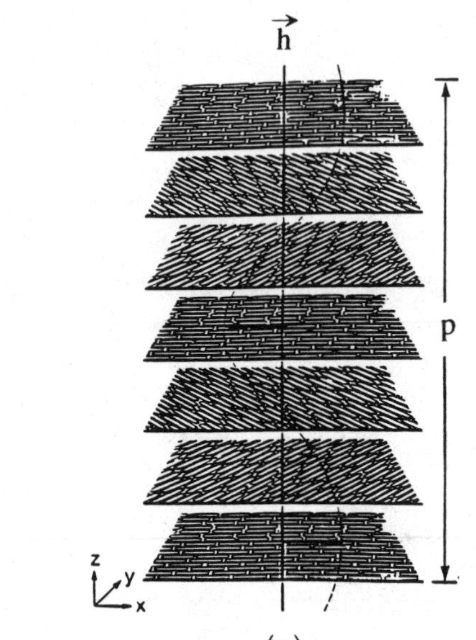


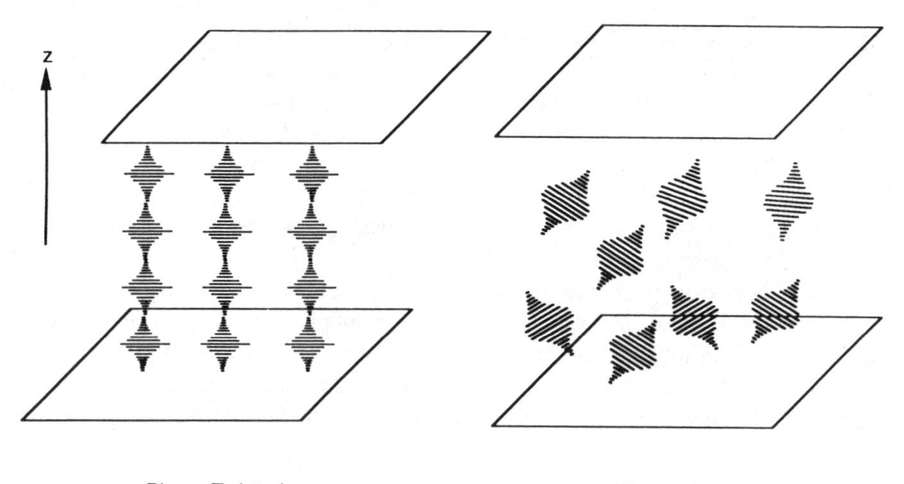


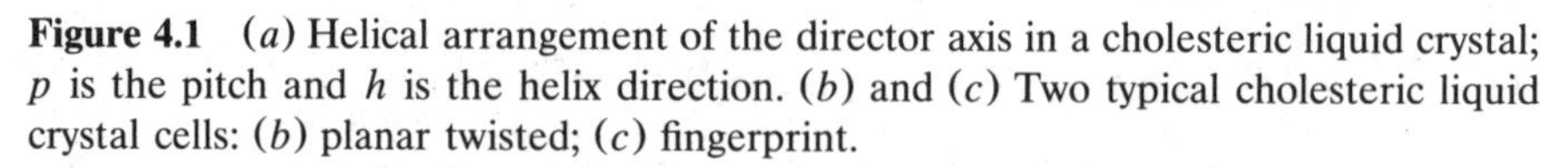

**Figure 4.1** (*a*) Helical arrangement of the director axis in a cholesteric liquid crystal; *p* is the pitch and *h* is the helix direction. (*b*) and (*c*) Two typical cholesteric liquid crystal cells: (*b*) planar twisted; (*c*) fingerprint.

structure (i.e., the twist deformation energy is involved). This is indeed rigorously demonstrated in Frank elastic theory (1), where consideration of the absence of mirror symmetry in cholesterics results in the addition of another factor to the twist deformation energy:

$$K_2(\hat{h}\cdot\nabla\times\hat{h})^2 \quad \rightarrow \quad K_2(\hat{n}\cdot\nabla\times\hat{n}+q_0)^2 \tag{4.1}$$

nematic cholesteric

With reference to Figure 4.1, the director axis is described by $\hat{n}=(n_x, n_y, n_z)$, where $n_x=\cos\theta(z)$, $n_y=\sin\theta(z)$, and $n_z=0$. Note that this configuration corresponds to a state of minimum free energy

$$F_2=\frac{1}{2}K_2(\hat{n}\cdot\nabla\times\hat{n})=\frac{1}{2}K_2\left(-\frac{\partial\theta}{\partial z}+q_0\right)^2=0 \tag{4.2}$$

if $q_0=\partial\theta/\partial z$, that is,

$$\theta=q_0 z \tag{4.3}$$

For a general distortion therefore the free energy of a cholesteric liquid crystal is given by

$$F_d=\tfrac{1}{2}K_1(\nabla\cdot\hat{n})+K_2(\hat{n}\cdot\nabla\hat{n}+q_0)^2+K_3(\hat{n}\times(\nabla\times\hat{n}))^2 \tag{4.4}$$

If the pitch of the cholesteric is not changed by an external probing field (electric, magnetic, or optical), the physical properties of cholesterics are those of locally uniaxial crystals. In other words, the anisotropies in the dielectric constant ($\Delta\varepsilon=\varepsilon_{\parallel}-\varepsilon_{\perp}$), electric conductivity ($\Delta\sigma=\sigma_{\parallel}-\sigma_{\perp}$), magnetic susceptibility ($\Delta\chi^m=\chi^m_{\parallel}-\chi^m_{\perp}$), and so on are defined with respect to the local director axis direction. On the other hand, if one refers to the helical axis (cf. Figure 4.1, $z$ direction), an applied probing field along $z$ will "see" the $\perp$ components (i.e., $\varepsilon_{\perp}$, $\sigma_{\perp}$, $\chi_{\perp}$, etc.). If the probing field is along a direction perpendicular to $z$, it will effectively "see" the average of the $\perp$ and $\parallel$ components [i.e., $\frac{1}{2}(\varepsilon_{\parallel}+\varepsilon_{\perp})$, $\frac{1}{2}(\sigma_{\parallel}+\sigma_{\perp})$, etc.].

Just as in the nematic case, the application of an applied magnetic or electric field gives rise to additional terms in the free energy given by

$$F_{\text{mag}}=-\tfrac{1}{2}\Delta\chi^m(\hat{n}\cdot\mathbf{H})^2 \tag{4.5a}$$

and

$$F_{\text{el}}=-\tfrac{1}{2}\Delta\varepsilon(\hat{n}\cdot\mathbf{E})^2 \tag{4.5b}$$

respectively, as well as some terms which are independent of the orientation of the director axis. $\Delta\chi^m = \chi_{\parallel}^m - \chi_{\perp}^m$ for cholesterics are usually quite small (about $10^{-9}$ in cgs units) in magnitude and negative in sign. In other words, the directors tend to align normal to the magnetic field. Cholesteric liquid crystals synthesized by adding chiral molecules to nematics such as MBBA tend to have positive $\Delta\chi^m$'s (1).

### 4.1.2 Field-Induced Effects and Dynamics

The realignment or alignment of a cholesteric liquid crystal in an applied electric or magnetic field, in the purely dielectric interaction picture (i.e., no current flow), results from the system's tendency to minimize its total free energy.

Clearly, the equilibrium configuration of the director axis depends on its initial orientation and the direction of the applied fields, as well as the signs of $\Delta\chi^m$ and $\Delta\varepsilon$. We will not delve into the various possible cases as they all involve the same basic mechanism; that is, the director axis tends to align parallel to the field for positive dielectric anisotropies, and normal to the field for negative dielectric anisotropies.

In the case of positive dielectric anisotropies, the field-induced reorientation process is analogous to that discussed for nematics (cf. Section 3.6). In cholesterics, however, the realignment of the director axis in the direction of the applied field will naturally affect the helical structure.

Figure 4.2$a$ shows the unperturbed director axis configuration in the bulk of an ideal cholesteric liquid crystal. Upon the application of a magnetic field, some molecules situated in the bulk regions $A$, $A'$,, and so on are preferentially aligned along the direction of the field; others, situated in regions $B$, $B'$, and so forth are not and would tend to reorient themselves along the field direction. As a result, the pitch of the helical structure will be increased; the helix is no longer of the ideal sinusoidal form. Finally, when

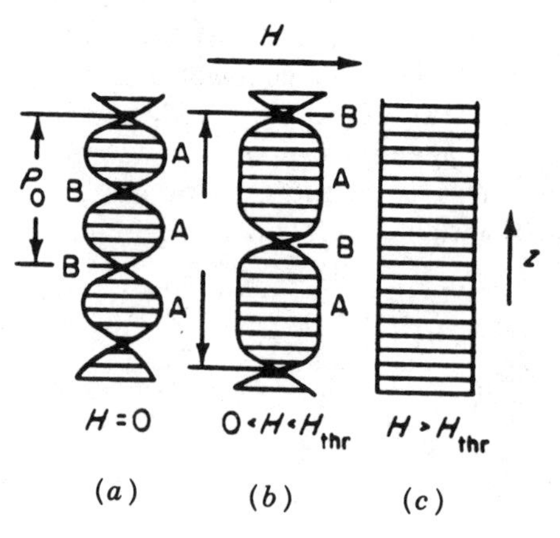


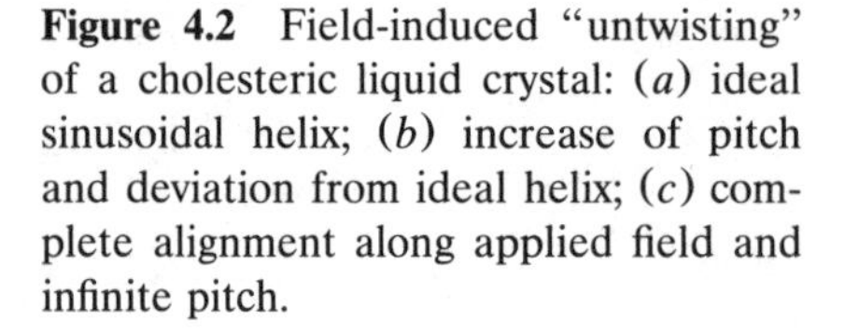

**Figure 4.2** Field-induced "untwisting" of a cholesteric liquid crystal: ($a$) ideal sinusoidal helix; ($b$) increase of pitch and deviation from ideal helix; ($c$) complete alignment along applied field and infinite pitch.

the field is sufficiently high, this "untwisting" effect is complete as the pitch approaches $\infty$, the cholesteric liquid crystal is said to have undergone a transition to the nematic phase (2).

This process can be described by the free-energy minimization process. The total free energy of the system is given by

$$F_{\text{total}} = \frac{1}{2} \int dz \left[ K_2 \left( \frac{\partial \phi}{\partial z} - q_0 \right)^2 - \Delta\chi^m H^2 \sin^2 \phi \right] \tag{4.6}$$

which, upon minimization, yields the Euler equation:

$$K_2 \frac{d^2\phi}{dZ^2} + \Delta\chi^m H^2 \sin\phi \cos\phi = 0 \tag{4.7}$$

which is analogous to (3.75b). One can define a coherence length $\xi_H$ by

$$\xi_H^2 = \frac{K_2}{\Delta\chi^m H^2} \tag{4.8}$$

Equation (4.8) thus becomes

$$\xi_H^2 \frac{d^2\phi}{dZ^2} = \sin\phi \cos\phi \tag{4.9}$$

Upon integration, it yields

$$\xi_H^2 \left( \frac{d\phi}{dZ} \right)^2 = \sin^2 \phi \tag{4.10}$$

The solution for $\phi(Z)$ is given by an elliptic function:

$$\sin\phi(Z) = Sn\left[ \left( \frac{Z}{\xi k} \right), k \right] \tag{4.11}$$

where $u = Z/\xi k$ and $k$ are the argument and modulus of the elliptic function, respectively. The condition for the free-energy minimum is given by

$$q_0 \xi = \frac{2E(k)}{\pi k} \tag{4.12}$$

and the field-dependent pitch $p(H)$ is given by

$$p(H) = p_0\left(\frac{2}{\pi}\right)^2 F(k)E(k) = 4\xi kF(k) \tag{4.13}$$

where $F(k)$ and $E(k)$ are complete elliptic integrals of the first and second kind.

When $k \to 1$, $E(k) = 1$, and $F(k)$ diverges logarithmically [i.e., $p(H) \to \infty$]. When $k \to 1$, (4.12) becomes

$$q_0\xi = \frac{2}{\pi} \tag{4.14}$$

or

$$q_0 \frac{K_2}{(\Delta\chi^m H^2)} = \frac{2}{\pi} \tag{4.15}$$

This shows that above a critical field defined by

$$H_c = \frac{\pi}{2}\left(\frac{K_2}{\Delta\chi^m}\right)^{1/2} q_0 = \pi^2\left(\frac{K_2}{\Delta\chi^2}\right)^{1/2} \frac{1}{p_0} \tag{4.16}$$

(where $p_0$ is the unperturbed pitch), the helix will be completely untwisted ($q \to 0$, $p \to \infty$); the system is essentially nematic. For a typical value of $K_2 = 10^{-6}$ dyne, $\Delta\chi^2 = 10^{-7}$ cgs units, $p_0 = 20$ $\mu$m, and $H_c \sim 15{,}000$ G. For $H < H_c$, the variation of $p$ with $H$ is well approximated by the expressions:

$$p = p_0\left[1 + \frac{(\Delta\chi^m)^2 P_0^4 H^4}{32(2\pi)^4 K_2^2} + \cdots\right] \tag{4.17}$$

It should be noted here that the preceding treatment of field-induced pitch change assumes that the cholesteric liquid crystal cell is thick and in an initially ideally twisted arrangement, and there is negligible influence from the cell walls. For thin cells or other initial director axis arrangements (e.g., fingerprint or focal-conic texture, etc.), the process will be more complicated. Nevertheless, this example serves well for illustrating the field-induced director axis reorientation and pitch change effect in cholesteric liquid crystals. Experimental measurements (3) in such systems have shown very good agreement with the theories.

### 4.1.3 Twist and Conic Mode Relaxation Times

For situations where an applied field is abruptly turned off, the relaxation time constants depend on the kind of deformation involved. There are two distinct forms of deformation: the pure twist one discussed previously and the so-called "umbrella" or conic mode (1). The first form of deformation does not involve fluid motion, whereas the latter does.

The dynamics of the field-induced twist deformation in cholesterics is described by an equation analogous to the equation for nematics:

$$\gamma_1 \frac{\partial \phi}{\partial t} - K_2 \frac{\partial^2 \phi}{\partial Z^2} - \Delta\chi^m H^2 \sin\phi \cos\phi = 0 \tag{4.18}$$

The dynamical equation for the conic distortion is much more complicated (4,5) and involves the other two elastic constants $K_3$ and $K_1$.

The corresponding relaxation times are as follows:

$$\tau_{\text{twist}} = \frac{\gamma_1}{K_2 q^2} \tag{4.19}$$

for the twist mode

$$\tau_{\text{conic}} \cong \frac{\gamma_1}{K_3 q_0^2 + K_1 q^2} \tag{4.20}$$

for the conic mode, where $\mathbf{q}$ is the wave vector characterizing the wave vector of the distortion.

These relaxation times are modified if an externally applied bias field is present and depend on the field and director axis configurations (5).

## 4.2 LIGHT SCATTERING IN CHOLESTERICS

### 4.2.1 Bragg Regime (Optical Wavelength ≈ Pitch)

In terms of their optical properties, a prominent feature of cholesterics is the helical structure of their director axes. Such helicity gives rise to selective reflection and transmission of circularly polarized light. Qualitatively, the main results for light incident perpendicularly on a planar sample, whose pitch is on the order of the optical wavelength, are summarized in Figure 4.3.

Consider a right-handed helix. An incident right circularly polarized light will be reflected as a right circularly polarized light, as the optical field follows the director axis rotation; that is, it follows the helix (Figure 4.3*a*). (Note that the right-handed or left-handed circular polarization is defined by an observer looking at the incoming light.) Under the Bragg condition:

$$\frac{2\pi}{q_0} = \lambda \tag{4.21}$$

the reflection is total, and there is no transmission.

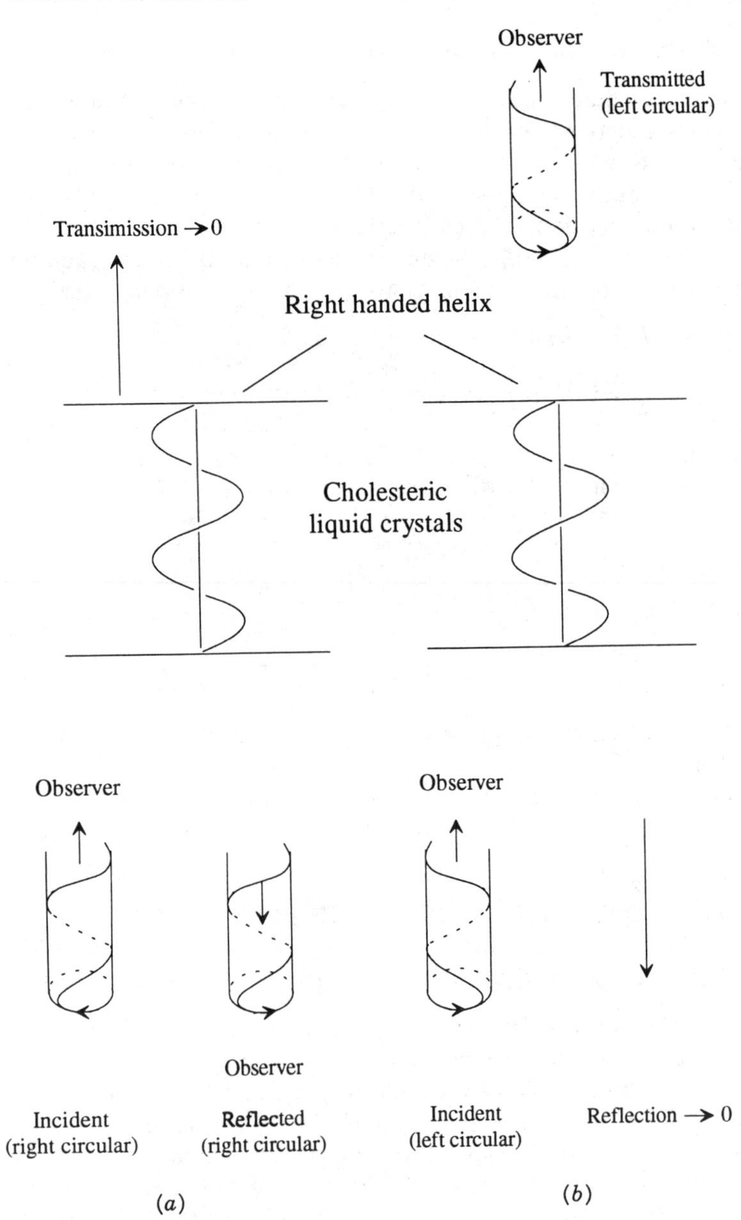


**Figure 4.3** Reflection and transmission of circularly polarized light in a cholestic liquid crystal with positive helix. (*a*) Incident light is right circularly; reflected light is right circularly polarized. (*b*) Left-circularly polarized light is totally transmitted.

Under the Bragg condition, an incident left circularly polarized light will be totally transmitted (Figure 4.3*b*).

The reflection property of a cholesteric liquid crystal is therefore opposite that of a conventional mirror, where an incident right circularly polarized light will be reflected as a left circularly polarized light and vice versa.

These results may be quantitatively analyzed, using the scattering theory (1) described in Chapter 5 and the electromagnetic approach given in the next section. The case of oblique incidence could also be analyzed by this approach, which shows that higher-order diffractions are possible. The Bragg diffraction becomes

$$p_0 \cos\gamma = m\lambda \qquad m = 1,2,3,\ldots \tag{4.22}$$

where $\gamma$ is the angle of the refracted light in the cholesteric liquid crystal. In this case the polarization states are elliptical. In general, these theories are very complicated and are outside the scope of the present treatment in this book. The interested reader is referred to the literature (6).

### 4.2.2 General Optical Propagation and Reflection: Normal Incidence

Consider a light wave propagating along the direction of the helix ($\hat{z}$ direction). For the locally uniaxial system, the electric displacement **D** and the electric field **E** are related by

$$\mathbf{D} = \hat{\varepsilon} : \mathbf{E} = \varepsilon_\perp \mathbf{E} + \Delta\varepsilon\, \hat{n}(\hat{n}\cdot\mathbf{E}) \tag{4.23}$$

where $\hat{\varepsilon}$ is the dielectric anisotropy tensor [equation (2.5)]. The two electric field components, $E_x$ and $E_y$, in plane wave form, are given by

$$E_x = Re\left[E_x(z)e^{-i\omega t}\right] \tag{4.24a}$$

$$E_y = Re\left[E_y(z)e^{-i\omega t}\right] \tag{4.24b}$$

and satisfy the Maxwell equation:

$$-\frac{d^2}{dz^2}\begin{pmatrix} E_x \\ E_y \end{pmatrix} = \left(\frac{\omega}{c}\right)^2 \hat{\varepsilon}(z)\begin{pmatrix} E_x \\ E_y \end{pmatrix} \tag{4.25}$$

Using $\hat{n} = (\cos\theta, \sin\theta, 0)$ and $\theta = q_0 Z$ in $\hat{\varepsilon}(z)$, we get

$$\hat{\varepsilon}(z) = \frac{\varepsilon_\parallel + \varepsilon_\perp}{2}\begin{pmatrix} 1 & 0 \\ 0 & 1 \end{pmatrix} + \frac{\Delta\varepsilon}{2}\begin{pmatrix} \cos 2q_0 & \sin 2q_0 z \\ \sin 2q_0 z & -\cos 2q_0 z \end{pmatrix} \tag{4.26}$$

In terms of circularly polarized waves polarized in the right (+) and left (−)

directions:

$$\hat{e}_{\pm} = \frac{\hat{x} \pm i\hat{y}}{\sqrt{2}} \tag{4.27}$$

(4.25) becomes

$$-\frac{d^2E^+}{dZ^2} = k_0^2 E^+ + k_1^2 \exp(2iq_0 z)E^- \tag{4.28}$$

$$-\frac{d^2E^-}{dZ^2} = k_0^2 E^- + k_1^2 \exp(-2iq_0 Z)E^+ \tag{4.29}$$

where $k_0^2 = (\omega/c)^2\langle\varepsilon\rangle$ and $k_1^2 = (\omega/c)^2\Delta\varepsilon$ and $\langle\varepsilon\rangle = [\varepsilon_{\parallel} + \varepsilon_{\perp}]/2$.

The solutions for $E^+$ and $E^-$ are of the form

$$E^+ = a\exp[i(l + q_0)Z] \tag{4.30}$$

$$E^- = b\exp[i(l - q_0)Z] \tag{4.31}$$

Substituting (4.30) and (4.31) into (4.28) and (4.29) gives a dispersion relationship between $\omega$ and $l$:

$$\left(-\left(\frac{\omega}{c}\right)^2\langle\varepsilon\rangle + l^2 + q_0^2\right)^2 - 4q_0^2 l^2 - \left(\frac{\omega}{c}\right)^4 \Delta\varepsilon^2 = 0 \tag{4.32}$$

A schematic plot of this relationship is given in Figure 4.4.

For $\omega_- < \omega < \omega_0$ (i.e., $cq_0/n_{\perp} < \omega < cq_0/n_{\parallel}$), only one wave with circular polarization may propagate. This is the Bragg reflection regime discussed in the previous section. The spectral width $\Delta\lambda$of this reflection band is proportional to the optical dielectric anisotropy:

$$\Delta\lambda = P_0\,\Delta n \tag{4.33}$$

Outside this selective reflection band, in general, there are two roots $l_1$ and $l_2$ with positive group velocity ($V_g = \partial\omega/\partial l > 0$); that is, there are two forward propagating waves and two backward propagating waves ($-l_1$ and $-l_2$). These mode structures are very sensitive to the parameter (1):

$$x = \frac{2q_0 l}{k_1^2} \tag{4.34}$$

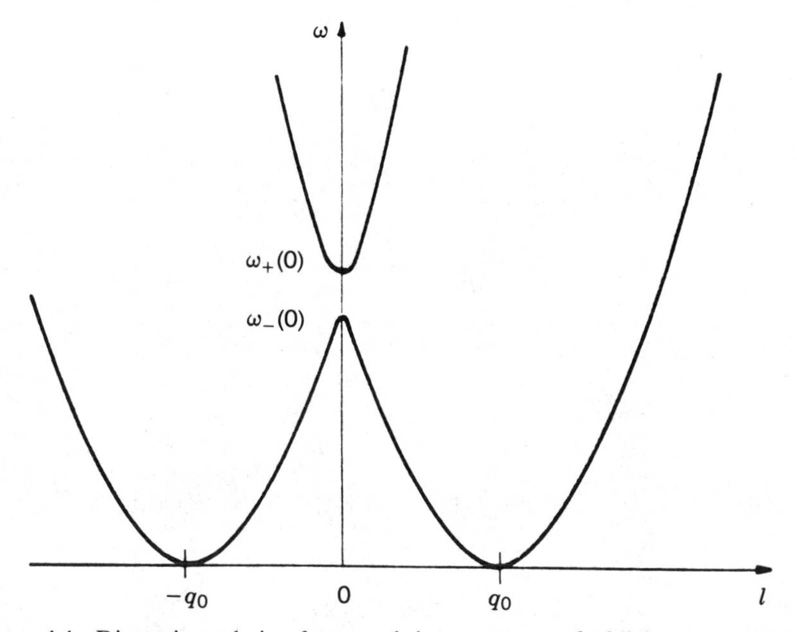


**Figure 4.4** Dispersion relation for $w$ and the wave vector $l$ of light propagating in a cholesteric liquid crystal.

For $x \ll 1$ (i.e., $\lambda \ll \Delta n P_0$), the optical wave is "guided" by the system; the electric vector of a linearly polarized wave (ordinary or extraordinary) follows the rotation of the director, and the angle of its rotation corresponds to the number of turns of the helix. This is also known as the Mauguin regime. (In Chapter 6 we will consider the case where the light intensity is high enough to create director axis distortion and therefore affects its propagation and other scattering characteristics in the context of nonlinear optics.)

For the case where $n_e$ and $n_O$ are close (i.e., $k_1^2$ is small compared to $q_0 l$, i.e., $x$ is large), the eigenmodes are nearly circular. In this case the two eigenmodes $l_1$ and $l_2$ are given by (1):

$$l_1 = k_0 + q_0 + \frac{k_1^4}{8k_0 q_0(k_0 + q_0)} + O(k_1^3) \tag{4.34}$$

$$l_2 = k_0 - q_0 + \frac{k_1^4}{8k_0 q_0(q_0 - k_0)} + O(k_1^3) \tag{4.36}$$

Writing $l_1 - q_0 = (\omega/c)n_1$ and $l_2 + q_0 = (\omega/c)n_2$, the optical rotation per unit length $\psi/d = (\omega/c)(n_1 - n_2)$ is given by the following two equivalent expressions:

$$\frac{\psi}{d} = \frac{k_1^4}{8q_0(k_0^2 - q_0^2)} \tag{4.37}$$

or

$$\frac{\psi}{d} = \frac{q_0}{32}\left(\frac{n_\varepsilon^2 - n_0^2}{n_\varepsilon^2 + n_0^2}\right)\frac{1}{\bar{\lambda}^{-2}}(1 - \bar{\lambda}^{-2}) \tag{4.38}$$

where $\bar{\lambda} = \lambda / P$. Note that in the vicinity of $\bar{\lambda} \approx 1$ for $(\lambda \approx P)$, that is, in the vicinity of the selective reflection band, the rotation per unit length can be very large. Also, the rotation changes sign at $\bar{\lambda} = 1$.

## 4.3 SMECTIC AND FERROELECTRIC LIQUID CRYSTALS: A QUICK SURVEY

Smectic liquid crystals process a higher degree of order than nematics; they exhibit both positional and directional ordering in their molecular arrangements. Long-range positional ordering in smectics is manifested in the form of layered structures, in which the director axis is aligned in various directions depending on the smectic phase. To date, at least nine distinct smectic phases, bearing the designation smectic-A, smectic-B, smectic-C through smectic-I have been identified, in chronological order (7). Some smectic liquid crystals, for example, smectic-C* materials, are ferroelectric; their molecules possess permanent dipole moments (8,9). A good example of a room-temperature smectic-A is OCB (4,4′-*n*-octylcyanobiphenyl), whose molecular structure is shown in Figure 4.5*a*. This material has also been studied in the context of nonlinear optical pulse propagation and optical wave mixing phenomena (10,11). The liquid crystal *n*CB ($n = 8$–$12$) also exhibits the smectic-A phase (9). The well-studied liquid crystal 4-*n*-octyloxy-4′-cyanobiphenyl (OOCBP), whose molecular structure is shown in Figure 4.5*b*, exhibits the smectic-C, smectic-A, and nematic phases as a function of temperature. Smectic liquid crystals that are ferroelectric include HOBACPC (9), DOBAMBC (12), XL13654 (13,14), and SCE9 (15). A well-studied ferroelectric liquid crystal is DOBAMBC; its molecular structure is shown in Figure 4.5*c*.

Because of these differences in the degree of order and molecular arrangement and the presence of a permanent dipole moment, the physical properties of smectic liquid crystals are quite different from those of the nematic phase. In this chapter we examine the pertinent physical theories and the optical properties of three exemplary types of smectics: smectic-A, smectic-C, and (ferroelectric) smectic-C*.

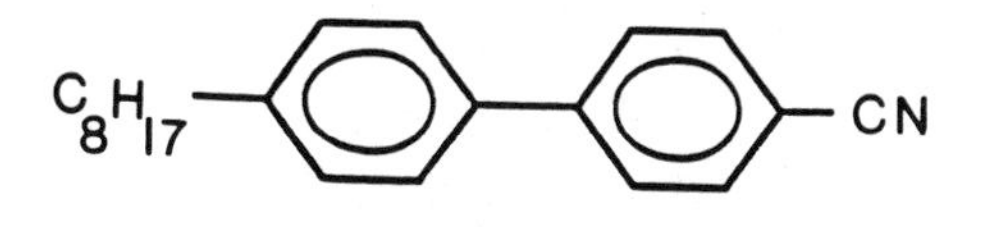


Sm A-nematic 21.5 - 33.5 nematic-isotropic 33.5 - 40.5

(*a*)

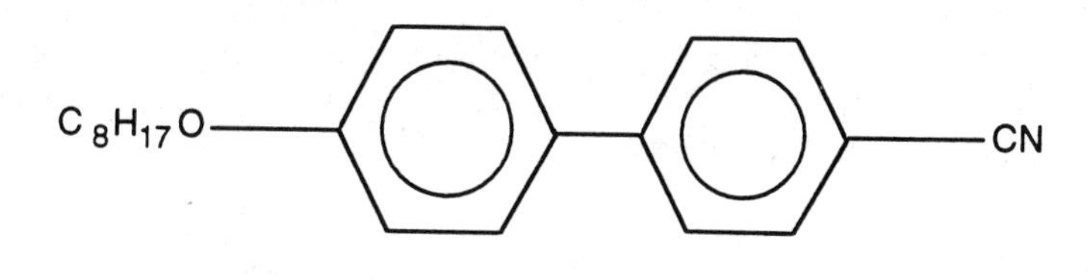


$C \xrightarrow{54.5^\circ C} S \xrightarrow{66.6^\circ C} N \xrightarrow{79.3^\circ C} I$

(*b*)

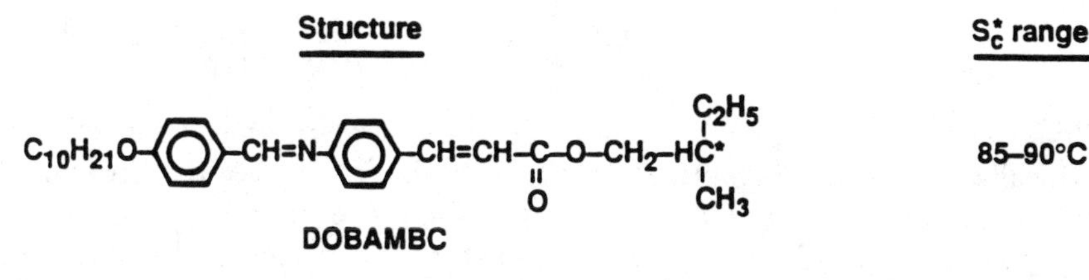


(*c*)

**Figure 4.5** (*a*) Smectic-A liquid crystal. (*b*) A liquid crystal compound that exhibits smectic-C, smectic-A, and nematic phases. (*c*) Molecular structure of a ferroelectric liquid crystal.

## 4.4 SMECTIC-A LIQUID CRYSTALS

### 4.4.1 Free Energies

The molecular arrangement of a smectic-A (SmA) liquid crystal is shown in Figure 1.11*a*. The physical properties of SmA are analogous to nematics in many ways. However, because of the existence of the layered structures, there are important differences in the dynamics and types of elastic deformations that could be induced by applied fields.

In an ideal single-domain SmA sample, in which the layers are parallel and equidistantly separated, the director axis components $n_x$ and $n_y$ are related to the layer displacement $u(x, y, z)$, in the limit of small distortion, by the following relationships:

$$n_x = -\frac{\partial u}{\partial x} \tag{4.39a}$$

$$n_y = \frac{\partial u}{\partial y} \tag{4.39b}$$

In the equilibrium case only $u(x, y, z)$ and its spatial derivatives are needed for describing elastic distortion in SmA. For example, a small director axis reorientation may be represented by

$$\theta(\mathbf{r}) \sim \frac{\partial u}{\partial z} \tag{4.40}$$

The energy associated with this distortion, which corresponds to a compression of the layer, is given by

$$F_{\text{comp}} = \frac{1}{2}\overline{B}\left(\frac{\partial u}{\partial z}\right)^2 \tag{4.41}$$

This process is analogous to compressibility of an isotropic liquid crystal. Typically, the compressibility $\overline{B}$ is on the order of $10^7$ to $10^8$ erg·cm$^{-3}$. Assuming further that: (i) there is no long-range transitional order in the plane, (ii) $z$ and $-z$ are equivalent (no ferroelectricity), and (iii) the deformation is small so that the molecules at any point remain perpendicular to the plane of the layer, the total free energy of the system can be derived (1):

$$F_{\text{total}} = F_0 + \frac{1}{2}\overline{B}\left(\frac{\partial u}{\partial z}\right)^2 + \frac{1}{2}K_1\left(\frac{\partial^2 u}{\partial x^2} + \frac{\partial^2 u}{\partial y^2}\right)^2 + \frac{1}{2}\Delta\chi^m H^2\left[\left(\frac{\partial u}{\partial x}\right)^2 + \left(\frac{\partial u}{\partial y}\right)^2\right] \tag{4.42}$$

The first term on the right-hand side is the unperturbed free energy. The second term is the energy associated with layer compression. The third term is the splay distortion energy which is identical in form to that in nematics, $\frac{1}{2}K_1(\nabla\cdot\hat{n})^2$. Similarly, the fourth term is the field-induced distortion energy, $\frac{1}{2}\Delta\chi^m(\hat{n}\cdot H)^2$, as in nematics.

Note that (4.42) contains only the splay distortion, which preserves the layer spacing (cf. Figure 4.6). The other two distortions, bend and twist, allowed in nematics are prohibited in the SmA phase as they involve extremely high distortion energy. This is also manifested in the form of a

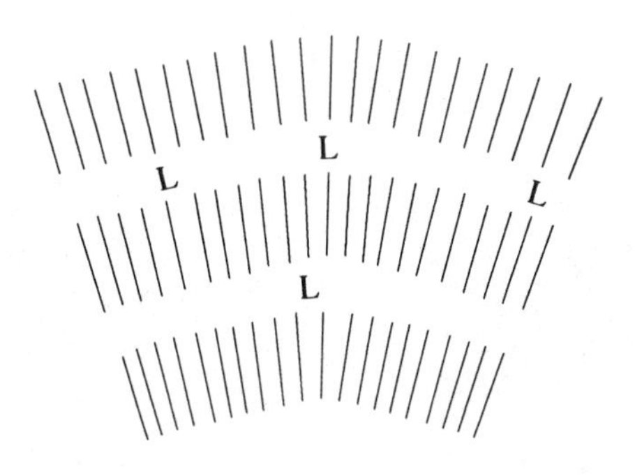


**Figure 4.6** Splay distortion in a SmA liquid crystal.

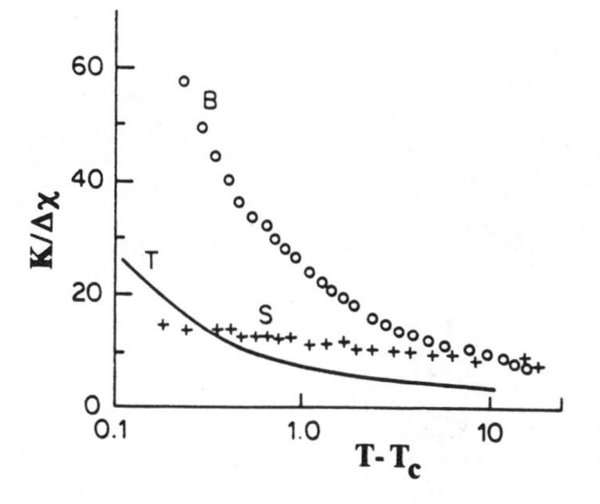


**Figure 4.7** Measured values of bend and splay elastic constant in the nematic phase as a function of the vicinity of nematic → SmA transition in CBOOA [after Cheung et al. (16)].

divergence in the values of the corresponding elastic constants $K_3$ and $K_2$ in the nematic phase as the temperature approaches the nematic → SmA transition (16), as shown in Figure 4.7.

### 4.4.2 Light Scattering in SmA Liquid Crystals

Light-scattering processes in SmA liquid crystals are governed by fluctuations in the layer displacement; they are analogous to scattering in nematics discussed in detail in the next chapter. In terms of the Fourier $U_q$ components of the layer displacement, the free energy in the absence of an external field is given by (1):

$$F = \sum_q \frac{1}{2}\left[\bar{B}q_z^2 + Kq_\perp^4\right]|u_q|^2 \tag{4.43}$$

where $q_z$ and $q_\perp$ are the scattering wave vector components parallel and perpendicular to the $z$ axis, respectively.

Following the derivation given in Chapter 5, the intensity of the scattered wave is given by

$$I(q) = \langle|\hat{i}\cdot\delta\varepsilon(q)\cdot\hat{f}|^2\rangle \tag{4.44}$$

where $\hat{i}$ and $\hat{f}$ are the polarization of the incoming and outgoing waves and $\delta\varepsilon(q)$ is the fluctuation in the dielectric constant tensor:

$$\delta\varepsilon = \Delta\varepsilon[\delta n : \hat{n} + \hat{n} : \delta n] \tag{4.45}$$

Using (4.39), this gives

$$\delta\varepsilon_{xz} = -\Delta\varepsilon \frac{\partial u}{\partial x} \tag{4.46a}$$

$$\delta\varepsilon_{yz} = -\Delta\varepsilon \frac{\partial u}{\partial y} \tag{4.46b}$$

Accordingly, the scattered intensity can be derived (1):

$$I = \frac{k_B T}{\overline{B}} \Delta\varepsilon^2 \frac{q^2}{q_z^2 + \lambda^2 q_\perp^4} \tag{4.47}$$

where $\lambda \equiv (K_1/B)^{1/2}$ and is on the order of the layer thickness. There are two distinct regimes: (1) $q_z \sim q_\perp \neq 0$ and (2) $q_z = 0$.

In case (1), note that $q\lambda \ll 1$ and we thus have

$$I = \Delta\varepsilon^2 \frac{k_B T}{\overline{B}} \frac{q_\perp^2}{q_z^2} \tag{4.48}$$

which is on the order of $k_B T/E_{\text{comp}}$, similar to isotropic liquid crystals (cf. Chapter 5). This is the usual case involving the transmission of light through a smectic film; the scattering is much smaller than in the nematic phase.

Case (2), however, corresponds to large scattering. If $q_z = 0$, equation (4.47) becomes

$$I = \Delta\varepsilon^2 \frac{k_B T}{\overline{B}\lambda^2 q^2} = \frac{\Delta\varepsilon^2 k_B T}{K_1 q^2} \tag{4.49}$$

which is analogous to scattering in the nematic phase. The mode of excitation causing large scattering corresponds to pure undulation for which the interlayer spacing is fixed.

## 4.5 SMECTIC-C LIQUID CRYSTALS

### 4.5.1 Free Energy

The finite tilt of the director axis from the layer normal (taken as the $\hat{z}$ axis) introduces a new degree of freedom, namely, a rotation around the $z$ axis, compared to the SmA phase. This rotation preserves the layer spacing and therefore does not require too much energy. Since $\partial u/\partial y$ and $\partial u/\partial x$ are equivalent to rotations around the $x$ and $y$ axes, respectively, we may express the free energy in Smectic-C (SmC) liquid crystals in terms of the rotation components:

$$\Omega_x = \frac{\partial u}{\partial y} \qquad \Omega_y = -\frac{\partial u}{\partial x} \qquad \Omega_z \tag{4.50}$$

Taking into account all the energy terms associated with director axis rotation, interlayer distortion, and possible coupling between them, the total free energy of the system is given by (1):

$$F = F_c + F_d + F_{cd} \tag{4.51}$$

where

$$F_c = \frac{1}{2}B_1\left(\frac{\partial\Omega_z}{\partial x}\right)^2 + \frac{1}{2}B_2\left(\frac{\partial\Omega_z}{\partial y}\right)^2 + \frac{1}{2}B_3\left(\frac{\partial\Omega_z}{\partial z}\right)^2 + B_{13}\frac{\partial\Omega_z}{\partial x}\frac{\partial\Omega_z}{\partial z} \tag{4.52}$$

$$F_d = \frac{1}{2}A\left(\frac{\partial\Omega_x}{\partial x}\right)^2 + \frac{1}{2}A_{12}\left(\frac{\partial\Omega_y}{\partial x}\right)^2 + \frac{1}{2}A_{21}\left(\frac{\partial\Omega_x}{\Omega y}\right)^2 + \frac{1}{2}\overline{B}\left(\frac{\partial u}{\partial z}\right)^2 \tag{4.53}$$

and

$$F_{cd} = C_1\frac{\partial\Omega_x}{\partial x}\frac{\partial\Omega_z}{\partial x} + C_2\frac{\partial\Omega_x}{\partial y}\frac{\partial\Omega_z}{\partial y} \tag{4.54}$$

Here $F_c$ is the free energy associated with director axis rotation without change of layer spacing, $F_d$ is due to layer distortions, and $F_{cd}$ is the cross term describing the coupling of these layer distortions and the free-rotation process.

### 4.5.2 Field-Induced Director Axis Rotation in SmC Liquid Crystals

In practical implemententations or switching devices, the logical thing to do is to involve only one or a small number of these distortions. If an external field is applied, the field-dependent terms [cf. equations (4.5*a*) and (4.5*b*)] should be added to the total free-energy expression.

The process of field-induced director axis distortion in SmC is analogous to the nematic case. For example, the first three terms on the right-hand side of (4.53) correspond to the splay term in nematics:

$$\frac{1}{2}K_1\left(\frac{\partial^2 u}{\partial x^2} + \frac{\partial^2 u}{\partial y^2}\right)^2 = \frac{1}{2}K_1\left(\frac{\partial\Omega_x}{\partial y} - \frac{\partial\Omega_y}{\partial x}\right)^2 \tag{4.55}$$

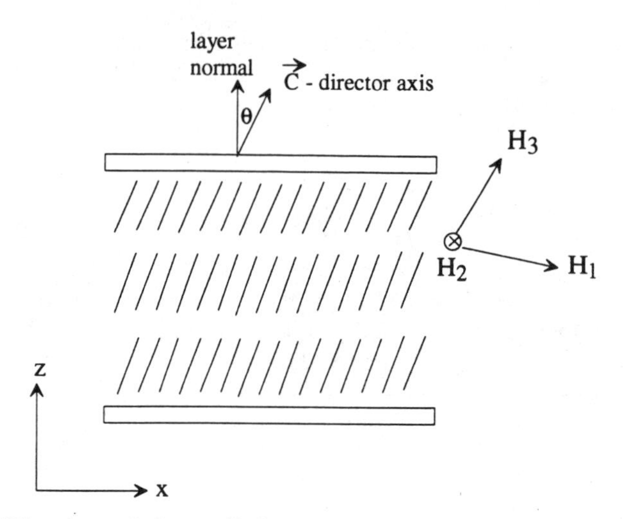


**Figure 4.8** Directions of the applied magnetic fields components $H_1$, $H_2$, and $H_3$ relative to the smectic $C$- axis.

Accordingly, if only such distortions (i.e., no layer displacement or coupling effects) are induced in a SmC sample by an applied field, Freedericksz transitions (discussed in the previous chapter for nematics) will occur.

Consider, for example, the effect caused by a magnetic field as depicted in Figure 4.8. The applied field has three components, $H_1$, $H_2$, and $H_3$, and the respective diamagnetic susceptibility components are $\chi_1^m$, $\chi_2^m$, and $\chi_3^m$; $\chi_3^m$ corresponds to the director axis, usually denoted as the $C$ axis; $\chi_2^m$ is along $y$, and $\chi_1^m$ is in a direction orthogonal to both the 3 and $z$ axes.

If the applied field is along the $C$ axis (i.e., $H_3$), a Freedericksz transition is possible for $\chi_2^m > \chi_3^m$. The director should rotate around the $z$ axis so that, in the strong field limit, its projection onto the smectic layer coincides with the $y$ axis. There is no change in the tilt angle $\theta_0$. The threshold field for this process is

$$H_{3C} = \frac{\pi \sin\theta}{d}\left(\frac{B}{\chi_2^m - \chi_3^m}\right)^{1/2} \tag{4.56}$$

where $B$ is the appropriate elastic constant defined by (4.52).

If the applied field is along the $y$ axis (i.e., $H_2$), a rotation of the $C$ axis around $z$, as depicted in Figure 4.9, will occur with a threshold field of

$$H_{2C} = \frac{\pi \sin\Omega}{d}\left[\frac{B}{\chi_1^m(\cos\Omega)^2 + \chi_3^m(\sin\Omega)^2 - \chi_2^m}\right]^{1/2} \tag{4.57}$$

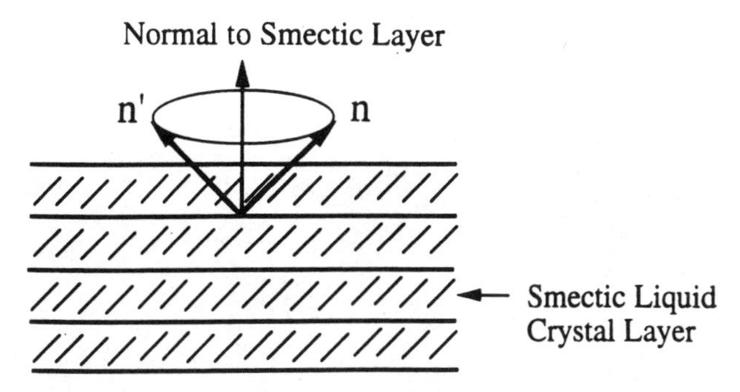


**Figure 4.9** Rotation of the $C$- axis around the layer normal; no change in layer spacing.

For an applied field along $H_1$ and $\chi_2^m > \chi_1^m$, the threshold field is given by

$$H_{1C} = \frac{\pi \sin \Omega}{d \cos \Omega} \left( \frac{B}{\chi_2^m - \chi_1^m} \right)^{1/2} \tag{4.58}$$

Usually $\chi_3^m > \chi_2^m \approx \chi_1^m$ and therefore case 2 involves the least field strength.

The case of optical field-induced director axis rotation is discussed in Chapter 6 in the context of nonlinear optics.

## 4.6 SMECTIC-C* AND FERROELECTRIC LIQUID CRYSTALS

Smectic-C*, or ferroelectric, liquid crystals, which possess nonzero spontaneous polarization **P**, may be classified into two categories. In the case of unwound SmC* liquid crystals, the director axis $\hat{n}$ is tilted at a fixed angle $\theta$ with respect to the layer normal. It follows that the direction of **P** is also fixed, as shown in Figure 1.11*d*. It is optically homogeneous. On the other hand, the director axis of helically modulated SmC* liquid crystals varies in a helical manner from layer to layer, as shown in Figure 4.10. The director axis precesses around the normal to the layer with a pitch that is much larger than the layer thickness. The magnitude of the pitch is on the order of the optical wavelength, and thus helically modulated SmC* liquid crystals are optically inhomogeneous.

In helically modulated SmC* liquid crystals, the bulk polarization is vanishingly small. The helicity can be unwound by an external field applied parallel to the smectic layers. It can also be unwound by surface effects if the samples are sufficiently thin (thickness < pitch), leading to the so-called surface-stabilized ferroelectric liquid crystal (SSFLC) with a nonvanishing macroscopic polarization.

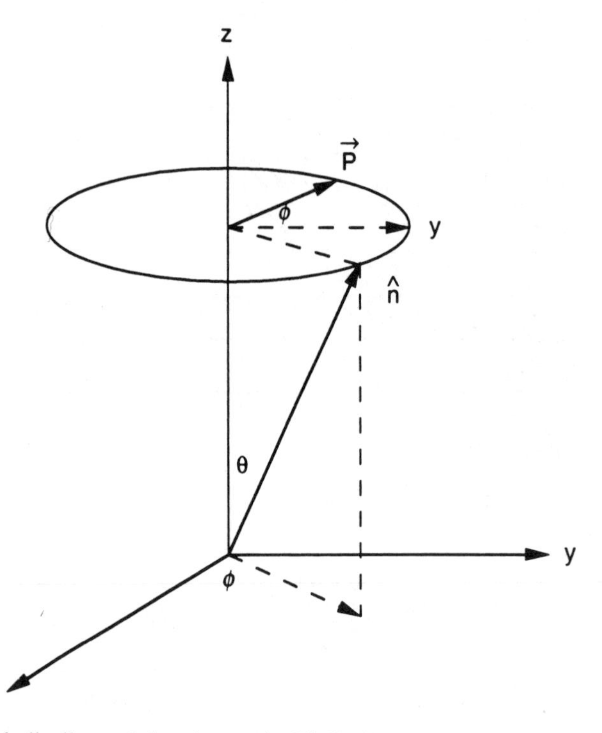


**Figure 4.10** A helically modulated smectic-C* liquid crystal; the directions of both $\hat{n}$ and **p** vary spatially in a helical manner.

Both the tilt angle $\theta$ and the spontaneous polarization **p** decrease in magnitude as the temperature of the system is increased. Above a critical temperature $T_{c^*A}$, a phase transition to the untilted SmA phase takes place (17).

Various optical effects arise as a result of the presence of the spontaneous polarization. Electrooptical effects are discussed in Khoo and Wu (9). In this section our attention will be focused on those physical and optical properties relevant for understanding the nonlinear optical responses of SmC* and related phenomena.

### 4.6.1 Free Energy of Ferroelectric Liquid Crystals

Recalling the free-energy expression for the nonferroelectric SmC liquid crystals discussed in the preceding section, one should not be too surprised to find that the free energy of ferroelectric liquid crystals is even more complicated. Besides the elastic energy $F_E$, several others play an equally important role in determining the response of the ferroelectric-liquid crystal to an external field. These include the surface energy density $F_s$, the spontaneous polarization density $F_p$, and the dielectric interaction energy density $F_{\text{diel}}$ with the applied field. These interactions have been studied by various workers; here we summarize the main results.

The elastic part of the free energy is analogous to the chiral nematic phase. It has been derived by Nakagawa et al. (18) and can be expressed as follows:

$$F_K = \tfrac{1}{2}\left[K_1(\nabla\cdot\hat{n})^2 + K_2(\hat{n}\cdot\nabla\times\hat{n} + Q_T)^2 + K_3(\hat{n}\times\nabla\times\hat{n} + Q_B b)^2\right] \quad (4.59)$$

where $Q_T$ and $Q_B$ are the inherent twist and bend wave numbers and $b = (\hat{n}\times\hat{k})$, where $\hat{k}$ is the layer normal unit vector.

Using the geometry shown in Figure 4.10, the director axis $\hat{n}$ becomes

$$\hat{n} = (\sin\theta\cos\phi, \sin\theta\sin\phi, \cos\theta) \quad (4.60)$$

Equation (4.59) becomes

$$F_E = F_0 + \frac{1}{2}A(1+\nu\sin^2\phi)\left(\frac{\partial\phi}{\partial y}\right)^2 + A(1+\nu)Q_0\sin\phi\left(\frac{\partial\phi}{\partial y}\right) \quad (4.61)$$

where

$$F_0 = \tfrac{1}{2}\left[K_2 Q_T^2 + K_3 Q_B^2\right] \quad (4.62a)$$

$$A = K_1\sin^2\theta \quad (4.62b)$$

$$B = \left(K_2\cos^2\theta + K_3\sin^2\theta\right)\sin^2\theta \quad (4.62c)$$

$$\nu = \frac{(B-A)}{A} \quad (4.62d)$$

$$Q^0 = \frac{\left[K_2 Q_T\sin\theta\cos\theta + K_3 Q_B\sin^2\theta\right]}{\left[A(1+\nu)\right]} \quad (4.62e)$$

The dielectric interaction energy density is simply given by $F_{\text{diel}} = -\tfrac{1}{2}\mathbf{D}\cdot\mathbf{E}$. For an applied electric field along the $y$ axis, for example, in the plane of the smectic layer, the interaction energy is given by

$$F_{\text{diel}} = -\tfrac{1}{2}\varepsilon_{yy}E_y^2 \quad (4.63)$$

where

$$\varepsilon_{yy} = \varepsilon_{\perp} \cos^2 \phi + \varepsilon_{\parallel} \sin^2 \phi \tag{4.64a}$$

$$\varepsilon'_{\parallel} = \varepsilon_{\perp} \cos^2 \theta + \varepsilon_{\parallel} \sin^2 \theta \tag{4.64b}$$

More generally, $F_{\text{diel}}$ may be written as

$$F_{\text{diel}} = -\tfrac{1}{2}\varepsilon_{\perp}(1 + \Delta\varepsilon' \sin^2 \phi)E^2 \tag{4.65}$$

where $\Delta\varepsilon' = (\varepsilon'_{\parallel} - \varepsilon_{\perp})/\varepsilon_{\perp}$.

The interaction of an electric field **E** with the permanent polarization **P** is simply $-\mathbf{P}\cdot\mathbf{E}$. For the configuration given in Figure 4.10, we have

$$F_p = -PEy \cos \phi \tag{4.66}$$

As in the other phases of liquid crystals, the free energy associated with the surface interaction is the most complicated one. It takes on further significance in the case of surface-stabilized ferroelectric liquid crystals. These surface interactions have been studied by various workers (19, 20). In their treatments the surface energy is expressed as

$$F_s = -g_1 \cos^2(\phi - \phi_s) - g_2 \cos(\phi - \phi_s) \tag{4.67}$$

where $\phi_s$ is the pretilt angle of the molecules and $g_1$ and $g_2$ are the respective coefficients for the nonpolar and polar surface interaction terms. Nakagawa et al. (18) later improved upon this expression:

$$F_s(\phi) = \sum_{1,2} h^{1,2}(\phi) \tag{4.68}$$

where

$$h^{1,2}(\phi) = -g^{1,2}\left\{C^{1,2} \exp\left[-\frac{\alpha'}{2}\sin^2(\phi - \phi^{1,2})\right] + [1 - C^{1,2}]\exp\left[-\frac{\alpha'}{2}\cos^2(\phi + \phi^{1,2})\right]\right\} \tag{4.69}$$

Here the superscripts 1 and 2 refer to the two cell boundary plates, $\alpha'$ is a parameter characterizing the anchoring potential, and $C^{1,2}$ relates the relative stability between the $\phi^{1,2}$ and the $\pi - \phi^{1,2}$ states. These expressions satisfy the symmetry requirement $h^{1,2}(\pi - \phi) = h^{1,2}(\phi)$ for $C^1 = C^2 = \frac{1}{2}$.

The total free energy of the ferroelectric liquid crystal system depicted in Figure 4.10 is therefore given by

$$G = \int_{\text{vol}} F\,dv \tag{4.70}$$

where $F = F_E + F_{\text{diel}} + F_p + F_s$.

The equilibrium configuration of the system is obtained by minimizing the total energy with respect to $\phi$:

$$\frac{\partial G}{\partial \phi} = 0 \tag{4.71}$$

This yields

$$\frac{\partial}{\partial y}(1 + \Delta\varepsilon' \sin^2\phi)\frac{\partial \phi}{\partial y} = \frac{-P_s}{\varepsilon_\perp} \sin\phi \frac{\partial \phi}{\partial y} \tag{4.72}$$

where $\phi$ is the electric field potential associated with the field $E = -\partial\phi/\partial y$ applied in the $y$ direction.

The dynamics of the molecular reorientation process is described by the torque balance equation:

$$\frac{\partial G}{\partial \phi} + \gamma_1 \frac{\partial \phi}{\partial t} = 0 \tag{4.73}$$

where $\gamma_1$ is the rotational viscosity coefficient. This yields

$$\gamma_1 \frac{\partial \phi}{\partial t} = A(1 + \nu \sin^2\phi)\left(\frac{\partial^2 \phi}{\partial y^2}\right) + \frac{\nu}{2} A \sin 2\phi \left(\frac{\partial \phi}{\partial y}\right)^2 + \frac{\Delta\varepsilon'}{2}\varepsilon_\perp E^2 \sin 2\phi + PE \sin\phi \tag{4.74}$$

This equation is greatly simplified under the one-constant approximation ($K_1 = K_2 = K_3 = K$). This gives

$$K \sin^2\theta \frac{\partial^2 \phi}{\partial y^2} + \frac{\Delta\varepsilon'}{2}\varepsilon_\perp E^2 \sin 2\phi + PE \sin\phi = \gamma_1 \frac{\partial \phi}{\partial t} \tag{4.75}$$

From this, one can see that the dynamics is controlled by the elastic torque (first term on the left-hand side), the optical dielectric torque (second term on the left-hand side), and the polarization torque (third term on the right-hand side).

The polarization part of the dynamics is governed by a time constant given by

$$\tau = \frac{\gamma_1}{PE} \tag{4.76}$$

Using typical values of $P$ and $\gamma_1$ for ferroelectric liquid crystals ($P \sim 10^{-5}$ cm$^{-2}$, $\gamma_1 \approx 10^{-2}$ mks units, $E = 10^7$ V·m$^{-1}$), $\tau$ is on the order of 100 $\mu$s. Ferroelectric liquid crystals of much faster response time have by now been developed by several research groups and industrial organizations (21).

### 4.6.2 Smectic-C* –Smectic-A Phase Transition

The principal parameter that distinguishes smectic-C* from smectic-A is the tilt angle $\theta_0$. Because of the chiral character of the molecule, the tilt precesses around the normal to the smectic layers, together with the transverse electric polarization **P** (cf. Figure 4.10). In the theories developed for describing the phase transition phenomena from the smectic-C* phase to the smectic-A phase, the tilt angle is treated as a primary order parameter of the system, very much as the director axis $\hat{n}$ in the nematic or cholesteric phase, while **P** is regarded as a secondary one.

Writing the two components of **P** and the tilt angle in the $x-y$ plane as $\mathbf{P} = (P_x, P_y)$ and $\boldsymbol{\theta} = (\theta_1, \theta_2)$, the free-energy density $f_0(z)$ of the system can be expressed as a Landau type of expansion in terms of $\theta_1$, $\theta_2$, $P_x$, and $P_y$ in the following form (22, 23):

$$\begin{aligned} f_0(z) = {} & \frac{1}{2}A(\theta_1^2 + \theta_2^2) + \frac{1}{4}B(\theta_1^2 + \theta_2^2)^2 - \Lambda\left(\theta_1 \frac{d\theta_2}{dz} - \theta_2 \frac{d\theta_1}{dz}\right) \\ & + \frac{1}{2}K_3\left[\left(\frac{d\theta_1}{dz}\right)^2 + \left(\frac{d\theta_2}{dz}\right)^2\right] + \frac{1}{2\varepsilon}(P_x^2 + P_y^2) \\ & - \mu\left(P_x \frac{d\theta_1}{dz} + P_y \frac{d\theta_2}{dz}\right) + C(P_x\theta_2 - P_y\theta_1) \end{aligned} \tag{4.77}$$

In this expression only the coefficient of the term quadratic in the primary parameter is temperature dependent, whereas the coefficient of the $P^2$ term is constant; this is so because it is not the interaction between the electric polarization that leads to a phase transition. The coefficient $A$ is of the form $A = A_0(T - T_{CA})$, where $T_{CA}$ is the smectic-C–smectic-A transition temperature, $K_3$ is the elastic constant, and $\Lambda$ is the coefficient of the so-called Lifshitz term responsible for the helicoidal structure. $\mu$ and $C$ are the coefficients of the flexoelectric and piezoelectric bilinear couplings between the tilt and the polarization.

The coefficients $\Lambda$ and $C$ are dependent on the chiral character of the molecules. For nonchiral molecules, $\Lambda$ and $C$ are zero; minimization of the free energy given in (4.77) yields a system where the director axis is homogeneously tilted below the transition temperature $T_c$. There is no linear coupling between the tilt and the polarization and thus $\mathbf{P} = 0$. For temperatures below $T_{CA}$, the smectic-C–smectic-A transition temperature, the magnitude of the tilt angle $\theta$ is given by a square root dependence:

$$\theta_0 = \sqrt{\frac{A_0}{B}(T - T_{CA})} \tag{4.78}$$

On the other hand, for chiral molecules, $\Lambda$ and $C$ are nonzero. In this case the free energy is minimized if $\theta$ and $P$ are described by the helical functions:

$$\theta_1 = \theta_0 \cos q_0 z \qquad \theta = \theta_0 \sin q_0 z \tag{4.79}$$

$$P_x = -P_0 \sin q_0 z \qquad P_y = P_0 \cos q_0 z \tag{4.80}$$

Note that $P$ is locally perpendicular to the tilt. Both process around $z$ with a pitch wave vector $q_0$ given by

$$q_0 = \frac{\Lambda + \varepsilon\mu C}{K_3 - \varepsilon\mu^2} \tag{4.81}$$

The magnitude of the spontaneous polarization $P_0$ is proportional to the tilt angle $\theta_0$:

$$P_0 = \varepsilon[C + \mu q_0]\theta_0$$

where $\theta_0$ is similar to that given in (4.78):

$$\theta_0 = \sqrt{\frac{A_0}{B}(T_{C*A} - T)} \tag{4.83}$$

The smectic-$C_*$ $\rightarrow$ smectic-A transition temperature $T_{C*A}$ is given by

$$T_{C*A} = T_{CA} + \frac{1}{A_0}\left[\varepsilon C^2 + K_3 q_0^2\right] \tag{4.84}$$

Above $T_{C*A}$, in the smectic-A phase, the two order parameters $P_0$ and $\theta_0$ vanish.

If an external field is present, as pointed out earlier, the helical structure can be unwound. From this point of view, the effect of the electric field is equivalent to "canceling" the elastic torque term $K_3 q_0^2$ in (4.84), resulting in a homogeneous (i.e., nonhelical C* system). The corresponding phase transition temperature for the unwound system is thus given by

$$T_{C^*A}^{\text{unwound}} = T_{CA} + \frac{1}{A_0}\varepsilon C^2 \tag{4.85}$$

Experimentally, $T_{CA}$, $T_{C^*A}$, and $T_{C^*A}^{\text{unwound}}$, and $T_{C^*A}^{\text{unwound}}$ are found to be very close to one another. This is expected as the chiral terms in the free-energy expression are basically small-perturbation terms. Their optical and electrooptical properties, however, are considerably modified by the presence of the chirality and spontaneous polarization. As we remarked earlier, ferroelectric liquid crystals provide a faster electrooptical switching mechanism (9). In the context of nonlinear optics, the non-centrosymmetry caused by the presence of **P** allows the generation of even harmonic light.

## REFERENCES


1. P. G. deGennes, "Physics of Liquid Crystals." Clarendon Press, Oxford, 1974.
2. R. B. Meyer, *Appl. Phys. Lett.* **14**, 208 (1968); P. G. deGennes, *Solid State Commun.* **6**, 163 (1968).
3. E. Sackmann, S. Meiboom, and L. C. Snyder, *J. Am. Chem. Soc.* **89**, 5982 (1967); J. Wysocki, J. Adams, and W. Haas, *Phys. Rev. Lett.* **20**, 1025 (1968); G. Durand, L. Leger, F. Rondele, and M. Veyssie, *ibid.* **22**, 227 (1969); R. B. Meyer, *Appl. Phys. Lett.* **14**, 208 (1969).
4. C. Fan, L. Kramer, and M. J. Stephen, *Phys. Rev. A* **2**, 2482 (1970).
5. J. D. Parson and C. F. Hayes, *Phys. Rev. A* **9**, 2652 (1974); see also deGennes (1).
6. See, for example, D. W. Berreman and T. J. Scheffer, *Phys. Rev. Lett.* **25**, 577 (1970); D. Taupin, *J. Phys.* (*Paris*) **30**, C4-C32 (1969).
7. G. W. Gray and J. Goodby, "Smectic Liquid Crystals: Textures and Structures." Leonard Hill, London, 1984.
8. R. B. Meyer, L. Liebert, L. Strzelecki, and P. J. Keller, *J. Phys. Lett. Orsay, Fr.* **36**, 69 (1975).
9. I. C. Khoo and S. T. Wu, "Optics and Nonlinear Optics of Liquid Crystals." World Scientific, Singapore, 1993; see also G. Anderson, I. Dahl, L. Komitov, S. T. Lagerwall, K. Sharp, and B. Stebler, *J. Appl. Phys.* **66**, 4983 (1989), for other chiral smectics.
10. I. C. Khoo and R. Normandin, *J. Appl. Phys.* **55**, 1416 (1984).
11. I. C. Khoo, R. R. Michael, and P. Y. Yan, *IEEE J. Quantum Electron.* **QE-23**, 1344 (1987).
12. N. M.Shtykov, M. I. Barnik, L. M. Blinov, and L. A. Beresnev, *Mol. Cryst. Liq. Cryst.* **124**, 379 (1985).
13. A. Taguchi, Y. Oucji, H. Takezoe, and A. Fukuda, *Jpn. J. Appl. Phys.* **28**, 997 (1989).

14. R. Macdonald, J. Schwartz, and H. J. Eichler, *Int. J. Nonlinear Opt. Phys.* **1**, 103 (1992).
15. J. Y. Liu, M. G. Robinson, K. M. Johnson, and D. Doroski, *Opt. Lett.* **15**, 267 (1990).
16. L. Cheung, R. B. Meyer, and H. Gruler, *Phys. Rev. Lett.* **31**, 349 (1973).
17. S. A. Pikin and V. L. Indenbom, *Ferroelectrics* **20**, 151 (1978).
18. M. Nakagawa, M. Ishikawa, and T. Akahane, *Jpn. J. Appl. Phys.* **27**, 456 (1988).
19. M. A. Handschy and N. A. Clark, *Ferroelectrics* **59**, 69 (1984).
20. Y. Yamada, T. Tsuge, N. Yamamoto, M. Yamawaki, H. Orihara, and Y. Ishibashi, *Jpn. J. Appl. Phys.* **26**, 1811 (1987).
21. See, for example, commercial information leaflets by the E. Merck (Germany) or BDH [UK] companies, see also Khoo and Wu (9).
22. S. A. Pikin and V. L. Indenbom, *Uspe. Fiz. Nauk* **125**, 251 (1978).
23. R. Blinc and B. Zeks, *Phys. Rev. A* **18**, 740 (1978).

# CHAPTER 5

# LIGHT SCATTERINGS

## 5.1 INTRODUCTION

In earlier chapters we discussed some specific light-scattering processes in the mesophases of liquid crystals. In particular, we found that the ability of the molecules to scatter light is very much dependent on the orientations and fluctuations of the director axis and their reconfiguration under applied fields. There are, however, light-scattering processes that occur on the molecular level that involve the electronic responses of the molecules. In this chapter we discuss the general approaches and techniques used for analyzing light-scattering processes in liquid crystals that are applicable in many respects to other media as well.

Approaches to the problems of light scattering in liquid crystals may be classified into two categories. In one category, such as Brillouin and Raman scatterings, knowledge of the actual molecular physical properties, such as resonances and energy level structures, is needed. On the other hand, in the electromagnetic formalism for light-scattering phenomena, one needs to invoke only the optical dielectric constants and their fluctuations. This latter approach is generally used for analyzing orientational fluctuations in liquid crystals.

The process of light scattering can also be divided into linear and nonlinear regimes. In linear optics the properties of liquid crystals are not affected by the incident light, which may be regarded as a probe or signal field. The resulting scattered or transmitted light, in terms of its spatial or temporal frequency spectrum and intensity, reflects the physical properties of the material. On the other hand, in the nonlinear optical regime the incident light interacts strongly with and modifies the properties of the liquid crystals. The resulting scattered or transmitted light will reflect these strong interactions.

In this and the preceding chapters, our attention is focused on linear optical scattering processes, which are nevertheless quite important in nonlinear optical phenomena. Nonlinear optics and the nonlinear optical properties of liquid crystals are presented in Chapters 6 to 10. We begin with a review of the electromagnetic theory of light-scattering terms associated with fluctuations of optical dielectric constants associated with temperature effects. Raman and Brillouin scatterings, which involve molecular energy levels and rovibrational excitations, are given later in this chapter.

## 5.2 GENERAL ELECTROMAGNETIC FORMALISM OF LIGHT SCATTERING IN LIQUID CRYSTALS

Scattering of light in a medium is caused by fluctuations of the optical dielectric constants, $\delta\varepsilon(\mathbf{r}, t)$. In isotropic liquids $\delta\varepsilon(\mathbf{r}, t)$ are mainly due to density fluctuations, caused by fluctuations in the temperature. In liquid crystals in their ordered phases, an additional and important contribution to $\delta\varepsilon(\mathbf{r}, t)$ arises from director axis fluctuations.

As a result of $\delta\varepsilon(\mathbf{r}, t)$, an incident light will be scattered. The direction, polarization, and spectrum of the scattered light depends on the optical–geometrical configuration.

In general, for a uniaxial birefringent medium such as a liquid crystal, the dielectric constant tensor may be written as

$$\varepsilon_{\alpha\beta} = \varepsilon_{\perp}\varepsilon_{\alpha\beta} + (\varepsilon_{\parallel} - \varepsilon_{\perp})n_{\alpha}n_{\beta} \tag{5.1}$$

where $n_\alpha$ and $n_\beta$ are the components of a unit vector $\hat{n}$ along the optical axis. (In liquid crystals $\hat{n}$ is the director axis.) Fluctuations in $\varepsilon_{\alpha\beta}$ come from changes in $\varepsilon_{\perp}$ and $\varepsilon_{\parallel}$ due to density and temperature fluctuations and from fluctuations in the directions of $\hat{n}$. Light scattering in liquid crystals was first quantitatively analyzed by deGennes (1), using classical electromagnetic theory and assuming that $\Delta\varepsilon$ is small.

Consider a small volume $dV$ located at a position $\mathbf{r}$ from the origin as shown in Figure 5.1. The induced polarization $\delta\mathbf{P}$ (dipole moment per unit volume) at this location by an incident field $\mathbf{E}_{\text{inc}} = \hat{i}E\exp[i(\mathbf{k}_i \cdot \mathbf{r} - \omega t)]$ is given by

$$\begin{aligned} \delta\mathbf{P} &= \varepsilon_0 \bar{\bar{\chi}} : \mathbf{E}_{\text{inc}} \\ &= \delta\bar{\bar{\varepsilon}} : \mathbf{E}_{\text{inc}} \end{aligned} \tag{5.2}$$

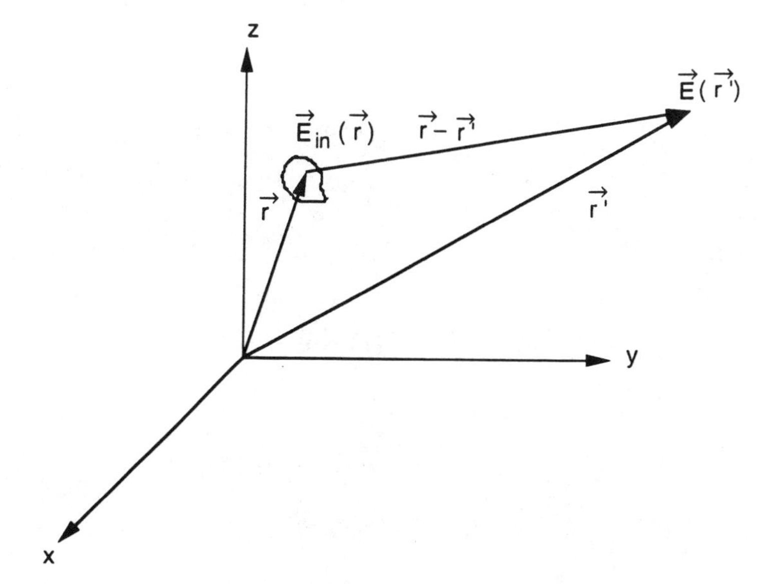


**Figure 5.1** Scattering of light from an elementary volume located at **r**.

where $\delta\bar{\bar{\varepsilon}}$ is the change in the dielectric constant tensor. This contributes an outgoing field at $\mathbf{r}'$ given by (2):

$$\mathbf{E}(\mathbf{r}') = \frac{\omega^2}{c^2 R}\left[\exp i(\mathbf{k}_f \cdot \mathbf{R} - \omega t)\right]\mathbf{P}_\omega(\mathbf{r}) \tag{5.3}$$

where $\mathbf{P}_\omega(\mathbf{r})$ is the component of $\delta\mathbf{P}$ normal to the direction $\mathbf{R} = \mathbf{r}' - \mathbf{r}$. The total outgoing field $\mathbf{E}_{\text{out}}$ at a position $\mathbf{r}$ in the so-called far-field zone ($|r' - r| \gg \lambda$) is given by the integral of all the radiated contributions from the volume.

Let $\hat{f}$ denote the polarization vector of the outgoing field and let $\mathbf{K}_f$ denote its wave vector; that is, writing

$$\mathbf{E}_{\text{out}} = \hat{f}E_{\text{out}} \exp i(\mathbf{k}_f \cdot \mathbf{r}' - \omega t) \tag{5.4}$$

we have

$$\hat{f} \cdot \mathbf{E}_{\text{out}} = \frac{\omega^2 E}{c^2 R} \exp(i\mathbf{k}_f \cdot \mathbf{r}') \int_{\text{vol}} \left[\hat{f} \cdot \delta\bar{\bar{\varepsilon}} : \hat{i}\right] \exp - i(\mathbf{q} \cdot \mathbf{r})\, dV \tag{5.5}$$

where $\mathbf{q} = \mathbf{k}_f - \mathbf{k}_i$. The scattering amplitude $\alpha_{fi}$ is defined by

$$\hat{f} \cdot \mathbf{E}_{\text{out}} = \alpha_{fi}\left[\frac{E}{R} \exp(i\mathbf{k}_f \cdot \mathbf{r}')\right] \tag{5.6a}$$

that is,

$$\alpha_{fi} = \frac{\omega^2}{c^2} \int_{\text{vol}} \delta\varepsilon_{fi}(\mathbf{r}) \exp(-i\mathbf{q}\cdot\mathbf{r})\, dV \tag{5.6b}$$

where $\delta\varepsilon_{fi}(\mathbf{r}) = \hat{f} : \delta\bar{\bar{\varepsilon}} : \hat{i}$.

Writing

$$\delta\varepsilon_{fi}(\mathbf{q}) = \frac{1}{V} \int_{\text{vol}} \delta\varepsilon_{fi}(\mathbf{r}) \exp(-i\mathbf{q}\cdot\mathbf{r})\, dV \tag{5.7}$$

we have

$$\alpha_{fi} = \frac{\omega^2}{c^2} V \delta\varepsilon_{fi}(\mathbf{q}) \tag{5.8}$$

The differential scattering cross section $d\sigma/d\Omega$ (per solid angle) is given by the *thermal* average of $|\alpha_{fi}|^2$, that is,

$$\frac{d\sigma}{d\Omega} = \langle |\alpha_{fi}|^2 \rangle \tag{5.9}$$

The thermal average is used because the fluctuations are attributed to the temperature.

## 5.3 SCATTERING FROM DIRECTOR AXIS FLUCTUATIONS IN NEMATIC LIQUID CRYSTALS

Scattering of light in nematic liquid crystals is a complicated problem, since so many vector fields are involved. The crucial parameters are the wave vectors $\mathbf{k}_i$ and $\mathbf{k}_f$, the scattering wave vector $\mathbf{q}$, the director axis orientation $\hat{n}$, and its fluctuations $\delta\mathbf{n}$ from its equilibrium direction $\hat{n}_0$. As developed by deGennes (1), the problem of analyzing light scattering in nematic liquid crystals can be greatly simplified if the coordinate system is properly defined in terms of the initial orientation of the director axis $\hat{n}_0$ with respect to the scattering wave vector $\mathbf{q}$.

As shown in Figure 5.2, the director axis fluctuation $\delta\mathbf{n}$ (which is normal to $\hat{n}_0$, since $|\hat{n}| = 1$) is decomposed into two orthogonal components $\delta\mathbf{n}_1$ and $\delta\mathbf{n}_2$, along the unit vectors $\hat{e}_1$ and $\hat{e}_2$, respectively. Note that one of them, $\delta\mathbf{n}_1$, is in the plane defined by $\mathbf{q}$ and $\hat{n}_0$ (taken as $\hat{z}$), and the other, $\delta\mathbf{n}_2$, is perpendicular to the $q-z$ plane.

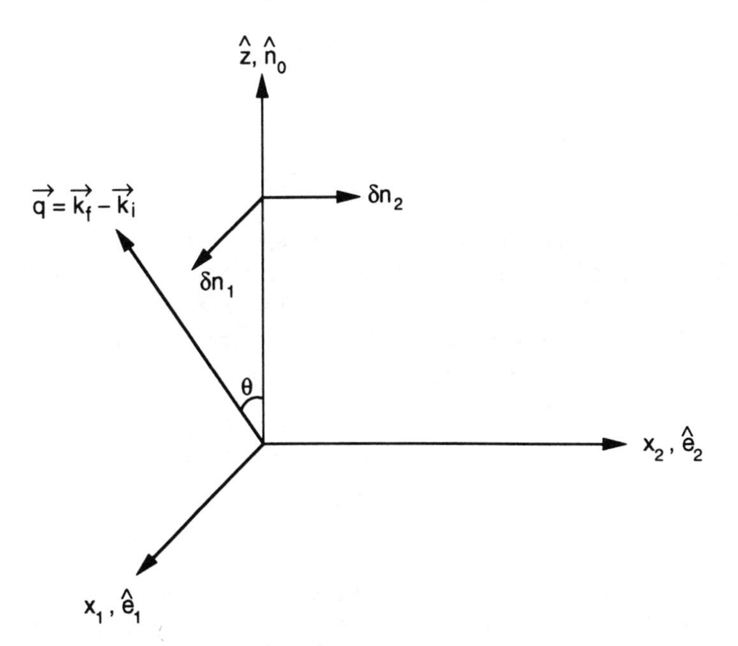


**Figure 5.2** Coordinate system for analyzing light scattering in nematic liquid crystals in terms of two normal modes.

In this case one can express $\delta n_1(\mathbf{r})$ and $\delta n_2(\mathbf{r})$ in terms of their Fourier components as

$$\delta n_{1,2}(\mathbf{r}) = \sum_q n_{1,2}(\mathbf{q}) \exp(i\mathbf{q}\cdot\mathbf{r}) \tag{5.10}$$

The inverse Fourier transform is given by

$$n_{1,2}(\mathbf{q}) = \frac{1}{V}\int n_{1,2}(\mathbf{r}) e^{-i\mathbf{q}\cdot\mathbf{r}}\, dV \tag{5.11}$$

The total free energy associated with this director axis deformation is, from (3.6), given by

$$F_{\text{total}} = \frac{1}{2}\int K_1\left(\frac{\partial n_1}{\partial x_1} + \frac{\partial n_2}{\partial x_2}\right)^2 + K_2\left(\frac{\partial n_1}{\partial x_2} - \frac{\partial n_2}{\partial x_1}\right)^2$$
$$+ K_3\left[\left(\frac{\partial n_1}{\partial x_3}\right)^2 + \left(\frac{\partial n_2}{\partial x_3}\right)^2\right] dV \tag{5.12}$$

Substituting (5.10) into (5.12), we thus have

$$F_{\text{total}} = \frac{V}{2} \sum_{q} \sum_{1,2} |n_{1,2}(q)|^2 [K_{1,2} q_1^2 + K_3 q_3^2] \tag{5.13}$$

An important feature of the total free energy is that it is the sum of the energies associated with the two normal modes $\delta n_1$ and $\delta n_2$; these modes are not coupled to each other. From classical mechanics, at thermal equilibrium, the thermally averaged energy of each normal mode is $\frac{1}{2}k_B T$, where $k_B$ is the Boltzmann constant. In other words, we have

$$\left\langle \frac{V}{2} |n_{1,2}(q)|^2 [K_{1,2} q_1^2] \right\rangle = \frac{1}{2} k_B T \tag{5.14}$$

that is,

$$\langle |n_{1,2}(q)|^2 \rangle = \frac{k_B T / V}{k_3 q_3^2 + k_{1,2} q_1^2} \tag{5.15}$$

From (5.1) the change in $\varepsilon_{\alpha\beta}$ associated with the director axis fluctuation comes from the second term:

$$\delta\varepsilon_{\alpha\beta} = \Delta\varepsilon [n_\alpha \delta n_\beta + \delta n_\alpha n_\beta] \tag{5.16}$$

This means

$$\delta\varepsilon_{fi} = \Delta\varepsilon [(\hat{n}_0 \cdot \hat{i})(\delta \mathbf{n} \cdot f) + (\hat{n}_0 \cdot \hat{f})(\delta\hat{n} \cdot i)] \tag{5.17}$$

where $\hat{i}$ and $\hat{f}$ are the incident and outgoing optical field polarization directions, assumed to be orthogonal to one another (cf. Figure 5.3).

If we express $\delta\mathbf{n}(\mathbf{r})$ in terms of its Fourier components, that is,

$$\delta n(\mathbf{q}) = \hat{e}_i n_i(\mathbf{q}) + \hat{e}_2 n_2(\mathbf{q}) \tag{5.18}$$

equation (5.17) becomes

$$\delta\varepsilon_{fi} = \Delta\varepsilon \sum_{\alpha=1,2} n_\alpha(q) [(\hat{i} \cdot \hat{n}_0)(f \cdot \hat{e}_\alpha) + (\hat{f} \cdot \hat{n}_0)(\hat{i} \cdot \hat{e}_\alpha)] \tag{5.19}$$

Using (5.8), (5.9) and $\delta\varepsilon_{fi}$ given previously and noting that when we square $\alpha_{fi}$ cross terms involving $n_1(q)$ and $n_2(q)$ disappear because the two modes

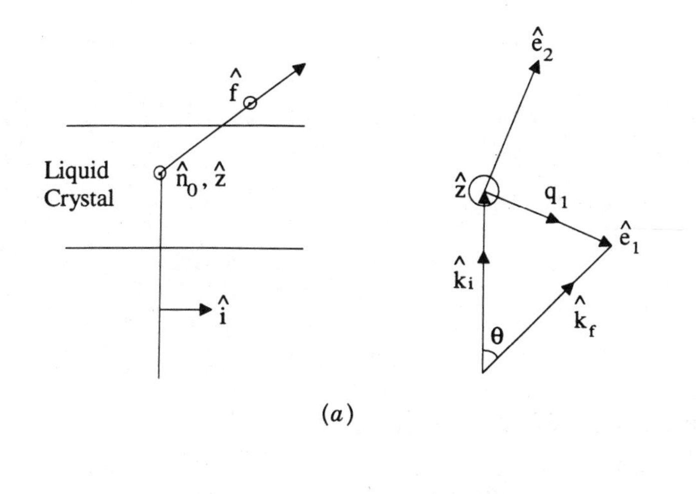


(a)

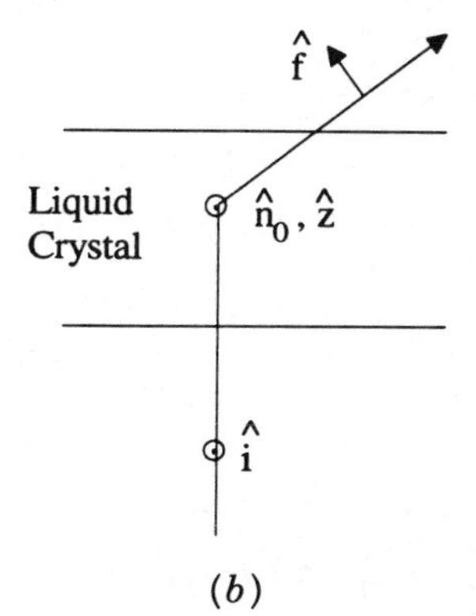


(b)

**Figure 5.3** Geometries for intense scattering of the incident light. Note that the incident and scattered light are cross polarized.

are decoupled, we finally have

$$\frac{d\sigma}{d\Omega} = \left(\frac{\Delta\varepsilon\,\omega^2}{c^2}\right)^2 V \sum_{\alpha=1,2} \langle |n_\alpha(q)|^2\rangle \left[(\hat{\imath}\cdot\hat{n}_0)(f\cdot\hat{e}_\alpha) + (\hat{f}\cdot\hat{n}_0)(\hat{\imath}\cdot\hat{e}_\alpha)\right]^2 \quad (5.20)$$

Using (5.15) for $\langle |n_{1,2}(q)|^2\rangle$, the differential cross section thus becomes

$$\frac{d\sigma}{d\Omega} = \left(\frac{\Delta\varepsilon\,\omega^2}{c^2}\right)^2 V \sum_{\alpha=1,2} \frac{k_B T}{\left[k_3 q_3^2 + k_\alpha q_1^2\right]} \left[f_\alpha(\hat{\imath}\cdot\hat{n}_0) + i_\alpha(\hat{f}\cdot\hat{n}_0)\right]^2 \quad (5.21)$$

Consider the scattering geometry shown in Figure 5.3*a*. Note that $q_1 \approx 2k\sin(\theta/2)$ (since $k_i \approx k_f \approx k$), $q_3 = 0$, $i_3 = f_1 = f_2 = 0$, $f_3 = 1$, $i_1 = \cos(\theta/2)$, and $i_2 = \sin(\theta/2)$. Inserting these parameters into (5.21) yields

$$\frac{d\sigma}{d\Omega} = \frac{\Delta\varepsilon^2 \omega^4}{c^4} V k_B T \left[ \frac{\cos^2(\theta/2)}{k_1 q_1^2} + \frac{\sin^2(\theta/2)}{k_2 q_1^2} \right] \tag{5.22}$$

Using $q_1 \sim 2k \sin(\theta/2)$, gives

$$\frac{d\sigma}{d\Omega} \sim \left[ \cot^2\left(\frac{\theta}{2}\right) + \frac{K_1}{K_2} \right] \tag{5.23}$$

One can similarly deduce the differential scattering cross sections for other geometries.

Equations (5.22) and (5.23) give good agreement with experimental observations (3, 3a):

1 Scattering is intense for crossed polarizations (i.e., when the incident and outgoing optical electric fields are orthogonal to each other).
2 Scattering is particularly strong at small scattering wave vector $\mathbf{q}$, small $\theta$.

If an external field is present (e.g., a magnetic field applied in the $z$ direction), the director axis fluctuations may be reduced. Quantitatively, this may be estimated by including in the free-energy equation (5.12) the magnetic interaction term

$$F_{\text{mag}} = \tfrac{1}{2} \int \Delta\chi^m H^2 (n_1^2 + n_2^2)\, dV \tag{5.24}$$

The reader can easily show that this will modify the result for $\langle |n_{1,2}(q)|^2 \rangle$ [cf. equation (5.15)] to yield

$$\left\langle |n_{1,2}(q)|^2 \right\rangle = \frac{k_B T / V}{k_3 q_3^2 + k_{1,2} q_1^2 + \Delta\chi H^2} \tag{5.25}$$

which shows that the fluctuation-induced scattering will be quenched at high field (4).

## 5.4 LIGHT SCATTERING IN THE ISOTROPIC PHASE OF LIQUID CRYSTALS

In the isotropic phase director axis orientations are random. The optical dielectric constant, a thermal average, is therefore a scalar parameter. The fluctuations in $\varepsilon$ in this case are due mainly to fluctuations in the density of the liquid, caused by temperature fluctuations.

Denoting the average dielectric constant in the isotropic phase by $\bar{\varepsilon}$ and denoting the local change in the volume by $u(\mathbf{r})$, the dielectric constant may be expressed as

$$\begin{aligned}\varepsilon &= \bar{\varepsilon} + \frac{d\varepsilon}{dV}u(\mathbf{r}) \\ &= \bar{\varepsilon} + \varepsilon' u(\mathbf{r})\end{aligned} \tag{5.26}$$

The compressional energy associated with the volume change is

$$F_u = \tfrac{1}{2}\int W|u(\mathbf{r})|^2\, dV \tag{5.27}$$

$$= \frac{VW}{2}\sum_q |u(\mathbf{q})|^2 \tag{5.28}$$

where $u(\mathbf{q})$ is the Fourier transform of $u(\mathbf{r})$, in analogy to our previous analysis of $n(\mathbf{r})$ and $n(\mathbf{q})$, and $W$ is the isothermal compressibility.

Applying the equipartition theorem, we get

$$\left\langle |u(\mathbf{q})|^2 \right\rangle = \frac{k_B T}{WV} \tag{5.29}$$

From (5.8) and noting that $\delta\varepsilon_{fi}(\mathbf{q}) \equiv \hat{f}\cdot\hat{i}\varepsilon' u(\mathbf{q})$, the scattering amplitude $\alpha_{fi}$ is given by

$$\alpha_{fi}^2 = \left[\frac{\omega^2}{c^2} V\hat{f}\cdot\hat{i}\varepsilon'\right]^2 |u(q)|^2 \tag{5.30}$$

This finally gives the differential cross section

$$\left(\frac{d\sigma}{d\Omega}\right)_{\text{iso}} = V\left(\frac{\varepsilon'\omega^2}{c^2}\hat{f}\cdot\hat{i}\right)^2 \frac{k_B T}{W} \tag{5.31}$$

Light scattering in the isotropic phase is considerably less than in the nematic phase, as one may see from the photographs of a transmitted laser beam through a nematic liquid crystal in Figures 5.4*a* and *b*, for temperatures below and above, respectively, the phase transition temperature $T_c$. Because of such a drastic change in the scattering loss, $T_c$ is sometimes called the clearing temperature.

More quantitatively, one can compare the corresponding scattering cross sections [cf. equations (5.21) and (5.31)]. The ratio $\sigma_R$ of $d\sigma/d\Omega$ (nematic)

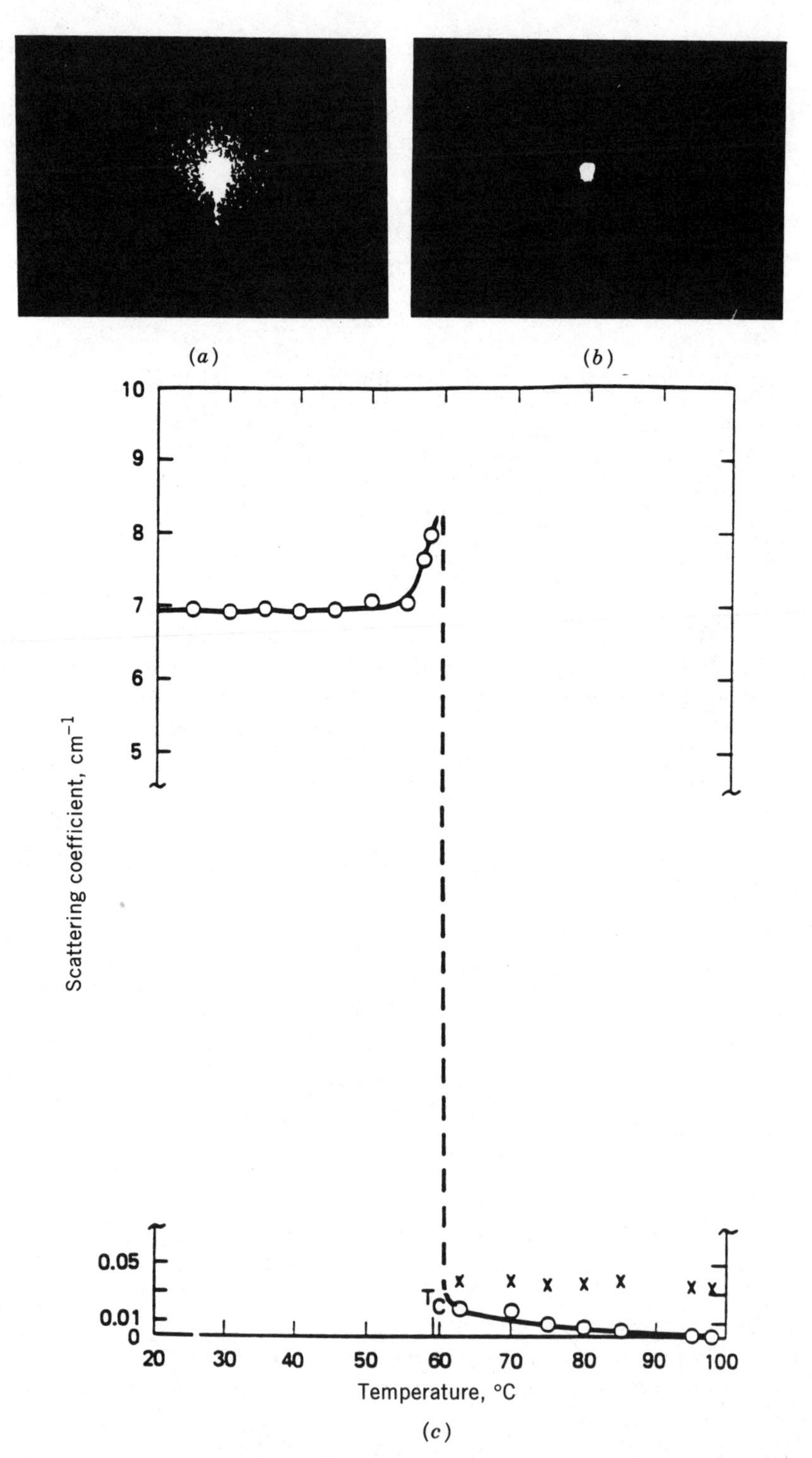


**Figure 5.4** Photograph of the transmitted laser beam through a nematic liquid crystal film (100 $\mu$m thick): (*a*) below and (*b*) above the nematic–isotropic phase transition temperature. (*c*) Experimentally measured scattering coefficient for below and above $T_c$ of an unaligned E7 sample [after Wu (7)].

over $d\sigma/d\Omega$ (isotropic) is on the order of

$$\sigma_R \sim \frac{\Delta\varepsilon}{\varepsilon'} \frac{W}{Kq^2} \tag{5.32}$$

Letting $a$ be the characteristic length of the binding energy of the molecule $u$, we have $K \sim u/a$, $W \sim u/a^3$, $1/q \sim \lambda$ (optical wavelength), and $\Delta\varepsilon/\varepsilon' \sim 1$. Therefore,

$$\sigma_R \sim \left(\frac{\lambda}{a}\right)^3 \tag{5.33}$$

Typically, $\lambda \sim 5000$ Å, $a \approx 20$ Å, so $\sigma_R \sim 10^6$; that is, scattering in the nematic phase is about six orders of magnitude larger than scattering in the isotropic phase (cf. Figure 5.4*c*). This is attributed to the fact that dilation leading to density change in the isotropic phase is characterized by $W$; this involves much more energy than rotation of the director axis, characterized by $K$, in the nematic phase.

As the temperature of the isotropic liquid crystal is lowered toward $T_c$, the molecular correlation becomes appreciable, and the scattering contributed by the orientational fluctuations will begin to dominate. The scattering cross section is proportional to $\langle|\delta\varepsilon_{fi}(\mathbf{q})|^2\rangle$ [cf. equation (5.9)]. Ignoring the tensorial nature of $\varepsilon$, the quantity $|\delta\varepsilon(\mathbf{q})|^2$ in the $\mathbf{q}$ variable may be expressed in terms of the correlation of $\varepsilon(\mathbf{r})$; that is, the scattering cross section is proportional to

$$\sigma_{\text{iso}} \sim \int \langle \varepsilon(\mathbf{r}_1)\varepsilon(\mathbf{r}_2) \rangle \exp(i\mathbf{q}\cdot(\mathbf{r}_1 - \mathbf{r}_2))\, dV_{1,2} \tag{5.34}$$

Using the Landau theory (1, 5),

$$\langle \varepsilon(0)\varepsilon(R) \rangle \approx \frac{e^{-R/\xi}}{R} \tag{5.35}$$

where $\xi$ is the so-called correlation length:

$$\xi = \xi_0 \left( \frac{T^*}{T - T^*} \right)^{1/2} \tag{5.36}$$

Substituting (5.35) into (5.34) gives

$$\sigma_{\text{iso}} \approx 4\pi\xi^2 \tag{5.37}$$

$$\approx (T - T^*)^{-1} \tag{5.38}$$

This is in agreement with the experimental observation of Lister and Stinson (6).

## 5.5 TEMPERATURE, WAVELENGTH, AND CELL GEOMETRY EFFECTS ON SCATTERING

While the ability to scatter light easily is a special advantage of nematics for various optical and electrooptical applications, it also imposes a severe limitation on the optical path length through the liquid crystal. The actual amount of scattering experienced by a light beam depends on a variety of factors, including the cell thickness, geometry, boundary surface conditions, external fields, and optical wavelength. Here we summarize some of the general observations.

As experimentally observed by Chatelain (3), light scattering in the nematic phase is remarkably independent of the temperature vicinity to $T_c$ (i.e., independent of $T_c - T$), in spite of the dependence of the scattering cross section on several highly temperature-dependent parameters such as the dielectric anisotropy $\Delta\varepsilon$ and the elastic constants $K_1$, $K_2$, and $K_3$. This is actually due to the fortuitous cancelation of the temperature dependences of $K$ and $\Delta\varepsilon$. In the Landau–deGennes theory, $K$ is proportional to the square of the order parameter, $S^2$, whereas $\Delta\varepsilon$ is proportional to $S$. From (5.13) note that $d\sigma/d\Omega \sim \Delta\varepsilon^2/K$ and is therefore not dependent on $S$. In other words, the scattering cross section $d\sigma/d\Omega$ is relatively constant with respect to variation in $T_c - T$.

This temperature independence is also noted in nonlinear scattering (3a), involving the mixing of two coherent laser beams in nematic liquid crystals. The coherently scattered wave intensity is also shown to be proportional to $\Delta\varepsilon^2/K$ and is thus independent of $T_c - T$.

The differential scattering cross sections given in (5.21) and (5.31) allow us to deduce the optical wavelength dependence. Note that, basically, for nematics

$$\frac{d\sigma}{d\Omega} \sim \left(\frac{\omega^2}{c^2}\right)^2 \frac{1}{q^2} \sim \frac{1}{\lambda^2} \tag{5.39}$$

since $\omega/c = K = 2\pi/\lambda$ and $q \sim 1/\lambda$ for a fixed scattering angle. For isotropic liquid crystals

$$\left(\frac{d\sigma}{d\Omega}\right)_{\text{iso}} \sim \left(\frac{\omega^2}{c^2}\right)^2 \sim \frac{1}{\lambda^4} \tag{5.40}$$

which is a well-known dependence for an isotropic medium.

The wavelength dependence of the scattering cross section of nematic liquid crystals, in terms of losses experienced by a laser beam in traversing an aligned cell, has been measured in our laboratory by Liu (7). Figure 5.5 shows a plot of the loss constant $\alpha$ in an aligned E46 sample (defined by $I_{\text{transmitted}} = I_{\text{incident}} e^{-\alpha d}$, where $d$ is the cell thickness) for several argon and He–Ne laser lines. In general, a dependence of the form $\lambda^{-n}$ ($n = 2.39$) is observed.

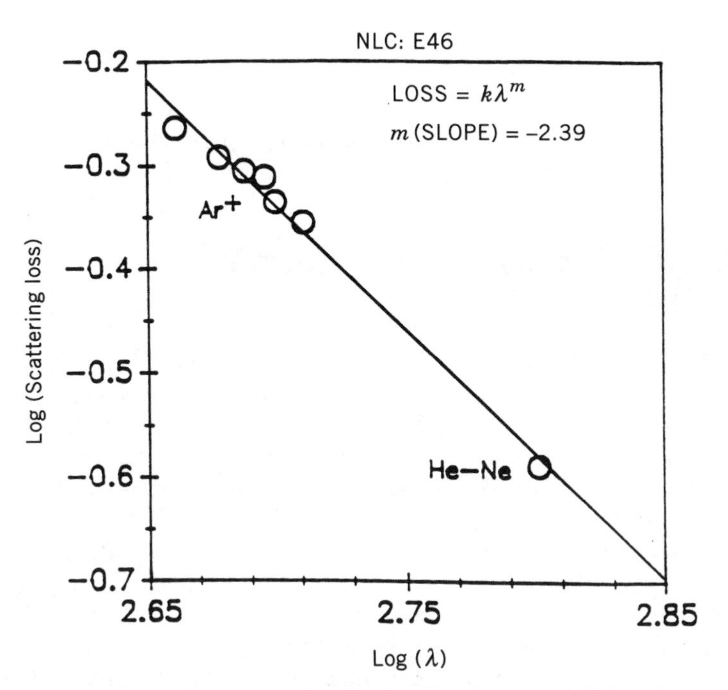


**Figure 5.5** Experimentally measured scattering loss in a nematic liquid crystal (E46) that is homeotropically aligned. In a 90-μm-thick sample, the typical scattering loss was measured to be about 18 $\text{cm}^{-1}$.

Scattering of light in nematic cells is highly dependent on the cell geometry, especially the cell thickness. This may be seen by including the boundary elastic restoring energy term, on the order of $K/d^2$ ($d$ is the cell thickness) into the calculation for the director axis fluctuations. For $d > q^{-1}$, this boundary energy term is not significant. On the other hand, for $d \ll q$, one would expect that the scattering would be reduced by these boundary effects.

An important manifestation of the dependence of light scattering on the liquid crystal cell geometry is seen in studies on nematic optical fibers (8). The latter are made by filling a microcapillary tube with nematic liquid crystals. It is found that if the core diameter of the nematic fiber is on the order of 10 μm or less, the scattering loss is dramatically reduced to 1 or 2 dB/cm, compared to the usual value of about 20 dB/cm for a flat cell.

A rigorous calculation for the director axis fluctuations in a cell of finite thickness (as opposed to the deGennes theory developed previously for bulk film) has been performed by Zeldovich and Tabiryan (9). Furthermore, their theory also treats the problem of phase coherence of the light just after passing the cell (i.e., in the so-called near zone). It is found that, in general, very strong transverse phase fluctuations, caused by thermal fluctuations of the director axis, are experienced only by an obliquely incident extraordinary wave; on the other hand, an ordinary wave undergoes no phase fluctuation for any alignment of the nematic liquid crystal cell (homeotropic, planar, or hybrid) at normal or oblique incidence. These phase fluctuations across the beam's cross section are manifested in strong speckled scatterings of the transmitted beam.

## 5.6 SPECTRUM OF LIGHT AND ORIENTATION FLUCTUATION DYNAMICS

Just as the scattered light is distributed over a spectrum of wave vectors $\mathbf{k}_f$, for a given incident wave vector $\mathbf{k}_i$, the frequency of the scattered light is distributed over a spectrum of frequencies $\omega_f$, for a given incident light frequency $\omega_i$. This spread in frequency $\Delta\omega = \omega_f - \omega_i$ is inherently related to the fact that the director axis fluctuations are characterized by finite relaxation time constants.

Generally, if the relaxation dynamics is in an exponential form [e.g., $\Delta n(t) = \Delta n(0)e^{-t/\tau}$], the frequency spectrum associated with this process, which is obtainable by the Fourier transform of $\Delta n(t)$, is a Lorentzian. The half-width of the Lorentzian $\Delta\omega$ is related to $\tau$ by $\Delta\omega = 2/\tau$.

As shown in Chapter 3, the dynamics of the orientational fluctuations depends on the distortion modes involved and the corresponding viscosity coefficient.

For a pure twist-type distortion, the equation of motion governing the elastic restoring energy and the viscous force is of the form

$$K_2 \frac{\partial^2 \Delta n(\mathbf{q})}{\partial x^2} - \gamma_1 \frac{\partial \Delta n(\mathbf{q})}{\partial t} = 0 \tag{5.41}$$

where $\mathbf{x}$ is the direction of the wave vector $\mathbf{q}$ of the fluctuation. We may write

$$\Delta n(\mathbf{q}) = \Delta n(t) \exp(iqx) \tag{5.42}$$

Equation (5.41) could be solved to give

$$\Delta n(q) \sim \Delta n(0) \exp\left(-\frac{q^2 K_2}{\gamma_1} t\right) \exp(iqx) \tag{5.43}$$

The time constant for the twist deformation is thus

$$\tau_{\text{twist}} = \frac{\gamma_1}{K_2 q^2} \tag{5.44}$$

and the scattered light will possess a frequency spectrum, centered at $\omega_i$ with a half-width

$$\Delta\omega_{\text{twist}} = \frac{2K_2q^2}{\gamma_1} \tag{5.45}$$

The dynamics of the other two types of orientation distortions, bend and splay deformations, are more complicated because such distortions are necessarily accompanied by flow (i.e., physical translation motion of the liquid crystal); this phenomenon is sometimes called the backflow effect, which may be regarded as the reverse effect of the flow-induced reorientation effect discussed in Chapter 3. A quantitative analysis of these processes (10) shows that the half-widths of the spectra associated with pure splay and pure bend deformations are given by, respectively,

$$\Delta\omega_{\text{splay}} = \frac{2K_1q^2}{\gamma_1 - \alpha_3^2/\eta_2} \tag{5.46a}$$

and

$$\Delta\omega_{\text{bend}} = \frac{2K_3q^2}{\gamma_1 - \alpha_2^2/\eta_1} \tag{5.46b}$$

where $\alpha_2$ and $\alpha_3$ are Leslie coefficients and $\eta_1$ and $\eta_2$ are viscosity coefficients defined in Chapter 3.

Using typical values [in c.g.s. units] of $K_1 \sim 6\times10^{-7}$, $K_2 \sim 3\times10^{-7}$, $K_3 \sim 8\times10^{-7}$, $\gamma_1 \sim 76\times10^{-2}$, $\eta_2 \sim 103\times10^{-2}$, $\eta_1 \sim 41\times10^{-2}$ for MBBA, $\gamma_1 - \alpha_3^2/\eta_2 = 19\times10^{-2}$, and $\gamma_1 - \alpha_3^2/\eta_1 = 126\times10^{-2}$, these line widths are estimated to be on the order of

$$\begin{aligned} \Delta w_{\text{twist}} &\sim 10^{-6}q^2 \\ \Delta w_{\text{splay}} &\sim 10^{-5}q^2 \\ \Delta w_{\text{bend}} &\sim 10^{-6}q^2 \end{aligned} \tag{5.47}$$

Depending on the value of $q$, the scattering wave vector, these frequency bandwidths are on the order of a few to about $10^2$ Hz. For example, if the wavelength of light $\lambda = 0.5$ $\mu$m $= 0.5\times10^{-4}$ cm and the scattering angle is $10^{-2}$, we have $q^2 \approx 1.6\times10^6$ and so $\Delta w_{\text{splay}} \sim 16$, whereas $\Delta w_{\text{twist}} \sim \Delta w_{\text{bend}} \sim 1.6$. These are usually referred to as the "slow" mode spectrum.

On the other hand, the relaxation process associated with the backflow is a comparatively faster one. This flow process is characterized by a time constant (11):

$$\tau_f \approx \frac{\rho}{\eta_2 q^2} \tag{5.48}$$

and thus a frequency bandwidth $\Delta\omega_f$ of

$$\Delta\omega_f \sim \frac{2\eta_2 q^2}{\rho} \tag{5.49}$$

Using $\eta_2 \sim 1\times10^{-2}$ and $\rho \sim 1$, we have $\Delta\omega_f \sim 2\times10^{-2}q^2$, which is about four or five orders of magnitude larger than the slow-mode spectrum.

## 5.7 RAMAN SCATTERINGS

### 5.7.1 Introduction

Raman scatterings provide very useful spectroscopic and molecular structural information (12). The process involves the interaction of the incident lasers with the vibrational or rotational excitations of the material. These Raman transitions involve two energy levels of the material connected by a two-photon process, as shown in Figure 5.6. Actually, of course, there are many more levels involved, but the following two-level model is sufficient for describing the basic physics of the interaction.

When an incident laser of frequency $\omega_1$ propagates through the material, two distinct Raman processes could happen. In the so-called Stokes scatterings, the electron in level 1 will be excited to level 2, and light emission at a frequency $\omega_s = \omega_1 - \omega_R$ will occur. In anti-Stokes scatterings, the electron in level 2 makes a transition to level 1, owing to its interaction with the incident laser, and light emission at $\omega_a = \omega_1 + \omega_R$ will take place.

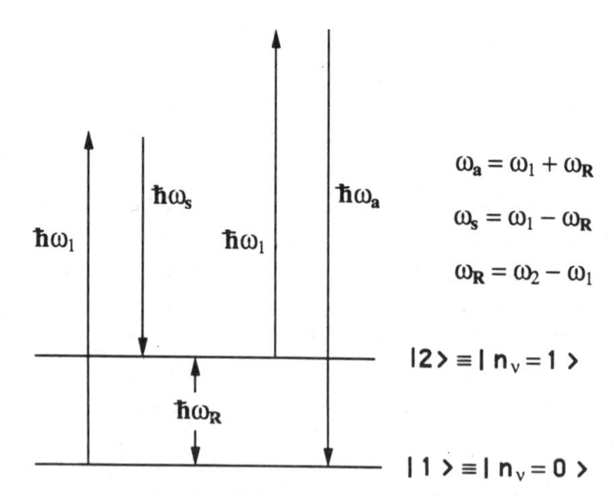


**Figure 5.6** Raman scattering involving the generation of Stokes and anti-Stokes light.

The population of level 1 is usually greater than that of level 2, and Stokes scattering will occur more efficiently than anti-Stokes scattering. However, if level 2 is unusually populated (e.g., as a result of a laser-induced transition from level 1), this notion regarding Stokes and anti-Stokes scattering efficiency is not valid, especially when intense laser fields are involved.

Since the process involves the emission of light by the material, it may occur in either spontaneous or stimulated fashion. In Chapter 9 stimulated Raman (as well as Brillouin) scattering processes, where the generated radiations are phase coherent with respect to the incident wave, are discussed in the context of nonlinear optics. Here we present a discussion of spontaneous and stimulated Raman scattering processes in terms of quantum mechanics.

### 5.7.2 Quantum Theory of Raman Scattering: Scattering Cross Section

Raman scatterings are due to the interaction of light with material vibrations. Let $q$ be the normal vibrational coordinate and let $U(q)$ be the potential energy associated with this mode. Because of the optical field displacement $q_v$ on $q$, the polarizability $\alpha(q)$ of the atom (or molecules comprising the nonlinear material) becomes

$$\alpha(q) = \alpha(q_0) + \left(\frac{\partial \alpha}{\partial q}\right)_{q_0} q_v \tag{5.50}$$

This gives an induced dipole moment $\mathbf{d}_{\text{ind}}$:

$$\mathbf{d}_{\text{ind}} = \left[\varepsilon_0 \left(\frac{\partial \alpha}{\partial q}\right)_{q_0} q_v\right]\mathbf{E} \tag{5.51}$$

and an interaction Hamiltonian $H_R$:

$$H_R = -\mathbf{d}_{\text{ind}} \cdot \mathbf{E} = -\varepsilon_0 \left(\frac{\partial \alpha}{\partial q}\right)_{q_0} q_v \mathbf{E} \cdot \mathbf{E} \tag{5.52}$$

In a quantized harmonic oscillator representation [cf. Yariv (13)], $q_v$ and $E$ can be expressed in terms of annihilation $(a)$ and creation operators $(a^+)$:

$$q_v \sim (a_v^+ + a_v) \tag{5.53a}$$

$$E \sim \sqrt{\omega_1}\,(a_1^+ - a_1) + \sqrt{\omega_s}\,(a_s^+ - a_s) \tag{5.53b}$$

The first term on the right-hand side of (5.53*b*) comes from the incident laser field $E_1$, and the second term comes from the Stokes field $E_s$.

The action of the annihilation $a$ and creation operators $a^+$ is as follows: Let $|n_1\rangle$, $|n_s\rangle$, and $|n_v\rangle$ denote the energy states of the laser, the Stokes, and the molecular vibrations, respectively. Then we have $a_1|n_1\rangle \to |n_1 - 1\rangle$; $a_1^+|n_1\rangle \to |n_1 + 1\rangle$; $a_s|n_s\rangle \to |n_s - 1\rangle$; $a_s^+|n_s\rangle \to |n_s + 1\rangle$; $a_v|n_v\rangle \to |n_v - 1\rangle$; $a_v^+|n_v\rangle \to |n_v + 1\rangle$. In other words, the action of the operators $a$ and $a^+$ is to decrease or to increase, respectively, the number of excitations in the corresponding energy state.

In this picture the total wave function of the system is represented by a product of the laser, the Stokes, and the vibrational states, that is,

$$|\psi\rangle = |n_1\rangle |n_s\rangle |n_v\rangle \tag{5.54}$$

From (5.52) and (5.53) we can see that, in general, the interaction Hamiltonian $H_R$ for Raman scattering consists of a triple product in the $a$'s and $a^+$'s. In particular, Stokes scattering is associated with the action of a term $a_1 a_s^+ a_v^+$ on the initial total wave function of the system $|i\rangle_s = |n_1\rangle|n_s\rangle|0\rangle_v$, which yields a final state where the vibration state (initially a vacuum state) is increased by one unit, while a photon is removed from the incident laser and a Stokes photon is created:

$$\underset{\{|i_s\rangle\}}{a_1 a_s^+ a_v^+ \{|n_1\rangle|n_s\rangle|n_v = 1\rangle\}} \to \underset{\{|f_s\rangle\}}{\{|n_1 - 1\rangle|n_s + 1\rangle|n_v = 1\rangle\}} \tag{5.55}$$

Similarly, the inverse scattering process corresponds to the action of $a_1^+ a_s a_v$ on an initial state $|i\rangle_a = |n_1\rangle|n_s\rangle|1\rangle$, that is,

$$\underset{\{|i_a\rangle\}}{a_1^+ a_s a_v \{|n_1\rangle|n_s\rangle|n_v = 1\rangle\}} \to \underset{\{|f_a\rangle\}}{\{|n_1 + 1\rangle|n_s - 1\rangle|n_v = 0\}} \tag{5.56}$$

In accordance with perturbation theory (13), the probability of these transitions is proportional to the square moduli of the matrix elements for these processes. For Stokes scattering,

$$W_s \propto |\langle f_s|a_1 a_s^+ a_v^+|i_s\rangle^2 = Dn_1(n_s + 1) \tag{5.57}$$

For inverse scattering,

$$W_a \propto |\langle f_a|a_1^+ a_s a_v|i_a\rangle|^2 = D(n_1 + 1)n_s \tag{5.58}$$

where $D$ is the proportional constant which may be shown by a rigorous quantum mechanical calculation (13) to be the same for these two processes.

Considering the growth in time of the Stokes photons and using (5.57) and (5.58), we have

$$\frac{dn_s}{dt} = DN_1 n_1(n_s + 1) - DN_2 n_s(n_1 + 1) \tag{5.59}$$

where $N_1$ and $N_2$ are the respective probabilities of occupation of the ground state $|1\rangle$ $(n_v = 0)$ and the excited state $|2\rangle$ $(n_v = 1)$.

In ordinary (spontaneous) Raman scattering, $n_s \ll 1$ and $N_2 \ll N_1$. We thus have

$$\frac{dn_s}{dt} \sim DN_1 n_1 \tag{5.60}$$

Since the number of photons is conserved, $dn_s/dt = -dn_1/dt$, (5.60) yields

$$\frac{dn_1}{dt} = -DN_1 n_1 \tag{5.61}$$

or

$$\frac{dn_1}{dz} = -\frac{D}{c/n(\omega_1)} N_1 n_1 \tag{5.62}$$

where $n(\omega_1)$ is the refractive index. This gives

$$n_1(z) = n_1(0) e^{-\beta_1 z} \tag{5.63}$$

where

$$\beta_1 = \frac{DN_1 n(\omega_1)}{c} \tag{5.64}$$

is the attenuation constant for the incident laser at $\omega_1$. Note that $\beta_1$ is *not dependent on the incident laser intensity*. Also, from (5.60), the number of Stokes photons scattered is proportional to the incident laser intensity at any given time. By considering the number of Stokes photons scattered into a differential solid angle $d\Omega$, the proportional constant $D$ in (5.64) is shown in Yariv (13) to be related to the differential cross section $d\sigma/d\Omega$ by

$$\left.\frac{d\sigma}{d\Omega}\right|_{90°,\phi} = \frac{3\nu_s^3 n^3(\omega_s) n(\omega_1) DV \Delta\nu N_1}{\nu_1 N c^4} \tag{5.65}$$

where $N$ is the density of the molecules, $V$ is the volume, $n(\omega_s)$ and $n(\omega_1)$ are the respective refractive indices at Stokes and incident laser frequencies, and $\Delta\nu$ is the natural line width of the transition.

From (5.60) we can see that the rate of production of Stokes photons is proportional to the incident laser intensity (i.e., proportional to $n_1$). Hence, if the incident laser is intense enough, the Stokes photon number $n_s$ can be substantial ($\gg 1$). In this case the scattering process will take on a totally different form, and the Stokes wave will grow exponentially. To see this, consider (5.59) for the case $n_1, n_s \gg 1$. We have

$$\frac{dn_s}{dt} = D(N_1 - N_2)n_1 n_s \tag{5.66}$$

This gives

$$\frac{dn_s}{dz} = \frac{Dn(\omega_s)}{c}(N_1 - N_2)n_1 n_s \tag{5.67}$$

or, for the Stokes intensity,

$$I_s(z) = I_s(0)e^{g_s z} \tag{5.68}$$

where

$$g_s = \frac{Dn(\omega_s)}{c}(N_1 - N_2)n_1 \tag{5.69}$$

Since $n_1$ is related to $I_1$ by

$$\frac{n_1}{v} = \frac{n(\omega_1)}{h\nu_1 c} I_1$$

we have

$$g_s = \frac{Dn(\omega_s)n(\omega_1)}{h\nu_1 c^2}(N_1 - N_2)I_1 \tag{5.70}$$

In other words, the Stokes wave experiences an exponential gain (if $N_1 > N_2$) constant that is proportional to the incident laser intensity. If this gain is larger than the loss $\alpha$ owing to absorption, random scatterings, and so on experienced by the Stokes wave, it is possible to have laser oscillations (i.e., the generation of Stokes lasers). Such a process is called stimulated Raman scattering.

For a more detailed discussion of stimulated scatterings as nonlinear optical processes, the reader is referred to Chapter 9.

## 5.8 BRILLOUIN AND RAYLEIGH SCATTERINGS

In general, for pure nonabsorbing media, the spectrum of spontaneously scattered light from an incident laser centered at $\omega_1$ is of the form given in Figure 5.7. The Raman–Stokes radiations (at $\omega = \omega_1 - \omega_R$) and the anti-Stokes radiations (at $\omega = \omega_1 + \omega_R$) are far to the side of the central frequency.

Around the central frequency $\omega_1$, in general, the spectrum consists of a Rayleigh scattering peak centered at $\omega_1$, two Brillouin scattering peaks at $(\omega_1 \pm \omega_B)$, and a broad background that is generally referred to as the Rayleigh wing.

Rayleigh scattering is due to the entropy or temperature fluctuations, which cause *nonpropagating* density (and therefore dielectric constant) fluctuations. On the other hand, Brillouin scattering is due to the propagating pressure (i.e., sound) waves and is often referred to as the electrostrictive effect. An incident laser can generate copropagating or counterpropagating sound waves at a frequency $\omega_B$. Hence, the spectrum of the scattered light consists of a doublet centered at $\omega_1 \pm \omega_B$.

Rayleigh wing scattering is due to the orientational fluctuations of the anisotropic molecules. For typical liquids these orientational fluctuations are characteristic of the *individual* molecules' movements and occur in a very short time scale ($\lesssim 10^{-12}$ s). Consequently, the spectrum is quite broad. In liquid crystals studies of individual molecular orientation dynamics have shown that the relaxation time scale is on the order of picoseconds (14), and thus the Rayleigh wing spectrum for liquid crystals is also quite broad.

From the discussion given in earlier chapters, we note that in liquid crystals the main scattering is due to collective or correlated orientational fluctuations. These, of course, are much slower processes than individual molecular motions. The spectrum from these correlated or collective orientational fluctuations is therefore very sharp and is embedded in the central Rayleigh peak region.

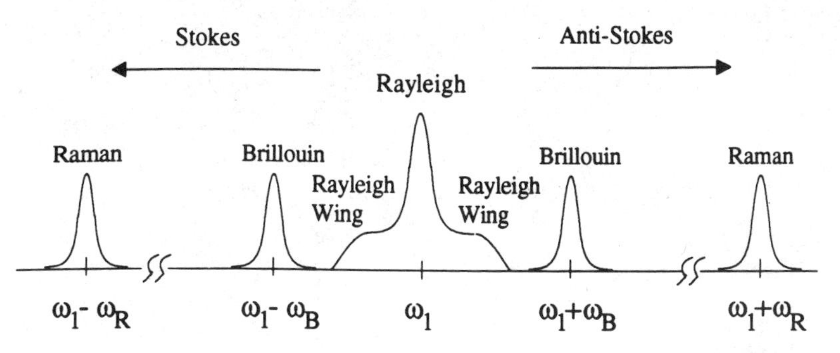


**Figure 5.7** A typical spectrum of light spontaneously scattered from a liquid or liquid crystal.

The dynamics of orientational scattering has been discussed on various occasions in the preceding chapter. We shall discuss here the mechanism of Brillouin and Rayleigh scatterings.

The origin of these scattering processes is the dielectric constant change $\Delta\varepsilon$ associated with a change $\Delta\rho$ in the density, that is,

$$\Delta\varepsilon = \frac{\partial\varepsilon}{\partial\rho}\,\Delta\rho \tag{5.71}$$

The fundamental independent variables for the density are the pressure $P$ and the entropy $S$. Therefore, we can express $\Delta\rho$ as

$$\Delta\rho = \left(\frac{\partial\rho}{\partial P}\right)_S \Delta P + \left(\frac{\partial\rho}{\partial S}\right)_P \Delta S \tag{5.72}$$

where the terms on the right-hand side of (5.72) correspond to the adiabatic and isobaric density fluctuations, respectively.

### 5.8.1 Brillouin Scattering

The pressure wave $\Delta P$ obeys an equation of motion of the form (15):

$$\frac{\partial^2 \Delta P}{\partial t^2} - \Gamma\,\nabla^2 \frac{\partial \Delta P}{\partial t} - v_s^2\,\nabla^2 \Delta P = 0 \tag{5.73}$$

where $v_s$ is the velocity of sound and $\Gamma$ is the damping constant.

The velocity of sound can be expressed in terms of the compressibility $C_s$ or the bulk modulus $\beta$:

$$C_s = \frac{1}{\beta} = -\frac{1}{v}\frac{\partial V}{\partial P} = \frac{1}{\rho}\left(\frac{\partial\rho}{\partial P}\right)_S \tag{5.74}$$

Then

$$v_s^2 = \left(\frac{\partial P}{\partial\rho}\right)_S = \frac{1}{C_s\rho} \tag{5.75}$$

The damping constant can be further expressed as

$$\Gamma = \frac{1}{\rho}\left[\frac{4}{3}\eta_s + \eta_b\right] \tag{5.76}$$

where $\eta_s$ is the shear viscosity coefficient and $\eta_b$ is the bulk viscosity coefficient.

The presence of the Brillouin doublet in the scattered wave from the medium in which there is an acoustic wave may be seen as follows. For simplicity, we will ignore all tensor characteristics of the problem. Let the pressure wave be of the form

$$\Delta P = \Delta P_0 \, e^{i(qz - \omega_B t)} + \text{c.c.} \tag{5.77}$$

As a result of this change in pressure, a dielectric constant change occurs and is given by

$$\begin{aligned} \Delta\varepsilon &= \left(\frac{\partial \varepsilon}{\partial \rho}\right) \Delta\rho \\ &= \left(\frac{\partial \varepsilon}{\partial \rho}\right)\left(\frac{\partial \rho}{\partial P}\right)_S \Delta P \end{aligned} \tag{5.78}$$

Using the definition for the electrostrictive coefficient $\gamma^e = \rho(\partial\varepsilon/\partial\rho)$ and $C_s = 1/\rho(\partial\rho/\partial P)$, we thus have

$$\Delta\varepsilon = C_s \Gamma^e \Delta P \tag{5.79}$$

Writing the incident optical electric field as

$$\mathbf{E}_{\text{inc}} = \mathbf{E}_0 (e^{i(\mathbf{k}\cdot\mathbf{r} - \omega t)} + \text{c.c.}) \tag{5.80}$$

we can see that the dielectric constant will give rise to an induced polarization

$$\begin{aligned} \mathbf{P} &= \Delta\varepsilon \, \mathbf{E}_{\text{inc}} \\ &= C_s \gamma^e \Delta P \, \mathbf{E}_{\text{inc}} \end{aligned} \tag{5.81}$$

From (5.77) for $\Delta P$ and (5.80) for $\mathbf{E}_{\text{inc}}$, the reader can easily verify that the scattered waves contain two components. One component possesses a wave vector $\mathbf{k}_- = \mathbf{k} - \mathbf{q}$ oscillating at a frequency $\omega - \omega_B$. This corresponds to the scattering of light from a retreating sound wave. The other scattered component is characterized by a wave vector $\mathbf{k}_+ = \mathbf{k} + \mathbf{q}$ oscillating at a frequency $\omega + \omega_B$. This corresponds to the scattering of light from an oncoming sound wave. These two components are very close to each other in magnitude.

The wave vectors of the incident $\mathbf{k}_1$ and the scattered $\mathbf{k}_2$ light and the acoustic wave vector $\mathbf{q}$ are related by the wave vector addition rule shown in Figure 5.8 for both processes.

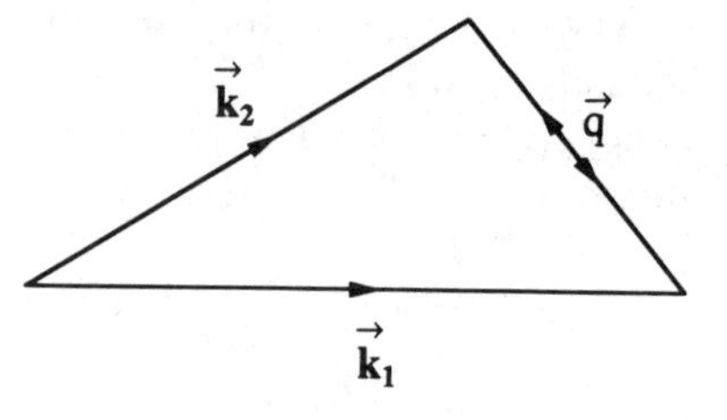


**Figure 5.8** Wave vector addition rule for the Brillouin scattering processes. For Stokes scattering, $\omega_2 = \omega_1 - \omega_B$ and $\mathbf{k}_2 = \mathbf{k}_1 - q$; for anti-Stokes scattering, $\omega_2 = \omega_1 + \omega_B$ and $\mathbf{k}_2 = \mathbf{k}_1 + \mathbf{q}$.

Since $|\mathbf{k}_1| \approx |\mathbf{k}_2|$ and $|q| = 2|k_1|\sin(\theta/2)$, the frequency $\omega_B$ of the acoustic wave is thus given by

$$\omega_B = |q|v_s = 2n\omega\frac{v_s}{c}\sin\left(\frac{\theta}{2}\right) \tag{5.82}$$

The spectral width of the scattered line is related to the acoustic damping constant $\Gamma$. This may be obtained by substituting $\Delta P$ in (5.77) into the acoustic equation (5.73). This gives

$$\omega_B^2 = q^2\left(v_s^2 - i\omega_B\Gamma\right) \tag{5.83}$$

or

$$q^2 = \frac{\omega_B^2}{v_s^2 - i\omega_B\Gamma} \simeq \frac{\omega_B^2}{v_s^2}\left(1 + \frac{i\omega_B\Gamma}{v_s^2}\right) \tag{5.84}$$

$$q \sim \frac{\omega_B}{v} + \frac{i\Gamma q^2}{2v_s} \tag{5.85}$$

if $\omega_B\Gamma \ll v_s^2$ (which is usually the case). This shows that the acoustic wave is characterized by a damping constant

$$\Gamma_B = \frac{\Gamma q^2}{2} \tag{5.86}$$

The inverse of this, $\tau_p = 1/\Gamma_B$, is often referred to as the phonon lifetime. As a result, the spectrum of the Brillouin doublet is broadened by an amount

$$\delta\omega = \frac{1}{\tau_p} = \Gamma_B \tag{5.87}$$

Using $q = 2|k_1|\sin(\theta/2)$, we get

$$\delta\omega = 2n^2\Gamma\frac{\omega^2}{c^2}\sin^2\left(\frac{\theta}{2}\right) \tag{5.88}$$

### 5.82 Rayleigh Scattering

The entropy fluctuation $\Delta S$ obeys a diffusion equation similar to that for the temperature fluctuation:

$$\rho C_p \frac{\partial \Delta S}{\partial t} - \kappa \nabla^2 S = 0 \tag{5.89}$$

where $C_p$ is the specific heat at constant pressure and $\kappa$ is the thermal conductivity. A solution of the diffusion equation is a *nonpropagative* function of the form

$$\Delta S = \Delta S_0 \, e^{-\Gamma_T t_e - i\mathbf{q}\cdot\mathbf{r}} \tag{5.90}$$

where the thermal damping constant $\Gamma_T$ is given by

$$\Gamma_T = \frac{\kappa}{\rho C_p} q^2 \tag{5.91}$$

Following the preceding analysis, we can see that the scattering caused by the entropy fluctuation does not shift the frequency. Instead, because of the exponentially decaying dependence, $e^{-\Gamma_T t}$, it broadens the light by an amount $\delta\omega = \Gamma_T$. Again, since $q = 2|\mathbf{k}_1|\sin(\theta/2)$, we have

$$\delta\omega = \frac{4\kappa}{\rho C_p}|k_1|^2 \sin^2\left(\frac{\theta}{2}\right) \tag{5.92}$$

## 5.9 THERMAL BRILLOUIN AND RAYLEIGH SCATTERING

The preceding section dealt with the scattering of light caused by density fluctuations in a nonabsorbing medium; the density variations originate from the electrostrictive effect. In an absorptive medium density variations can be induced simply by the heating effect accompanying the photoabsorption and subsequent thermalization processes. The resulting temperature rise creates both adiabatic and isobaric density fluctuations associated with the changes in pressure and entropy, respectively [cf. equation (5.72)]. Accordingly, both propagating and nonpropagating density variations will be created. The corresponding light-scattering processes are called thermal Brillouin scattering and thermal Rayleigh scattering.

Therefore, in a medium with a finite absorption constant, we expect two sets of Brillouin and Rayleigh scattering events to take place. One is associated with the electrostrictive component, and the other with the absorptive component. These processes are usually studied in the context of nonlinear optics involving high-power pulsed lasers. Also, the theoretical framework for these processes is usually developed using the density $\rho$ and the temperature $T$ as the thermodynamic variables. Since $\rho$ and $T$ are both dependent on the pressure $P$ and the entropy $S$, these equations for $\rho$ and $T$ are coupled. This is the subject of the discussion in Chapter 7.

## REFERENCES


1. P. G. deGennes, "The Physics of Liquid Crystals." Clarendon, Oxford Press, 1974.
2. L. D. Landau and E. M. Lifshitz, "Electrodynamics of Continuous Media." Pergamon, London, 1960; see also J. D. Jackson, "Classical Electrodynamics." Wiley, New York, 1963.
3. P. Chatelain, *Acta Crystallogr.* **1**, 315 (1948).

3a. I. C. Khoo, *Phys. Rev. A* **27**, 2747 (1983).

4. J. C. Filippini, *Phys. Rev. Lett.* **39**, 150 (1977). B. Malraison, Y. Poggi, and E. Guyon, *Phys. Rev. A* **21**, 1012 (1980).
5. L. D. Landau and E. M. Lifshitz, "Statistical Physics." Pergamon, London, 1975.
6. J. D. Litster and T. Stinson, *Phys. Rev. Lett.* **30**, 688 (1973); see also T. W. Stinson and J. D. Litster, *ibid.* **25**, 503 (1970).
7. T. H. Liu, Ph.D. Thesis, Pennsylvania State University, University Park (1987); see also I. C. Khoo and S. T. Wu, "Optics and Nonlinear Optics of Liquid Crystals." World-Scientific, Singapore, 1993.
8. M. Green and S. J. Madden, *Appl. Opt.* **28**, 5202 (1989).
9. B. Ya. Zeldovich and N. V. Tabiryan, *Sov. Phys.—JETP (Engl. Transl.)* **54**, 922 (1981).
10. A. Saupe, *Annu. Rev. Phys. Chem.* **24**, 441 (1973).
11. Orsay Liquid Crystal Group, *J. Chem. Phys.* **51**, 816 (1969).
12. See, for example, Shen Jen, N. A. Clark, and P. S. Pershan, *J. Chem. Phys.* **66**, 4635 (1977).
13. A. Yariv, "Quantum Electronics." Wiley, New York, 1985.
14. See, for example, J. R. Lalanne, B. Martin, and B. Pouligny, *Mol. Cryst. Liq. Cryst.* **42**, 153 (1977); see also C. Flytzanis and Y. R. Shen, *Phys. Rev. Lett.* **33**, 14 (1974).
15. I. L. Fabelinskii, "Molecular Scattering of Light." Plenum, New York, 1968.

# CHAPTER 6

# LASER-INDUCED NONELECTRONIC OPTICAL NONLINEARITIES IN LIQUID CRYSTALS

## 6.1 GENERAL OVERVIEW OF LIQUID CRYSTAL NONLINEARITIES

In linear optical processes the physical properties of the liquid crystal, such as its molecular structure, individual or collective molecular orientation, temperature, density, population of electronic levels, and so forth, are not affected by the optical fields. The direction, amplitude, intensity, and phase of the optical fields are affected in a unidirectional way (i.e., by the liquid crystal's physical parameters). The optical properties of liquid crystals may, of course, be controlled by some externally applied dc or low-frequency fields; this gives rise to a variety of electrooptical effects which are widely used in many electrooptical display and image-processing applications.

Liquid crystals are also optically highly nonlinear materials in that their physical properties (temperature, molecular orientation, density, electronic structure, etc.) are easily perturbed by an applied optical field (1–7). Nonlinear optical processes associated with electronic mechanisms are discussed in Chapter 8. In this section and the next chapter, we discuss the principal nonelectronic mechanisms for the nonlinear optical responses of liquid crystals.

Since liquid crystalline molecules are anisotropic, a polarized light from a laser source can induce an alignment or ordering in the isotropic phase, or a realignment of the molecules in the ordered phase. These result in a change in the refractive index.

Other commonly occurring mechanisms that give rise to refractive index changes are laser-induced changes in the temperature, $\Delta T$, and the density, $\Delta\rho$. These changes could arise from several mechanisms. A rise in temperature is a natural consequence of photoabsorptions and the subsequent inter- and intramolecular thermalization or nonradiative energy relaxation processes. In the isotropic phase the change in the refractive index is due to the density change following a rise in temperature. In the nematic phase the refractive indices are highly dependent on the temperature through their dependence on the order parameters, as well as on the density, as discussed in Chapter 3.

Temperature changes inevitably lead to density changes via the thermoelastic coupling (i.e., thermal expansion). However, density changes can also be due to the electrostrictive effect (i.e., the liquid crystal molecules' movement toward a region of high laser field). To gain a first-order understanding of the electrostrictive effect, let us ignore for the moment the anisotropy of the liquid crystalline parameters. Consider a molecule situated in a region illuminated by an optical field $\mathbf{E}$. The field induces a polarization $\mathbf{p} = \alpha\mathbf{E}$, where $\alpha$ is the molecular polarizability. The electromagnetic energy expended on the molecule is thus

$$u_E = -\int_0^E \mathbf{p}\cdot d\mathbf{E} = -\frac{1}{2}\alpha(\mathbf{E}\cdot\mathbf{E}) \tag{6.1}$$

If the electric field is spatially varying (i.e., there are regions of high and low energy density $E^2$), there will be a force acting on the molecule given by

$$\mathbf{F}_{\text{molecule}} = -\nabla u_E = \tfrac{1}{2}\alpha\nabla(\mathbf{E}\cdot\mathbf{E}) \tag{6.2}$$

that is, the molecule is pulled into the region of increasing field strength (cf. Figure 6.1). As a result, the density in the high field region is increased by an amount $\Delta\rho$. This increase in density gives rise to an increase in the dielectric constant (and therefore the refractive index) by an amount

$$\begin{aligned}\Delta\varepsilon &= \rho\frac{\partial\varepsilon}{\partial\rho}\left(\frac{\Delta\rho}{\rho}\right)\\ &= \gamma^e\left(\frac{\Delta\rho}{\rho}\right)\end{aligned} \tag{6.3}$$

where

$$\gamma^e = \rho\left(\frac{\partial\varepsilon}{\partial\rho}\right) \tag{6.4}$$

is the electrostrictive constant.

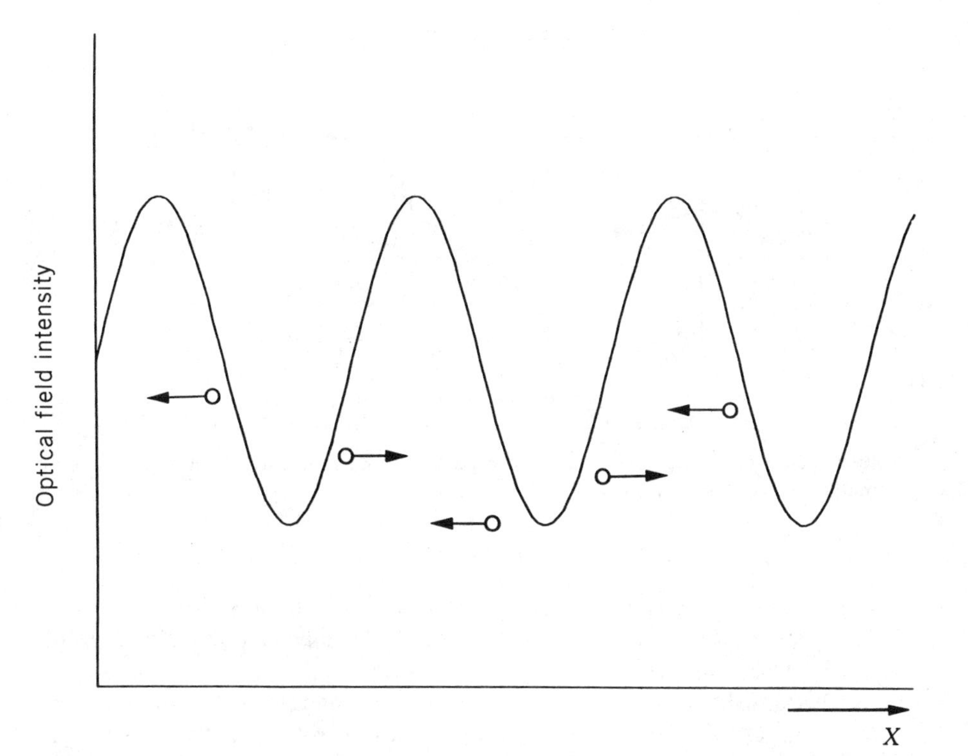


**Figure 6.1** Molecules are pulled toward regions of higher field strength by electrostrictive forces.

Large density and temperature changes could also give rise to flows and director axis reorientation. An intense laser field induces flows in liquid crystals via the pressure it exerts on the system (8, 9). The pressure $p(\mathbf{r}, t)$ creating the flow may originate from the thermoelastic or electrostrictive effects mentioned previously in conjunction with thermal and density changes. In nonabsorbing liquid crystals the flow is due mainly to electrostrictive forces of the type shown in (6.2) which are derived generally from the so-called Maxwell stress (9):

$$\mathbf{F} = (\mathbf{D} \cdot \nabla)\mathbf{E}^* - \tfrac{1}{2}\nabla(\mathbf{E} \cdot \mathbf{D}^*) \tag{6.5}$$

Electrostrictive effects are highly dependent on the gradient of the electromagnetic fields, which are naturally present in intensely focused or spatially highly modulated pulsed lasers.

Flows also give rise to director axis realignment (cf. Figure 6.2). In the extreme case of flow, the liquid crystal is forced to vacate the site it occupied

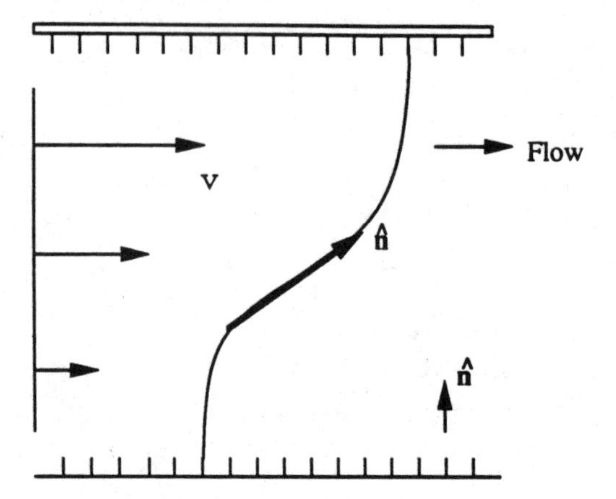


**Figure 6.2** An example of the flow-induced director axis reorientation effect in nematics. $v$ is the flow velocity.

(i.e., an empty space is left). Laser-induced flow effects thus give rise to large index changes, as observed in several studies involving nanosecond or picosecond laser pulses (8–10).

In this and the next chapters, laser-induced changes in the director axis orientation $\theta(\mathbf{r}, t)$, density $\rho(\mathbf{r}, t)$, temperature $T(\mathbf{r}, t)$, and flows are separately discussed for all the principal mesophases of liquid crystals. An intense laser pulse can also generate electronic nonlinearities in liquid crystals. This is treated separately in Chapter 8.

## 6.2 LASER-INDUCED MOLECULAR REORIENTATIONS IN THE ISOTROPIC PHASE

### 6.2.1 Individual Molecular Reorientations in Anisotropic Liquids

In the isotropic phase the liquid crystal molecules are randomly oriented owing to thermal motion, just as in conventionally anisotropic liquids. An intense laser field will force the anisotropic molecules to align themselves in the direction of the optical field through the dipolar interaction (cf. Figure 6.3), in order to minimize the energy. Such a process is often called laser-induced ordering; that is, the laser induced some degree of preferred orientation in an otherwise random system. Because the molecules are birefringent, this partial alignment gives rise to a change in the effective optical dielectric constant (i.e., an optical field intensity-dependent refractive index change).

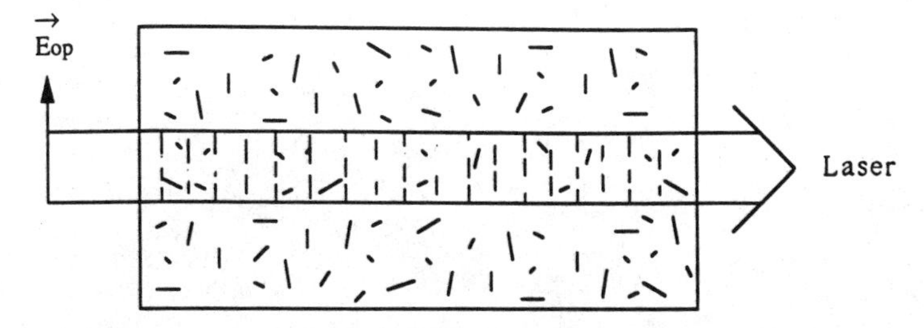


**Figure 6.3** Laser-induced ordering in an anisotropic liquid.

If the laser is polarized in the $\hat{x}$ direction, as shown in Figure 6.4, the induced polarization in the $\hat{x}$ direction is given by

$$P_x = \varepsilon_0 \, \Delta\chi_{xx}^{\mathrm{op}} \, E_x \tag{6.6}$$

where $\Delta\chi^{\mathrm{op}}$ is the optically induced change in the susceptibility. In terms of the principal axes 1 and 2,

$$P_x = P_2 \cos\theta + P_1 \sin\theta \tag{6.7}$$

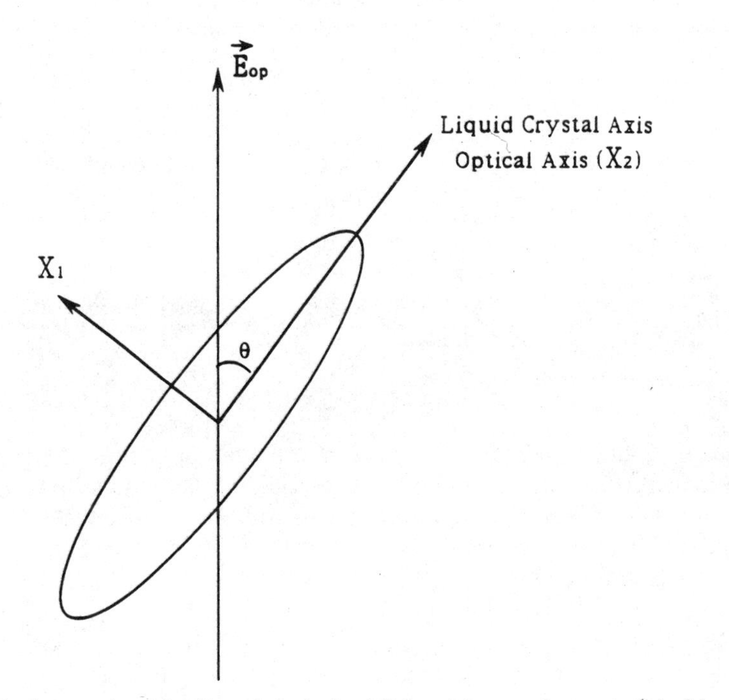


**Figure 6.4** Interaction of a linearly polarized light with an anisotropic (birefringent) molecule.

where

$$P_1 = \varepsilon_0 \chi_{11} E_1 = \varepsilon_0 \chi_{11} E_x \sin\theta \tag{6.8}$$

$$P_2 = \varepsilon_0 \chi_{22} E_2 = \varepsilon_0 \chi_{22} E_x \cos\theta \tag{6.9}$$

We therefore have

$$\Delta\chi_{xx}^{\mathrm{op}} = \chi_{22}\cos^2\theta + \chi_{11}\sin^2\theta \tag{6.10}$$

This may be expressed in terms of the average susceptibility:

$$\bar{\chi} = \tfrac{1}{3}(\chi_{22} + 2\chi_{11}) \tag{6.11}$$

and the susceptibility anisotropy:

$$\Delta\chi = \chi_{22} - \chi_{11} \tag{6.12}$$

by

$$\Delta\chi^{\mathrm{op}} = \bar{\chi} + \Delta\chi\left\langle\left(\cos^2\theta - \tfrac{1}{3}\right)\right\rangle \tag{6.13}$$

In (6.13) the angle brackets containing the factor $(\cos^2\theta - \frac{1}{3})$ signify that $\Delta\chi^{\mathrm{op}}$ is a macroscopic parameter and an ensemble average; it can be expressed in terms of the (induced) order parameter $Q \equiv \langle\frac{3}{2}\cos^2\theta - \frac{1}{2}\rangle$, by

$$\Delta\chi_{xx}^{\mathrm{op}} = \bar{x} + \tfrac{2}{3}\Delta\chi Q \tag{6.14}$$

The total polarization $P_x$ therefore becomes

$$\begin{aligned} P_x &= \left[\varepsilon_0\bar{\chi} + \varepsilon_0 \tfrac{2}{3}\Delta\chi Q\right] E_x \\ &= p_x^{\mathrm{L}} + P_x^{\mathrm{NL}} \end{aligned} \tag{6.15}$$

where the linear polarization $P_x^{\mathrm{L}} = \varepsilon_0 \bar{\chi} E$ is the contribution from the unperturbed system, and the nonlinear polarization

$$P_x^{\mathrm{NL}} = \tfrac{2}{3}\varepsilon_0 \Delta\chi QE \tag{6.16}$$

arises from the laser-induced molecular orientation, or ordering.

From (6.16) one can see that the molecular orientational nonlinearity in the isotropic phase of a liquid crystal is directly proportional to the laser-induced order parameter $Q$.

In typical anisotropic liquids (e.g., $CS_2$ or liquid crystals at temperatures far above $T_c$), the value of $Q$ may be obtained by a statistical mechanics approach. In the completely random system, the average orientation is described by a distribution function $f(\theta)$:

$$Q = \left\langle \tfrac{3}{2}\cos^2\theta - \tfrac{1}{2}\right\rangle = \int_0^{\pi} f(\theta)\left(\tfrac{3}{2}{}^2\theta - \tfrac{1}{2}\right) d\theta \tag{6.17}$$

In a steady state the equilibrium value for $f(\theta)$ is given by

$$f(\theta) = \frac{e^{-\mathscr{E}/K_B T}}{\int_0^{\pi} f(\theta)\, d\cos\theta} \tag{6.18}$$

where $\mathscr{E}$ is the interaction energy of a molecule $[\mathscr{E} = \varepsilon_0(\Delta\alpha/2)|E|^2]$ and $\Delta\alpha$ is the molecular polarization anisotropy ($\Delta\alpha = \Delta\chi/N$). $K_B$ is the Boltzmann constant.

Laser-induced individual molecular orientations in liquid crystalline systems have been studied by several groups (10–13). In the study by Lalanne et al. (13), both the uncorrelated individual molecular reorientations and the correlated reorientation effects described in the next section induced by picosecond laser pulses have been measured. Typically, the individual molecular motion is characterized by a response time on the order of a few picoseconds, and fluctuations in these individual molecular motions give rise to a broad central peak in the Rayleigh scattering measurement; this is usually referred to as the Rayleigh wing scattering component, and it always exists in ordinary liquids (cf. Chapter 5). On the other hand, the correlated molecular reorientational effect discussed in the next section is characterized by a response time on the order of $10^1$ to $10^2$ ns. This gives rise to a narrow central component in the Rayleigh scattering spectrum and could also be called a Rayleigh wing component because it also originates from orientational fluctuations.

Besides these molecular motions, an optical field could also induce other types of orientation effects in liquid crystalline systems, for example, nuclear reorientation caused by the field-induced nuclear orientational anisotropy. This process is sometimes referred to as the nuclear optical Kerr effect (10) as it results in an optical intensity-dependent change in the optical dielectric constant. Such effects in liquid crystals have been investigated by Fayer et al. (10). In general, these nuclear motions are characterized by a rise time of a few picoseconds and a decay time of about $10^2$ ps, in experiments involving subpicosecond laser pulses. In general, the dynamics of these nuclear motions are also more complex than the individual molecular reorientational effects discussed previously (10).

### 6.2.2 Correlated Molecular Reorientation Dynamics

For liquid crystals, owing to pretransitional effects near $T_c$, the induced ordering $Q$ exhibits interesting correlated dynamic and temperature-dependent effects.

In general, very intense laser pulses are required to create appreciable molecular alignment in liquids. To quantitatively describe the pulsed laser-induced effect, a time-dependent approach is needed. In this regime $f(\theta)$ obeys a Debye rotational diffusion equation (10):

$$\eta \frac{\partial f}{\partial t} = \frac{1}{\sin\theta} \frac{\partial}{\partial\theta} \left[ \sin\theta \left( \frac{\partial f}{\partial\theta} + 2\,\Delta\alpha \frac{|E|^2}{K_B T} \sin\theta \cos\theta\, f \right) \right] \tag{6.19}$$

where $\eta$ is the viscosity coefficient. Substituting (6.17) into (6.19), one obtains a dynamical equation for $Q$:

$$\frac{\partial Q}{\partial t} = -\frac{Q}{\tau_D} + \frac{2\,\Delta\alpha |E|^2}{3\eta} \tag{6.20}$$

where the relaxation time $\tau_D$ is given by

$$\tau_D = \frac{\eta}{5KT} \tag{6.21}$$

Equation (6.20) shows that $Q \sim |E|^2$, and thus the nonlinear polarization from (6.16) becomes

$$P_x^{\mathrm{NL}} \sim |E|^2 E \tag{6.22}$$

that is, a third-order nonlinear susceptibility which is proportional to $|E|^2$. This is equivalent to an intensity-dependent refractive index change (cf. Chapter 9).

In the vicinity of the phase transition temperature $T_c$, molecular correlations in liquid crystals give rise to interesting so-called pretransitional phenomena. This is manifested in the critical dependences of the laser-induced index change and the response time on the temperature. These critical dependences are described by Landau's (14) theory of second-order phase transition advanced by deGennes (15), as explained in Chapter 2.

The free energy per unit volume in the isotropic phase of a liquid crystal, in terms of a general order parameter tensor $Q_{ij}$, is given by

$$F = F_0 + \tfrac{1}{2} A Q_{ij} Q_{ji} - \tfrac{1}{4} \chi_{ij} E_i^* E_j \tag{6.23}$$

$$A = a(T - T^*) \tag{6.24}$$

where $A$ and $T^*$ are constants defined in Chapter 2.

From (6.23) the dynamical equation for $Q_{ij}$ becomes (14):

$$\eta \frac{\partial Q_{ij}}{\partial t} + A Q_{ij} = f_{ij} \tag{6.25}$$

where

$$f_{ij} = \tfrac{1}{6}\Delta\chi\left(E_i^* E_j - \tfrac{1}{3}|E|^2\delta_{ij}\right) \tag{6.26}$$

The solution for $Q_{ij}$ is

$$Q_{ij}(t) = \int_{-\infty}^{t} \left[\frac{f_{ij}(t')}{\eta} e^{-(t-t')/\tau}\right] dt' \tag{6.27}$$

where

$$\tau = \frac{\eta}{A} = \frac{\eta}{a(T - T^*)} \tag{6.28}$$

The exact form of $Q_{ij}(t)$, of course, depends on the temporal characteristics of the laser field $E(t')$.

For simplicity, we assume that the incident laser pulse is polarized in the $\hat{i}$ direction, for example. Furthermore, we assume that the laser is a square pulse of duration $\tau_p$. We thus have $f_{ij} = f_{ii} = \frac{1}{9}\Delta\chi E^2$. For $0 < t < \tau_p$, we have

$$\begin{aligned} Q_{ii} &= \frac{1}{9}\left(\frac{\Delta\chi}{\eta}\right)\int_0^t E^2 e^{-(t-t')/\tau}\, dt' \\ &= \frac{\tau}{9}\left(\frac{\Delta\chi}{\eta}\right) E^2(1 - e^{-t/\tau}) \end{aligned} \tag{6.29}$$

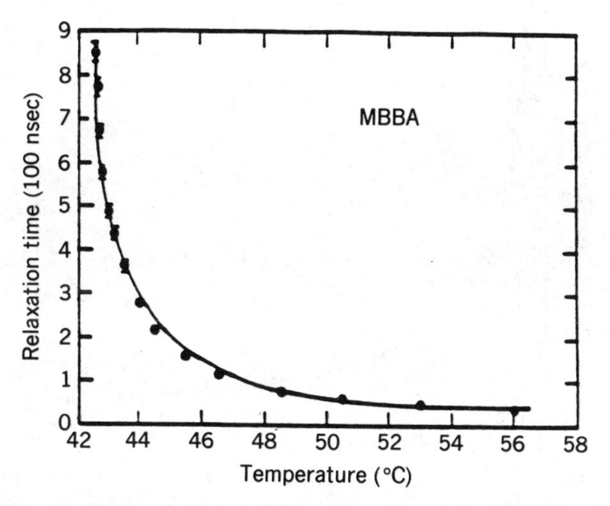


**Figure 6.5** Observed independence of the orientation relaxation time as a function of the temperature above $T_c$ [cf. Wong and Shen (10)].

For $t > \tau_p$, the order parameter freely relaxes and is described by an exponential function:

$$Q_{ii}(t) = \frac{\tau}{9}\frac{\Delta\chi}{\eta}E^2[1 - e^{-\tau_p/\tau}]e^{-t/\tau} \tag{6.30}$$

The subscript $ii$ on $Q$ denotes that we are evaluating the $i$th component of the polarization. Clearly, $Q_{ii}$ relaxes with a time constant $\tau$, which is given in (6.28). The dependence of $\tau$ on $(T - T^*)^{-1}$ shows that there is a critical slowing down as the system approaches $T^*$.

Since the linear polarization $P^{NL}$ is proportional to $Q$ and $Q$ is proportional to $\tau$ [cf. equation (6.29)], $P^{NL}$ is therefore proportional to $\tau$ [i.e., proportional to $(T - T^*)^{-1}$]. In other words, the nonlinearity of the isotropic phase will be greatly enhanced as one approaches $T^*$, just as its response time is greatly lengthened.

These phenomena, the critical slowing down of the relaxation and the enhancement of the optical nonlinearity near $T_c$, have been experimentally observed by Wong and Shen (10). In MBBA, for example, the observed relaxation times vary from about 100 ns at $T - T_c^* > 10°$ to 900 ns to $T - T^* \lesssim 1°$ (cf. Figure 6.5); the nonlinearity $\chi_{1122}$, for example, varies as $(2.2 \times 10^{-10})$ esu$/(T - T^*)$ (cf. Figure 6.6).

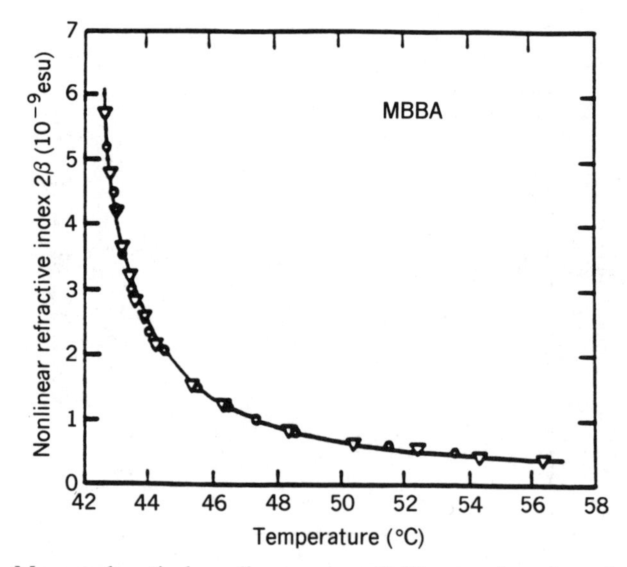


**Figure 6.6** Measured optical nonlinear susceptibility as a function of the temperature above $T_c$ [cf. Wong and Shen (10)].

### 6.2.3 Influence of Molecular Structure on Isotropic Phase Reorientational Nonlinearities

As explained in Chapter 1, molecular structures dictate the inter- and intramolecular fields which in turn influence all the physical properties of the liquid crystals. Molecular structures therefore are expected to influence the reorientational optical nonlinearities in both their magnitude and response time. The study by Madden et al. (16) has shed some light on this topic.

In the transient degenerate four-wave mixing study (cf. Chapter 9) conducted by these workers, the observed optical nonlinearity associated with molecular reorientation can be expressed (16) in terms of the parameter $C(t,k)$:

$$C(t,k) = \left\langle \left(\frac{\mathscr{L}}{K_B T}\right)\left(\sum_{\substack{i \\ \text{over all} \\ \text{molecules}}} \alpha^i_{xz}(t)e^{i\mathbf{k}\cdot\mathbf{r}^i(t)}\right)\left(\sum_j \alpha^j_{xz}(0)e^{i\mathbf{k}\cdot\mathbf{r}^j(0)}\right)^*\right\rangle \tag{6.31}$$

where $\alpha^i$ and $\mathbf{r}^i$ are the polarizability and position of molecule $i$ and $\mathscr{L}$ is a local field correction factor. $K_B$ is the Boltzmann constant and $T$ is the temperature. $C(t,k)$ is a measure of the time correlation function of a Fourier component of the polarizability of the liquid crystals.

Note that in (6.31) the $i \neq j$ terms contribute only when there is a correlation between the orientation of different molecules. Molecular orientational correlation in liquid crystals affects both the amplitude of $C(0,k)$ and its relaxation behavior.

In the case of weak coupling between molecular orientations and the transverse components of the momentum density of the liquid crystal (17, 18), the preceding expression reduces to

$$C(t,\mathbf{k}) \equiv C(t) = \rho\mathscr{L}\frac{2\gamma^2}{15K_B T}ge^{-t/\tau} \tag{6.32}$$

where $\gamma$ is the molecular polarizability anisotropy, $\rho$ is the number density, $\tau$ is the relaxation time, and $g$ is the static orientational correlation:

$$g \equiv \left\langle \alpha^i_{xz}(0) \sum_{\substack{j \\ \text{over all} \\ \text{molecules}}} \alpha^j_{xz}(0)\right\rangle \Big/ \langle \alpha^i_{xz}(0)\alpha^i_{xz}(0)\rangle \tag{6.33}$$

In an ordinary isotropic fluid, $g \approx 1$. On the other hand, for liquid crystal molecules near $T_c$, $g \approx K_B T/A(T - T^*)$, where $A$ comes from the Landau expansion [cf. equation (6.24)] of the free energy discussed in the preceding

**TABLE 6.1 Orientational Nonlinearity of Liquid Crystals Near $T_c$**

| Liquid Crystal | $\mathscr{L}\gamma^2/A$ | Observed Nonlinearity $(T-T^*)\chi^{(3)}$(esu) |
|---|---|---|
| 10CB | 0.49 | $1.36\times10^{-9}$ |
| 5CB | 0.84 | $2.71\times10^{-9}$ |
| 3CB | 1.51 | $5.0\ \times10^{-9}$ |

section. Using $g$ for a liquid crystal and (6.32), we have $C(t)=\frac{2}{15}\rho e^{-t/\tau}(\mathscr{L}\gamma^2/A)$.

From (6.32) for $C(t)$, one may conclude that the most important factor influencing the optical nonlinearity is $\gamma$, the molecular polarizability anisotropy. If everything else about the molecule remains constant, obviously a larger $\gamma$ means a larger optical nonlinearity. Consider, for example, cyanobiphenyls ($n$CBs), which are stable liquid crystals well known for their large polarizability anisotropy. In general, going to heavier members of a homologous series (i.e., larger-number $n$) increases $\gamma$, by increasing the size and anisotropy of the molecule (18). One would expect therefore a corresponding trend in the observed nonlinearity $C(0)$.

However, this is contradicted by the experimental measurements (17). Table 6.1 shows the observed results. As $n$ is increased from 3 to 10, where $\gamma$ increases, the factor $\mathscr{L}\gamma^2/A$ and the optical nonlinearity actually drop.

This "deviation" from the preceding notion of how molecular structures should influence optical nonlinearities is explained by the fact that, in general, many other physical parameters, besides $\gamma$, of the liquid crystals are modified as heavier (larger $n$) liquid crystals are synthesized. In the present case the other parameters are $\mathscr{L}$ and $A$. As seen in Table 6.1, $\mathscr{L}$ and $A$ together lead to a reverse trend on the optical nonlinearity as $\gamma$ is increased. Other parameters, such as the viscosity, shape parameter, and molecular volume, are also greatly changed as we go to heavier liquid crystal molecules in the $nC$B series, and they could adversely affect the resulting nonlinear optical response.

## 6.3 MOLECULAR REORIENTATIONS IN THE NEMATIC PHASE

In the nematic phase field-induced reorientation of the director axis arises as a result of the total system's tendency to assume a new configuration with the minimum free energy (1–7, 19). The total free energy of the system consists of the distortion energy $F_d$ and the optical dipolar interaction energy $F_{\text{op}}$, which are given by, respectively,

$$F_d=\tfrac{1}{2}K_1(\nabla\cdot\hat{n})^2+\tfrac{1}{2}K_2(\hat{n}\cdot\nabla\times\hat{n})^2+\tfrac{1}{2}K_3(\hat{n}\times\nabla\times\hat{n}) \qquad (6.34)$$

and

$$F_{\mathrm{op}} = -\frac{1}{4\pi}\int \mathbf{D}\cdot d\mathbf{E} = -\frac{\varepsilon_{\perp}}{8\pi}E^2 - \frac{\varepsilon_a\langle(\hat{n}\cdot\mathbf{E})^2\rangle}{8\pi} \tag{6.35}$$

where the angle brackets $\langle\ \rangle$ denote a time average. If the optical field $\mathbf{E}_{\mathrm{op}}$ is a plane wave [i.e., $\mathbf{E}_{\mathrm{op}} = \hat{p}|E|\cos(\omega t - kz)$, where $\hat{p}$ is a unit vector along the polarization direction], then $\langle E_{\mathrm{op}}^2\rangle = |E|^2/2$. [Note that (6.35) is written in cgs units.]

The first term on the right-hand side of (6.35) is independent of the director axis orientation, and hence it may be ignored when we consider the reorientation process. The second term indicates that the system favors a realignment of the director axis along the optical field polarization. In analogy to the elastic torque, an optical torque

$$\mathbf{m}_{\mathrm{op}} = \frac{\Delta\varepsilon}{4\pi}\langle(\hat{n}\cdot\mathbf{E})(\hat{n}\times\mathbf{E})\rangle \tag{6.36}$$

is associated with this free-energy term. We now consider a greatly simplified view of how such an optical torque gives rise to director axis reorientation in the nematic phase.

### 6.3.1 Simplified Treatment of Optical Field-Induced Director Axis Reorientation

Consider, for example, the interaction geometry depicted in Figure 6.7 where a linearly polarized laser is incident on a homeotropically aligned nematic liquid crystal. The propagation vector $\mathbf{K}$ of the laser makes an angle $(\beta + \theta)$ with the perturbed director axis: $\theta$ is the reorientation angle. For this case, if the reorientation angle $\theta$ is small, then only one elastic constant $K_1$ (for splay distortion) is involved. A minimization of the total free energy of the system yields a torque balance equation:

$$K_1\frac{d^2\theta}{dz^2} + \frac{\Delta\varepsilon\langle E_{\mathrm{op}}^2\rangle}{8\pi}\sin 2(\beta+\theta) = 0 \tag{6.37}$$

In the small $\theta$ approximation, this may be written as

$$2\xi^2\frac{d^2\theta}{dz^2} + (2\cos 2\beta)\theta + \sin 2\beta = 0 \tag{6.38}$$

where $\xi^2 = 4\pi K_1/[\Delta\varepsilon\langle E_{\mathrm{op}}^2\rangle]$.

Since most observed processes in nematic liquid crystals involve small reorientation, we will continue our discussion based on the small $\theta$ limit. In

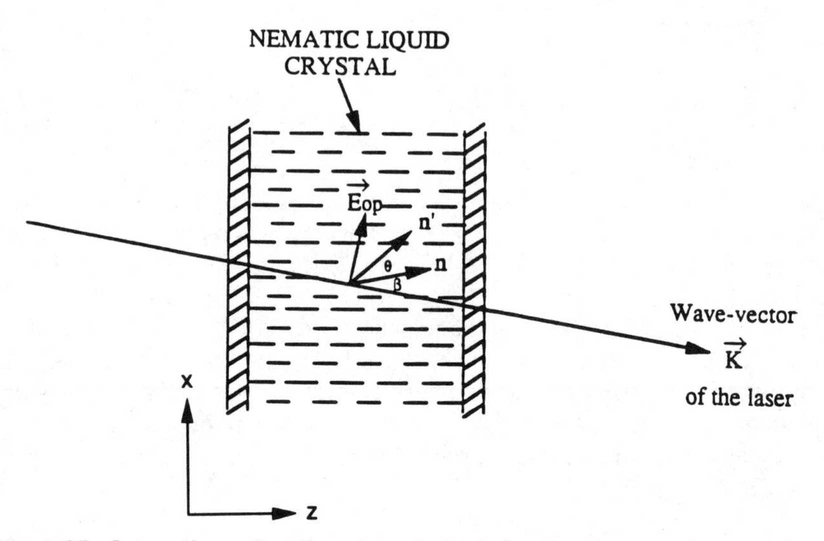


**Figure 6.7** Interaction of a linearly polarized (extraordinary ray) laser with a homeotropically aligned nematic liquid crystal film.

this case, using the so-called hard-boundary condition [i.e., the director axis is not perturbed at the boundary ($\theta = 0$ at $z = 0$ and at $z = d$)], the solution of (6.38) is

$$\theta = \frac{1}{4\xi^2} \sin 2\beta (dz - z^2) \tag{6.39}$$

that is, the reorientation is maximum at the center and vanishingly small at the boundary, as shown in Figure 6.8.

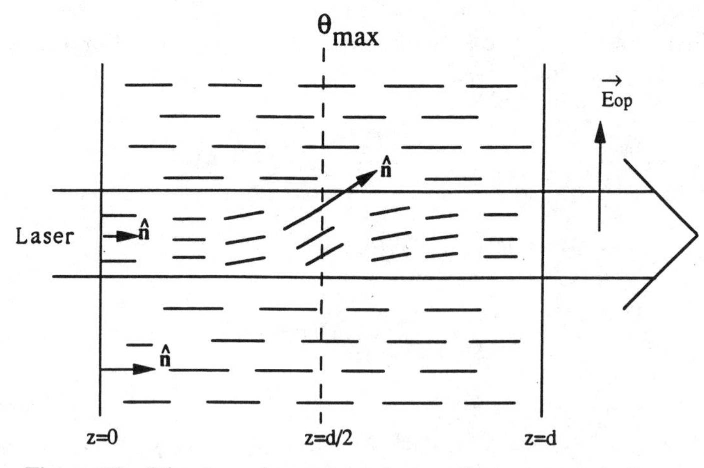


**Figure 6.8** Director axis reorientation profile in a nematic film.

As a result of this reorientation, the incident laser (an extraordinary wave) experiences a *z-dependent refractive index change* given by

$$\Delta n = n_e(\beta + \theta) - n_e(\beta) \tag{6.40}$$

where $n_e(\beta + \theta)$ is the extraordinary ray index

$$n_e(\beta + \theta) = \frac{n_{\parallel} n_{\perp}}{\left[ n_{\parallel}^2 \cos^2(\beta + \theta) + n_{\perp}^2 \sin^2(\beta + \theta) \right]^{1/2}} \tag{6.41}$$

For small $\theta$, the change in the refractive index $\Delta n$ is proportional to the square modulus of the optical electric field, that is,

$$\Delta n = n_2(z) \langle E_{\mathrm{op}}^2 \rangle \tag{6.42}$$

or

$$\Delta n = \alpha_2(z) I \tag{6.43}$$

with $\alpha_2(z)$ given by

$$\alpha_2(z) = \frac{(\Delta\varepsilon)^2 \sin^2(2\beta)}{4Kc} (dz - z^2) \tag{6.44}$$

Note that, when averaged over the sample thickness, the factor $(dz - z^2)$ gives $d^2/6$.

What is truly unique about nematic liquid crystals is the enormity of $\alpha_2$ compared to most nonlinear materials. Let us denote the average of the value of $\alpha_2$ over the film thickness by $\bar{\alpha}_2$. For a film thickness $d = 100\ \mu\mathrm{m}$, $\Delta\varepsilon \sim 0.6$, $K = 10^{-6}$, $\beta = 45°$, we have

$$\bar{\alpha}_2 = 5 \times 10^{-3}\ \mathrm{cm}^2/\mathrm{W} \tag{6.45}$$

In electrostatic units, this corresponds to a third-order susceptibility $\chi^{(3)}$ on the order of $9.54 \times 5 \times 10^{-3} \sim 5 \times 10^{-2}$ esu (cf. Chapter 9 on unit equivalence). This nonlinear coefficient is about eight orders of magnitude larger than that of $CS_2$ and about six orders of magnitude larger than the isotropic phase liquid crystal reorientational nonlinearity discussed in Section 6.2.3.

### 6.3.2 More Exact Treatment of Optical Field-Induced Director Axis Reorientation

If the anisotropies of the elastic constants and optical field propagation characteristics in the birefringent nematic film are taken into account, the resulting equations for the optical field, elastic and optical torques, and so on become very complicated (2–7) even for the simple interaction geometry depicted in Figure 6.4. The starting point of the analysis is the Euler–Lagrange equation associated with a small director axis reorientation angle $\theta(z)$ (20):

$$\frac{\partial F}{\partial \theta} - \frac{d}{dz}\frac{\partial F}{\partial \theta'} = 0 \tag{6.46}$$

where $F = F_d + F_{\text{op}}$ from (6.34) and (6.35). [Note that in a hybrid aligned nematic sample, one has to account for the surface anchoring energy as well (21).] In the present treatment we will assume the hard-boundary condition and ignore the surface interaction term. For reorientation in the $x-z$ plane, only the elastic energies associated with the splay ($K_1$) and bend ($K_3$) are involved.

Writing $\hat{n} = \sin\theta''\,\hat{x} + \cos\theta''\,\hat{z}$, we have

$$(\nabla\cdot\hat{n})^2 = \sin^2\theta''\left(\frac{\partial\theta''}{\partial z}\right)^2 \tag{6.47a}$$

$$(\hat{n}\times\nabla\times\hat{n}) = \cos^2\theta''\left(\frac{d\theta''}{dz}\right)^2 \tag{6.47b}$$

where $\theta'' = \theta + \beta$. Also, the optical field may be expressed as

$$\mathbf{E}_{\text{op}} = \left(E_x\hat{x} + E_z\hat{z}\right) \tag{6.48}$$

where $E_x$ and $E_z$ remain to be calculated from the Maxwell equation:

$$\nabla(\nabla\cdot\mathbf{E}_{\text{op}}) - \nabla^2\mathbf{E}_{\text{op}} - \frac{\omega^2}{c^2}\bar{\bar{\varepsilon}}\mathbf{E}_{\text{op}} = 0 \tag{6.49}$$

This is because the dielectric tensor $\bar{\bar{\varepsilon}}$ is dependent on the optical field $\mathbf{E}_{\text{op}}$ as a result of the optical field-induced director axis reorientation. In other words, (6.46) and (6.49) have to be solved in a self-consistent manner to yield $\theta(z)$ and $\mathbf{E}_{\text{op}}(z)$.

Using (6.47) and (6.48), (6.46) can be explicitly written as

$$\left(K_1\sin^2\theta'' + K_3\cos^2\theta''\right)\frac{d^2\theta}{dz^2} - (K_3 - K_1)\sin\theta''\cos\theta''\left(\frac{d\theta}{dz}\right)^2$$
$$+ \frac{\Delta\varepsilon}{16\pi}\left[\sin 2\theta''\left(|E_x|^2 - |E_y|^2\right) + \cos 2\theta''(E_xE_z^* + E_x^*E_z)\right] = 0 \tag{6.50}$$

The self-consistent solutions of $\mathbf{E}_{\text{op}}$ and $\theta(z)$ from these equations are quite complex and outside the scope of this chapter. We refer the reader to the literature (2–7, 20) for the details. It suffices to say that, in the limit of very small director axis reorientation, that is, $\theta \ll 1$, which is usually the case, the preceding equation reduces to the simplified one given in Section 6.3.1.

In the case where the optical field is incident perpendicularly to the sample, that is, $\beta = 0$, (6.50) becomes greatly simplified and yields an interesting result. Setting $\beta = 0$, we get

$$K_3 \frac{d^2\theta}{dz^2} + \frac{\Delta\varepsilon}{16\pi}\left[2\theta\left(|E_x|^2 - |E_z|^2\right) + |E_x E_z| + |E_x E_z|\right] = 0 \tag{6.51}$$

From $\nabla \cdot \mathbf{D} = 0$, we get

$$E_z \simeq - E_x \left(\frac{\Delta\varepsilon}{\varepsilon_{\parallel}}\right)\theta \tag{6.52}$$

Equation (6.51) therefore becomes

$$K_3 \frac{d^2\theta}{dz^2} + \frac{\Delta\varepsilon}{8\pi}\frac{\varepsilon_{\perp}}{\varepsilon_{\parallel}}|E_x|^2\theta = 0 \tag{6.53}$$

Equation (6.53) is analogous to the equation for the dc field-induced Freedericksz transition, recalling that we have made the approximation $\sin\theta = \theta$. We can thus define a so-called optical Freedericksz field $E_F$ given by

$$|E_F|^2 = \frac{8\pi^3 K_3}{\Delta\varepsilon}\frac{\varepsilon_{\parallel}}{\varepsilon_{\perp}}\frac{1}{d^2} \tag{6.54}$$

For $|E_x| < |E_F|$, $\theta = 0$. For $|E_x| > |E_F|$, director axis reorientation will take place.

The preceding discussion is based on the assumption that the incident laser is a plane wave. If the laser is a (focused) beam, with a beam size $\omega_0$ comparable to or smaller than the film thickness, transverse correlation effects will arise. Molecules situated "outside" the laser beam will exert torques on molecules "inside" the beam; conversely, molecules "inside" the beam could also exert torques on those on the "outside." The result is that the transverse dependence of the reorientation profile is not the same function as the incident laser beam's transverse profile. Put in another way, one may recognize that (6.53) is basically a diffusion equation, where the elastic term plays the role of the diffusive mechanism. As a result of this diffusive effect, as in many other physical processes, the spatial profile of the response is not the same as the excitation profile.

As shown in the detailed calculations given in Khoo et al. (22):

1 For $\beta = 0$, there is a threshold intensity for finite reorientation to occur. The threshold intensity depends on both the thickness of the film and the beam size $\omega_0$. For $\omega_0 \ll d$, the threshold intensity increases dramatically (compared with the value for a plane wave). There is no threshold intensity for field-induced reorientation in the $\beta \neq 0$ case.

2 For a Gaussian laser beam input, the reorientation profile is not Gaussian, although it is still a bell-shaped function with a half-width $\omega_\theta$ different from $\omega_0$. In general, for the $\beta \neq 0$ case, the half-width $\omega_\theta$ is always larger than $\omega_0$ for all values of $\omega_0$, approaching $\omega_0$ for large values of $\omega_0/d$ (cf. Figure 6.9*a*). On the other hand, for the $\beta = 0$ case, $\omega_\theta$ can be smaller than $\omega_0$, depending on whether $\omega_0$ is smaller or larger than $d$ (cf. Figure 6.9*b*).

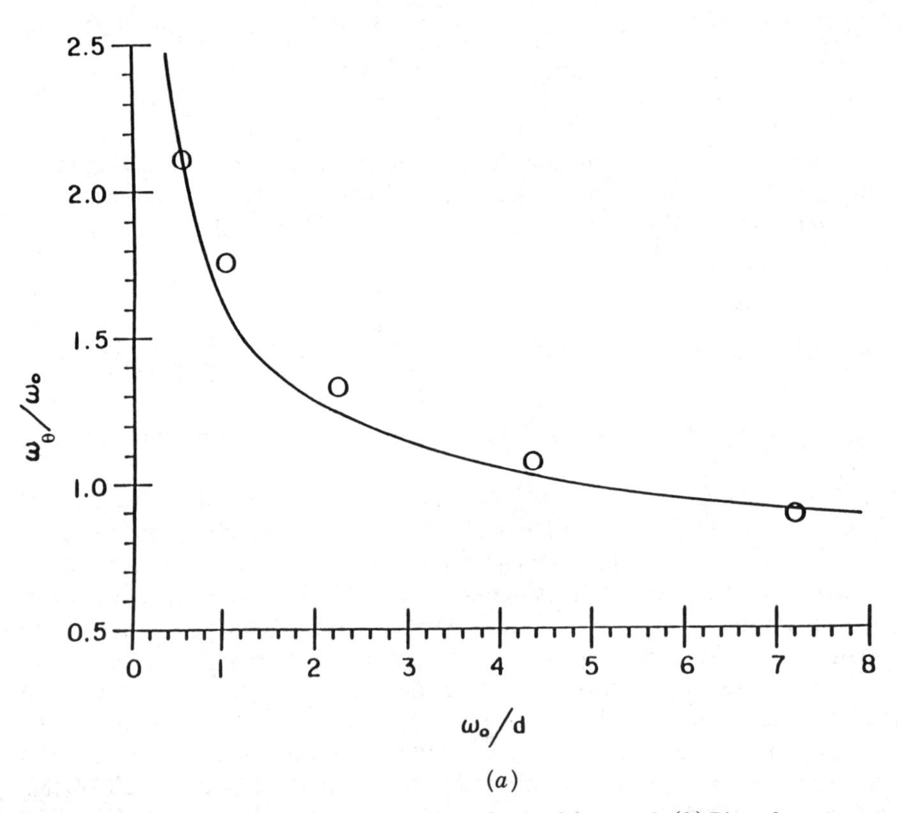


**Figure 6.9** (*a*) Plot of $\omega_\theta/\omega_0$ as a function of $\omega_0/d$ for $\beta \neq 0$. (*b*) Plot of $\omega_\theta/\omega_0$ as a function of $\omega_0/d$ for $\beta = 0$.

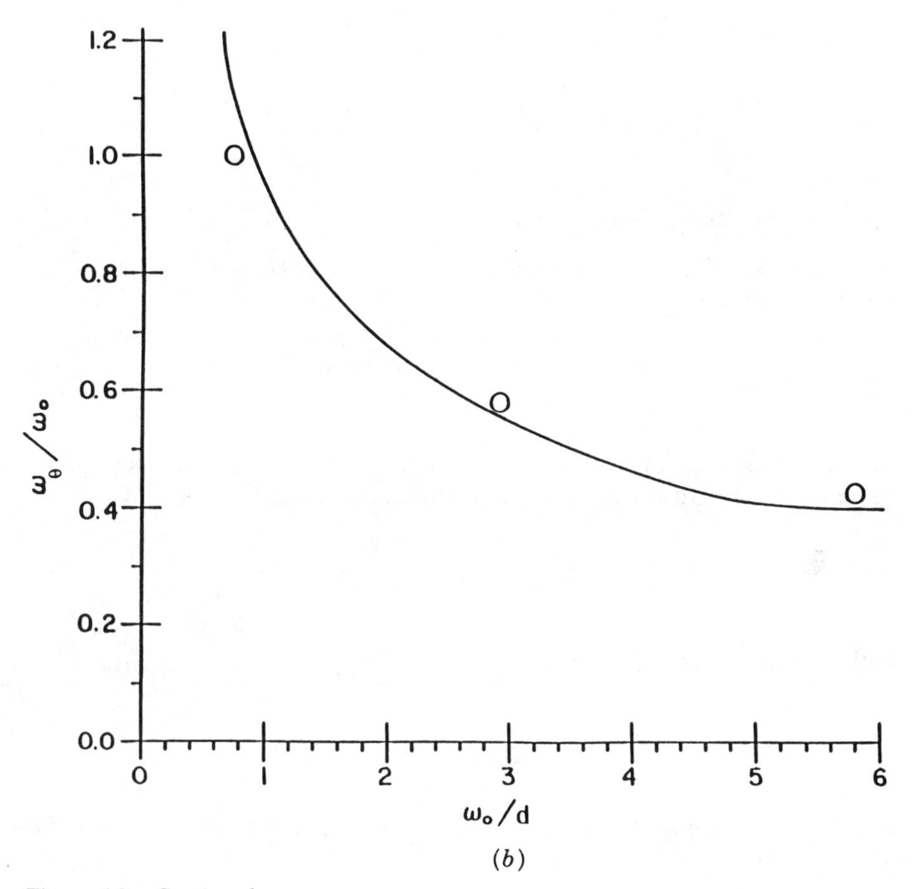


**Figure 6.9** *Continued*

## 6.4 NEMATIC PHASE REORIENTATION DYNAMICS

The dynamics of molecular reorientation are described by a torque balance equation. Balancing the viscous, elastic, and optical torques experienced by the molecule during the process of reorientation, the resulting equation is given by

$$\gamma \frac{\partial \theta}{\partial t} = K \frac{\partial^2 \theta}{\partial z^2} + \frac{\Delta\varepsilon \langle E_{\text{op}}^2 \rangle}{8\pi} \sin(2\beta + 2\theta) \tag{6.55}$$

for the interaction geometry given in Figure 6.4. For simplicity, we have again assumed that $\theta$ is small and that we can use the one-elastic-constant approximation. We have introduced an effective viscosity coefficient $\gamma$.

For small $\theta$, that is, $\sin\theta \approx \theta$, $\cos\theta \approx 1$, (6.55) becomes

$$\gamma\frac{\partial\theta}{\partial t} = K\frac{\partial^2\theta}{\partial z^2} + \frac{\Delta\varepsilon\langle E_{op}^2\rangle}{8\pi}\sin 2\beta + \theta\frac{\Delta\varepsilon\langle E_{op}^2\rangle}{4\pi}\cos 2\beta \tag{6.56}$$

### 6.4.1 Plane Wave Optical Field

Assuming that $E_{op}^2$ is a plane wave, we may write $\theta(t,z) = \theta(t)\sin(\pi z/d)$, and substituting it into (6.56), we get

$$\gamma\dot{\theta} = -\frac{K\pi^2}{d^2}\theta + \frac{\Delta\varepsilon\langle E_{op}^2\rangle\sin 2\beta}{8\pi} + \theta\frac{\Delta\varepsilon\langle E_{op}^2\rangle}{4\pi}\cos 2\beta \tag{6.57}$$

A good understanding of the reorientation dynamics can be obtained if we separate it into two regimes: (1) optical torque $\gg$ elastic torque

$$\left|\frac{\Delta\varepsilon\langle E_{op}^2\rangle}{8\pi}\sin(2\beta+2\theta)\right| \gg \left|K\frac{\partial^2\theta}{\partial z^2}\right|$$

and (2) optical torque $\ll$ elastic torque

$$\left|\frac{\Delta\varepsilon\langle E_{op}^2\rangle}{8\pi}\sin(2\beta+2\theta)\right| \ll \left|K\frac{\partial^2\theta}{\partial z^2}\right|$$

For case 1 we may ignore the elastic term in (6.57) and get an equation for $\theta$ of the form

$$\dot{\theta} = a + b\theta \tag{6.58}$$

where

$$a = \frac{\Delta\varepsilon\langle E_{op}^2\rangle}{8\pi\gamma}\sin 2\beta \tag{6.59}$$

$$b = \frac{\Delta\varepsilon\langle E_{op}^2\rangle}{4\pi\gamma}\cos 2\beta \tag{6.60}$$

If $E_{op}^2$ is associated with a square laser pulse (i.e., $E_{op}^2 = 0$ for $t < 0$; $E_{op}^2 = E_0^2$ for $0 < t < \tau_p$), the solution for $\theta$ is therefore, for $0 < t < \tau_p$,

$$\theta(t) = \frac{a}{b}(e^{bt} - 1) \tag{6.61}$$

From (6.61) we can see that $\theta(\tau_p)$ is appreciable only if $b\tau_p$ is appreciable. In other words, if the laser pulse duration is short (e.g., nanosecond), it has to be very intense in order to induce an appreciable reorientation effect. In this respect and because the surface elastic torque is not involved, the dynamical response of a nematic liquid crystal is quite similar to its isotropic phase counterpart. However, the dependence on the geometric factor $\sin 2\beta$ is a reminder that the nematic phase is nevertheless an (ordered) aligned phase, and its overall response is dependent on the direction of incidence and the polarization of the laser.

The effective optical nonlinearity in the transient case, compared to the steady-state value, can be estimated from (6.61).

In the short time limit (i.e., $bt \ll 1$), $\theta(t) \sim at$. Therefore, we have

$$\frac{\theta(t)}{\langle\theta\rangle} = \frac{12}{\pi^2}\left(\frac{K_1\pi^2}{\gamma d^2}\right)t \tag{6.62}$$

where $\langle\theta\rangle$ is the averaged (over the sample thickness) value of $\theta$ given in (6.39) for the steady-state case (e.g., cw laser). This may be expressed in another way:

$$\frac{\theta(t)}{\langle\theta\rangle} = \frac{12}{\pi^2}\frac{t}{\tau_r} \sim \frac{t}{\tau_r} \tag{6.63}$$

where $\tau_r$ is the nematic axis reorientation time, $\tau_r = \gamma d^2/K\pi^2$. Using typical values in the estimate of $\bar{\alpha}_2$ in (6.45): $\gamma = 0.1$ P, $d = 100$ $\mu$m, and $K = 10^{-6}$ dyne, we have $\tau_r \approx 1$ s. If we define a so-called effective nonlinear coefficient $\alpha_2^t$ for transient orientational nonlinearity, then

$$\alpha_2^t = \left(\frac{t}{\tau_r}\right)\bar{\alpha}_2 \tag{6.64}$$

For a 10-ns ($10^{-8}$) laser pulse, the effective nonlinearity is on the order of $5\times10^{-11}$ esu.

For case 2, which naturally occurs when the *laser pulse is over*, we have

$$\gamma\dot{\theta} = -\frac{K\pi^2}{d^2}\theta \tag{6.65}$$

that is,

$$\theta = \theta_{\max} e^{-t/\tau_r} \tag{6.66}$$

where

$$\alpha_{\max} = \frac{a}{b}(e^{b\tau_p} - 1) \tag{6.67}$$

The preceding discussion and results apply to the case where an extraordinary wave laser is obliquely incident on the (homeotropic) sample (i.e., $\beta \neq 0$). For the case where a laser is perpendicularly incident on the sample (i.e., its optical electric field is normal to the director axis), there will be a critical optical field $E_F$, the so-called Freedericksz transition field [cf. equation (6.54)], below which molecular reorientation will not take place. Second, the turn-on time of the molecular reorientation depends on the field strength above $E_F$ (i.e., on $E_{op} - E_F$). For small $E_{op} - E_F$, the turn-on time can approach many minutes! Studies with nanosecond and picosecond lasers (8, 9) have shown that under this perpendicularly incident (i.e., $\beta = 0$) geometry, it is very difficult to induce molecular reorientation through the mechanism discussed previously.

### 6.4.2 Sinusoidal Optical Intensity

In many nonlinear optical wave mixing processes involving the interference of two coherent beams, the resulting optical intensity imparted on the liquid crystal is a sinusoidal function. This naturally induces a spatially oscillatory director axis reorientational effect. Molecules situated at the intensity maxima will undergo reorientation, while those in the "dark" region (intensity minima) will stay relatively unperturbed. In analogy to the laser transverse intensity effect discussed at the end of the last section, molecules in these regions will exert torque on one another, and the resulting relaxation time constant will be governed by the characteristic length of these sinusoidal variations as well as the thickness $d$ of the nematic film.

Consider, for example, an optical intensity function of the form $\langle E_{op}^2 \rangle \sim E^2(1 + \cos qy)$. The induced reorientational angle will have a modulation of the form

$$\theta \sim \theta(t) \sin\left(\frac{\pi z}{d}\right) \cos qy \tag{6.68}$$

on top of a spatially uniform component. This will give rise to an extra elastic torque term given by

$$K \frac{\partial^2 \theta}{\partial y^2} = -Kq^2\theta \tag{6.69}$$

on the right-hand side of (6.56). Accordingly, the orientational relaxation dynamics when the optical field is turned off now becomes

$$\gamma\dot{\theta} = -K\left(\frac{\pi^2}{d^2} + q^2\right)\theta \tag{6.70}$$

This gives an orientational relaxation time constant:

$$\tau_r = \frac{\gamma}{K}\left[\frac{1}{\pi^2/d^2 + q^2}\right] \tag{6.71}$$

Writing the wave vector $q$ in terms of the grating constant $\Lambda = 2\pi/q$, we have

$$\tau_r = \frac{\gamma}{K}\left[\frac{1}{\pi^2/d^2 + 4\pi^2/\Lambda^2}\right] \tag{6.72}$$

From (6.72 we can see that if $\Lambda \ll d$, the orientational relaxation dynamics is dominated by $\Lambda$ (i.e., the intermolecular torques); conversely, if $d \ll \Lambda$, the dynamics is decided by the boundary elastic torques.

These influences of the intermolecular and the elastic torques in cw-laser-induced nonlinear diffraction effects in nematic films are reported in Khoo (23). There it is also noted that the optical nonlinearity associated with nematic director axis reorientation is proportional to the factor $\Delta\varepsilon^2/K$ [cf. equation (6.44).] Although both $\Delta\varepsilon$ and $K$ are strongly dependent on the temperature, the combination $\Delta\varepsilon^2/K$ is not. This is because $\Delta\varepsilon$ is propor-

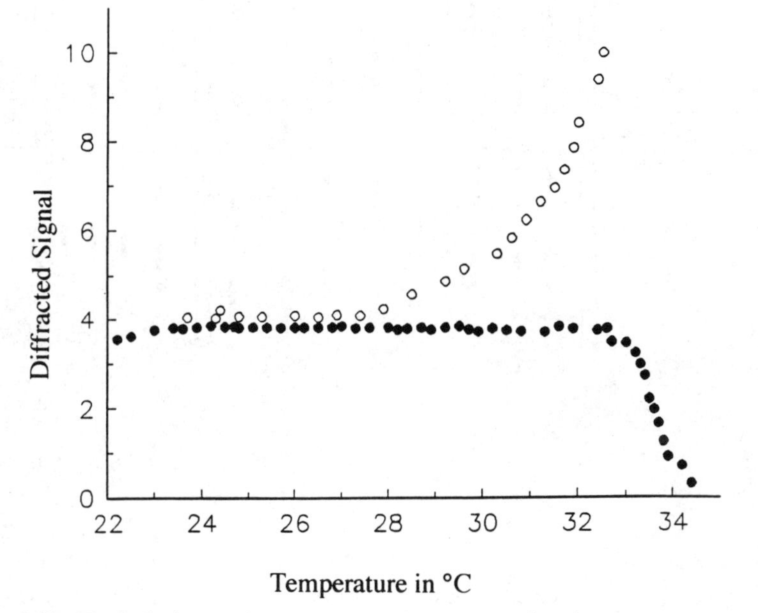


**Figure 6.10** Typical observed temperature dependence (dots) of the nonlinear side diffraction owing to the nematic axis reorientation effect [from Khoo (23)]. Circles are thermal index change effects discussed in the next chapter.

tional to the order parameter $S$, whereas $K$ is proportional to $S^2$. This is indeed verified in the experimental study of the temperature dependence of nonlinear diffraction (cf. Figure 6.10). The signal stays quite flat up to about 1° near $T_c$; near $T_c$, the nematic alignment begins to deteriorate and the signal diminishes. The factor $\Delta\varepsilon^2/K$ also appears in linear scattering associated with director axis fluctuations, as discussed in the previous chapter.

## 6.5 LASER-INDUCED DYE-ASSISTED MOLECULAR REORIENTATION

In nematic liquid crystals doped with some absorbing dye molecules, recent studies (24–26) have shown that the excited dye molecules could exert an intermolecular torque on the liquid crystal molecules that could be stronger than the optical torque. In particular, Janossy et al. (24) have observed that some classes of anthraquinone dye molecules, when photoexcited, will fur-

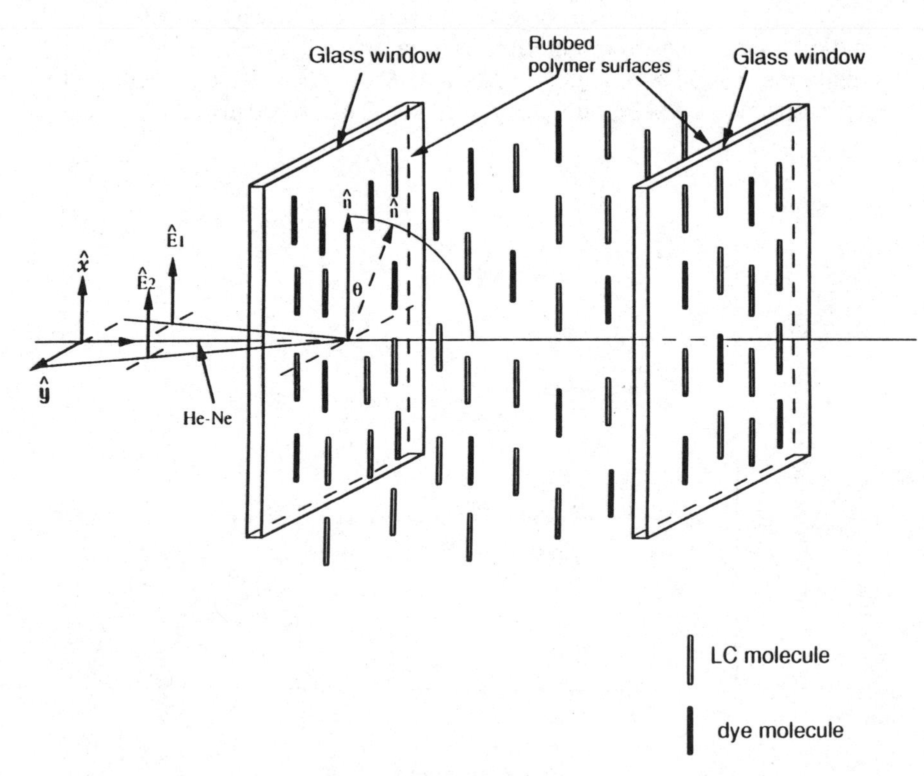


**Figure 6.11** Schematic diagram of the optical fields and their propagation direction in a planar aligned dye-doped liquid crystal. The two fields $\mathbf{E}_1$ and $\mathbf{E}_2$ are coherent beams derived from splitting a pump laser to induce dynamic grating in the liquid crystal sample. The He–Ne laser is the probe [cf. Khoo et al. (26)].

ther enhance such positive torque effects. On the other hand, Gibbons et al. and Chen and Brady (25) have observed that, under prolonged exposure, dye-doped liquid crystals (DDLCs) will align themselves in a direction orthogonal *to the optical electric field and the propagation wave vector* (i.e., in the $\hat{y}$ direction with reference to Figure 6.11). Under suitable surface treatment conditions, such reorientational effects can be made permanent (but erasable).

Khoo et al. (26) report the observation of a negative reorientational effect that occurs in the transient regime under short laser pulse illumination. Using a variety of polarization configurations between the pump and probe beam in the dynamic grating diffraction study, the authors have established that the negative change in the refractive index (i.e., negative nonlinearity) is associated with the *liquid crystal director axis realigning toward the z direction*. The efficiency of this reorientation process is governed by the types of dye molecules used as dopants (24, 26). An interesting and important point is that, at low optical power, the orientational effect actually is the dominating one, even in such highly absorptive material. The observed nonlinearity is orders of magnitude larger than the so-called giant optical nonlinearity of pure liquid crystals. In Chapter 10, we describe a very recent discovery of dc-field assisted photoreflective effect in dye-doped nematic liquid crystals.

## 6.6 REORIENTATION AND NONELECTRONIC NONLINEAR OPTICAL EFFECTS IN SMECTIC AND CHOLESTERIC PHASES

### 6.6.1 Smectic Phase

The basic physics of laser-induced molecular reorientations, as well as the thermal, density, and flow phenomena in the smectic phase, is similar to that occurring in the nematic phase; the main differences lie in the magnitude of the various physical parameters that distinguish the smectic from the nematic phase. In general, smectic liquid crystals are highly viscous (i.e., they do not flow as easily as nematics). Their tendency to have layered structures (i.e., positional ordering among molecules in a plane) also imposes further restrictions on the reorientation of the molecules by an external field. In the smectic phase the order parameter dependence on the temperature is less drastic than in the nematic phase, and thus the refractive indices $n_{\parallel}$ and $n_{\perp}$ are not sensitively dependent on the temperature. Smectic liquid crystals do possess one important intrinsic advantage over nematic liquid crystals. As a result of the presence of a higher degree of order and less molecular orientation fluctuations in this phase, light-scattering loss in the smectic phase is considerably less than in the nematic phase. This will be important for optical processes that may require longer interaction lengths.

Currently, most studies on nonlinear optics are conducted in nematic liquid crystals, because of their special properties discussed before. Since the

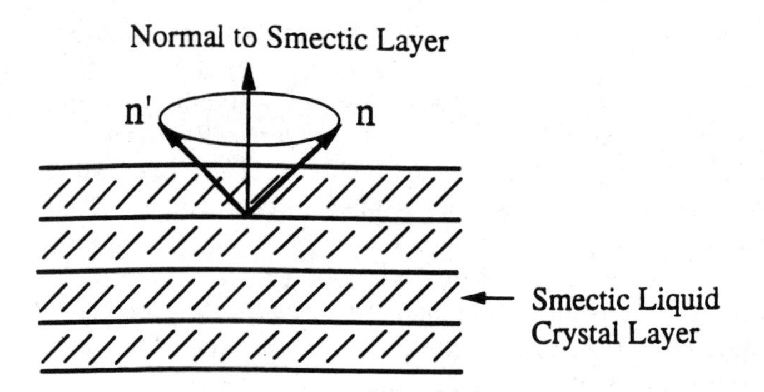


**Figure 6.12** Smectic-C liquid crystal axis azimuthal rotation by an external field.

basic physics in smectics is similar, we refer the reader to the literature quoted in the following discussion for the details. Nevertheless, there are some interesting studies worth special mention here.

The possibility of laser-induced director axis reorientations in the smectic phase was first theoretically studied by Tabiryan and Zeldovich (2) in 1981. For smectic-A and smectic-B, one can see that a reorientation of the director axis will involve a change in the layer spacing. As shown in Chapter 4, such a distortion will involve a tremendous amount of energy (13) and is therefore not observable under finite field strength. On the other hand, in smectic-C, it is possible to reorient the azimuthal component of the director axis of the molecule (cf. Figure 6.12). This transition involves only a rotation of the director about the normal to the layer and does not involve a distortion of the layer spacing.

The optical nonlinearity associated with such a reorientational process in the presence of an external orienting dc magnetic field is estimated by Tabiryan and Zeldovich (2) to be comparable to that in the nematic phase. In a later publication, Ong and Young (27) presented a detailed theory of a purely optically induced reorientation effect. Some preliminary observations of such a reorientation process in a freely suspended smectic-C firm were reported by Lippel and Young (28).

Recently, a detailed theory and experimental observation of laser-induced director axis reorientation in a chiral smectic-C* (ferroelectric) liquid crystal were reported (29). This study was conducted with a surface-stabilized ferroelectric liquid crystal in a planar oriented "bookshelf-like" configuration, where the director axis of the molecules is parallel, and the smectic layer perpendicular, to the cell walls (cf. Figure 6.13). Typically, the induced reorientation angle by a laser of intensity on the order of 3000 $W/cm^2$ is about 23° with a switching time measured to be on the order of a millisecond or less. The authors also noticed strong influence from laser heating of the sample.

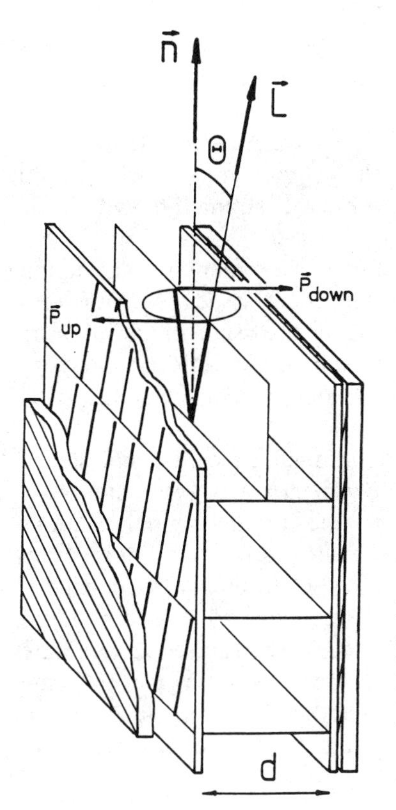


**Figure 6.13** "Bookshelf-like" configuration of a ferroelectric liquid crystal film used for the laser-induced director axis rotation effect [from Macdonald et al. (29)]. Here $\hat{n}$ is the normal to the smectic layer plane and $\vec{L}$ is the director axis direction. $\vec{P}$ is the spontaneous polarization.

### 6.6.2 Cholesteric Phase

Director axis reorientation in the cholesteric phase of liquid crystals was also first theoretically studied by Tabiryan and Zeldovich (2). The cholesteric phase is unusual in that the director axis is spatially spirally distributed with a well-defined pitch, resulting in selective reflection of light. The basic physics of optically induced director axis distortion in the cholesteric phase is analogous to its nematic counterpart. One writes down the free-energy density of the system which consists of the elastic distortion and the optical energy associated with the optical fields present in the material, that is,

$$F = \tfrac{1}{2}K_{11}(\nabla\cdot\hat{n})^2 + \tfrac{1}{2}K_{22}(\hat{n}\cdot\nabla\times\hat{n}+q)^2 + \frac{1}{2}K_{33}(\hat{n}\times\nabla\times\hat{n})^2 - \frac{\mathbf{D}\cdot\mathbf{E}}{8\pi} \tag{6.73}$$

where $q$ is the unperturbed wave vector of the helix with a pitch $p = 2\pi/q$.

By minimizing the free-energy density with respect to a director axis distortion, one arrives at the Euler–Lagrange equation for the director axis distortion angle.

If we compare (6.73) for cholesterics with (6.34) for nematics, the similarity is striking, though not surprising, since cholesterics are actually (chiral) nematics and are characterized by the same physical parameters and elastic distortion constants.

In Lee et al. (30), a theory and some qualitative experimental confirmation are presented on the optical retro-self-focusing effect associated with optically induced pitch dilation. The laser used has a Gaussian radial intensity distribution, which induces a radially varying pitch dilation effect, in analogy to the radially varying director axis reorientation induced by a Gaussian beam (cf. Section 6.3.2).

A quantitative study of laser-induced orientational effects in cholesterics was performed by Galstyan et al. (31). The director axis reorientation is induced by two counterpropagating waves of opposite circular polarizations (right- and left-handed). The signal beam originates as scattered noise from the pump beam and is amplified via stimulated scattering. In their experiment laser pulses on the order of 800 $\mu$s, with energy up to several hundred millijoules focused to a spot diameter of 35 $\mu$m, were required to generate observable stimulated scattering effects. This corresponds to a laser intensity on the order of a few megawatts per square centimeter, which seems to be the usual intensity level needed for observing stimulated scatterings in nematics as well (32).

## REFERENCES


1. I. C. Khoo, Nonlinear optics of liquid crystal. In "Progress in Optics" (E. Wolfe, ed.). North-Holland Publ., Amsterdam 1988.
2. N. V. Tabiryan and B. Ya. Zel'dovich, in a series of articles published in *Mol. Cryst. Liq. Cryst.* **62**, 637; **69**, 19, 31 (1981) on nematic, smectic and cholesteric liquid crystals.
3. I. C. Khoo and S. L. Zhuang, Nonlinear optical amplification in a nematic liquid crystal above the Freedericksz transition. *Appl. Phys. Lett.* **37**, 3 (1980).
4. A. S. Zolotko, V. F. Kitaeva, N. Kroo, N. N. Sobelev, and L. Chillag, *Pis'ma Eksp. Teor. Fiz.* **32**, 170 (1980); *JETP Lett.* (*Engl. Transl.*) **32**, 158 (1980).
5. I. C. Khoo, Optically induced molecular reorientation and third order nonlinear optical processes in nematic liquid crystal. *Phys. Rev.* A **23**, 2077 (1981).
6. S. D. Durbin, S. M. Arakelian, and Y. R. Shen, *Phys. Rev. Lett.* **47**, 1411 (1981).
7. R. M. Herman and R. J. Serinko, Nonlinear-optical processes in nematic liquid crystals near Freedericksz transitions. *Phys. Rev. A* **19**, 1757 (1979).

8. I. C. Khoo, R. G. Lindquist, R. R. Michael, R. J. Mansfield, and P. G. LoPresti, Dynamics of picosecond laser-induced density, temperature, and flow-reorientation effects in the mesophases of liquid crystals. *J. Appl. Phys.* **69**, 3853 (1991); "Flow" effect, where the liquid crystal molecules are displaced by the intense optical field, has also been observed in the smectic phase; see I. C. Khoo and R. Normandin, Nanosecond laser-induced transient and erasable permanent grating diffractions and ultrasonic waves in a nematic film. *J. Appl. Phys.* **55**, 1416 (1984); see also K. Kiyano and J. B. Ketterson, Ultrasonic study of liquid crystals. *Phys. Rev. A* **12**, 615 (1975).
9. H. J. Eichler and R. Macdonald, Flow alignment and inertial effects in picosecond laser-induced reorientation phenomena of nematic liquid crystals. *Phys. Rev. Lett.* **67**, 2666 (1991), see also J. D. Jackson, "Classical Electrodynamics" Wiley, New York, 1985.
10. See, however, F. W. Deeg and M. D. Fayer, Analysis of complex molecular dynamics in an organic liquid by polarization selective subpicosecond transient grating experiments. *J. Chem. Phys.* **91**, 2269 (1989), and references therein, where the observed individual molecular reorientation effect is associated with nuclear reorientation whose dynamics is more complex than the Debye relaxation process described in G. K. L. Wong and Y. R. Shen, Study of pretransitional behavior of laser-field-induced molecular alignment in isotropic nematic substances *Phys. Rev. A* **10**, 1277 (1974).
11. C. Flytzanis and Y. R. Shen, Molecular theory of orientational fluctuations and optical Kerr effect in the isotropic phase of a liquid crystal. *Phys. Rev. Lett.* **33**, 14 (1974).
12. D. V. G. L. N. Rao and S. Jayaraman, Pretransitional behaviour of self-focusing in nematic liquid crystals. *Phys. Rev. A* **10**, 2457 (1974).
13. J. Prost and J. R. Lalanne, Laser-induced optical Kerr effect and the dynamics of orientational order in the isotropic phase of a nematogen *Phys. Rev. A* **8**, 2090 (1973); see also J. R. Lalanne, B. Martin, and B. Pouligny, Direct observation of picosecond reorientation of molecules in the isotropic phases of nematogens. *Mol. Cryst. Liq. Cryst.* **42** 153 (1977).
14. L. D. Landau, *in* "Collected Papers of L. D. Landau" (D. Ter Haar, ed.). Gordon & Breach, New York, 1965.
15. P. G. deGennes, "The Physics of Liquid Crystals." Clarendon Press, Oxford, 1974.
16. P. A. Madden, F. C. Saunders, and A. M. Scott, Degenerate four-wave mixing in the isotropic phase of liquid crystals: The influence of molecular structure. *IEEE J. Quantum Electron.* **QE-22**, 1287 (1986).
17. J. Berne and R. Pecora, "Dynamic Light Scattering with Applications to Chemistry, Biology, and Physics." Wiley (Interscience), New York, 1976.
18. I. H. Ibrahim and W. Haase, On the molecular polarizability of nematic liquid crystals. *Mol. Cryst. Liq. Cryst.* **66**, 189 (1981); see also G. R. Luckhurst and G. W. Gray, "The Molecular Physics of Liquid Crystals." Academic Press, London, 1979.
19. N. V. Tabiryan, A. V. Sukhov, and B. Ya. Zel'dovich, The orientational optical nonlinearity of liquid crystals. *Mol. Cryst. Liq. Cryst.* **136** 1 (1986). In this review

article, various optical field-nematic axis configurations and the associated nonlinearities are discussed.

20. F. Simoni, Nonlinear optical phenomena in nematics. *In* "Physics of Liquid Crystalline Materials" (I. C. Khoo and F. Simoni, eds.). Gordon & Breach, Philadelphia, 1991.
21. G. Barbero, F. Simoni, and P. Aiello, Nonlinear optical reorientation in hybrid aligned nematics. *J. Appl. Phys.* **55**, 304 (1984).
22. I. C. Khoo, P. Y. Yan, and T. H. Liu, Nonlinear transverse dependence of optically induced director axis reorientation of a nematic liquid crystal film—theory and experiment. *J. Opt. Soc. Am. B* **4**, 115 (1987), and references therein.
23. I. C. Khoo, Reexamination of the theory and experimental results of optically induced molecular reorientation and nonlinear diffraction in nematic liquid crystals: Spatial frequency and temperature dependence. *Phys. Rev. A* **27**, 2747 (1983).
24. I. Janossy, L. Csillag and A. D. Lloyd, Temperature dependence of the optical Freedericksz transition in dyed nematic liquid crystals. *Phys. Rev. A* **44**, 8410–8413 (1991). I. Janossy and T. Kosa, Influence of anthraquinone dyes on optical reorientation of nematic liquid crystals. *Opt. Lett.* **17**, 1183–1185 (1992).
25. W. M. Gibbons, P. J. Shannon, S.-T. Sun, and B. J. Swetlin, Surface-mediated alignment of nematic liquid crystals with polarized laser light. *Nature (London)* **351**, 49–50 (1991); see also A. G.-S. Chen and D. J. Brady, Surface-stabilized holography in an azo-dye-doped liquid crystal. *Opt. Lett.* **17**, 1231–1233 (1992).
26. I. C. Khoo, H. Li, and Yu Liang, Optically induced extraordinarily large negative orientational nonlinearity in dye-doped liquid crystal. *IEEE J. Quantum Electron.* **QE-29**, 1444 (1993).
27. H. L. Ong and C. Y. Young, Optically induced molecular reorientation in smectic-C liquid crystal. *Phys. Rev.* **A 29**, 297 (1984).
28. P. H. Lippel and C. Y. Young, Observation of optically induced molecular reorientation in films of smectic-C liquid crystal. *Appl. Phys. Lett.* **43**, 909 (1983).
29. R. Macdonald, J. Schwartz, and H. J. Eichler, Laser-induced optical switching of a ferroelectric liquid crystal. *Int. J. Nonlinear Opt. Phys.* **1**, 119 (1992).
30. J.-C. Lee, S. D. Jacobs, and A. Schmid, Retro-self-focusing and pinholing effect in a cholesteric liquid crystal. *Mol. Cryst. Liq. Cryst.* **150B**, 617 (1987).
31. T. V. Galstyan, A. V. Sukhov, and R. V. Timashev, Energy exchange between optical waves counterpropagating in a cholesteric liquid crystal. *Sov. Phys.—JETP (Engl. Transl.)* **68**(5), 1001 (1989).
32. I. C. Khoo, R. R. Michael, and P. Y. Yan, Simultaneous occurrence of phase conjugation and pulse compression in simulated scatterings in liquid crystal mesophases. *IEEE J. Quantum Electron.* **QE23**, 1344 (1987); see also D. N. Ghosh Roy and D. V. G. L. N. Rao, Optical pulse narrowing by backward, transient stimulated Brillouin scattering. *J. Appl. Phys.* **59**, 332 (1986).

# CHAPTER 7

# THERMAL, DENSITY, AND OTHER NONELECTRONIC NONLINEAR MECHANISMS

## 7.1 INTRODUCTION

In Chapter 5 we discussed how the electrostrictive effect gives rise to density fluctuations in liquid crystals, which are manifested in frequency shifts and broadening in the spectra of the scattered light. In absorbing media density fluctuations are also created through the temperature rise following the absorption of the laser. The nature of optical absorption in liquid crystals, as in any other material, depends on the laser wavelength.

The (linear) transmission spectrum of a typical liquid crystal is shown in Figures 7.1 and 7.2, where dips in the curve correspond to strong single-photon absorption. The linear absorption constant $\alpha$ is quite high at wavelengths near or shorter than the ultraviolet ($\alpha \approx 10^2$–$10^3$ cm$^{-1}$), where the absorption band begins; in the visible and near-infrared regions, the absorption constant is typically small (with $\alpha \ll 1$ cm$^{-1}$ and $\alpha \lesssim 10$ cm$^{-1}$, respectively); in the mid-infrared and longer regions, the absorption constant is higher ($\alpha \approx 10$–$10^2$ cm$^{-1}$).

Under intense laser illumination, two- and multiphoton absorption processes will occur, as depicted in Figure 7.3. In this case a so-called nonabsorbing material in the single-photon picture outlined previously could actually be quite absorptive, if the two- or multiphoton process corresponds to a real transition to an excited state.

From the standpoint of understanding laser-induced temperature and density changes in liquid crystals, these photoabsorption processes may be

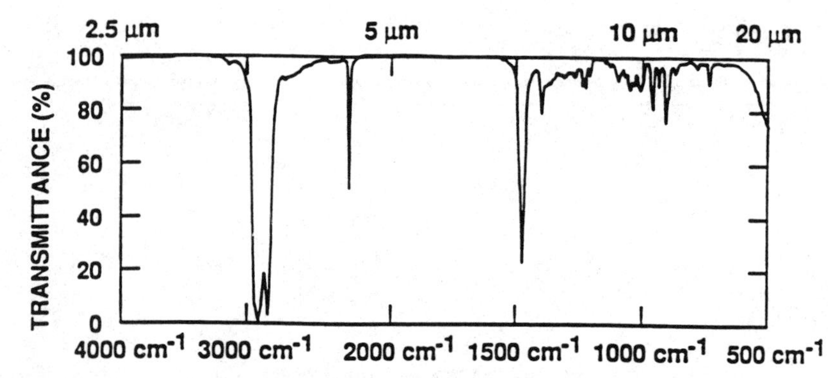


**Figure 7.1** Transmission spectrum of a nematic liquid crystal (3 CCH) in the 2.5 to 20 $\mu$m regime [after Khoo and Wu (4)].

simply represented as a means of transferring energy to the molecule. Figure 2.5 schematically depicts the scenario following the absorption of the incoming photons by the liquid crystalline molecules.

The equation describing the rate of change of energy density $u$ is given by

$$\frac{\partial u}{\partial t} = \frac{nc\alpha E^2}{4\pi} + D\nabla^2 u - \frac{u}{\tau} \tag{7.1}$$

The first term on the right-hand side denotes the rate of absorption of the light energy, the second term denotes the rate of energy diffusion, and the third term describes the thermalization via inter- and intramolecular relaxation processes—usually collisions.

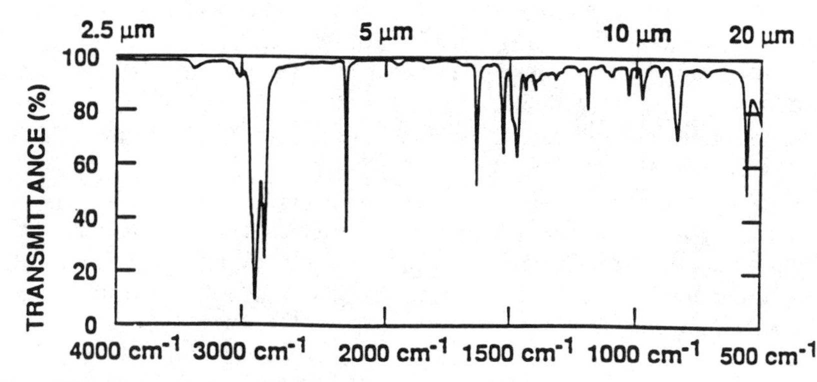


**Figure 7.2** Transmission spectrum of a nematic liquid crystal (7 PCH) in the 2.5 to 20 $\mu$m regime [after Khoo and Wu (4)].

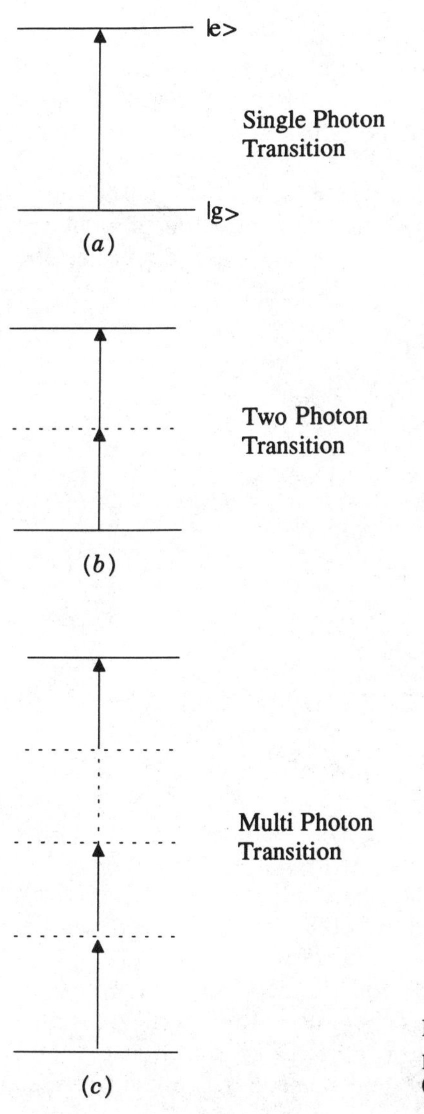


**Figure 7.3** (*a*) Single-photon absorption process; (*b*) Two-photon absorption process; (*c*) multiphoton absorption process.

Liquid crystal molecules are large and complex. The inter- and intramolecular relaxation processes following photoabsorption are extremely complicated. However, it is well known that *individual* molecular motions in both liquid crystalline and isotropic phases are all characterized by very fast picosecond relaxation times. One can therefore argue that, following photoabsorption, the thermalization time it takes to convert the absorbed energy into heat is quite short, typically on the order of picoseconds. Accordingly, if our attention is on the nonlinear optical responses of liquid crystals that are characterized by relaxation times in the nanosecond and longer time scale, we may adopt a formalism that ignores the thermalization process. Therefore, we may assume that the rate of energy density transfer reaches a steady state very quickly (i.e., $\partial u/\partial t \to 0$), and the diffusion process has no time to act. In this case we have $u/\tau = nc\alpha E^2/4\pi$.

The equations describing the laser-induced temperature $T$ and density $\rho$ changes are coupled. This is because both $\rho$ and $T$ are functions of the entropy $S$ and the pressure $P$ (1). Writing the temperature $T(\mathbf{r},t)$ and the density $\rho(\mathbf{r},t)$ as

$$T(\mathbf{r},t) = T_0 + \Delta T(\mathbf{r},t) \tag{7.2}$$

$$\rho(\mathbf{r},t) = \rho_0 + \Delta\rho(\mathbf{r},t) \tag{7.3}$$

the coupled hydrodynamical equations are given by (1,2):

$$-\frac{\partial^2}{\partial t^2}(\Delta\rho) + v^2\nabla^2(\Delta\rho) + v^2\beta_T\rho_0\nabla^2(\Delta T) + \frac{\eta}{\rho_0}\frac{\partial}{\partial t}\nabla^2(\Delta\rho) = \frac{\gamma^e}{8\pi}\nabla^2(E^2) \tag{7.4}$$

and

$$\rho_0 C_v \frac{\partial}{\partial t}(\Delta T) - \lambda_T\nabla^2(\Delta T) - \frac{(C_p - C_v)}{\beta_T}\frac{\partial}{\partial t}(\Delta\rho) = \frac{u}{\tau} = \frac{\alpha nc}{4\pi}E^2 \tag{7.5}$$

where $\rho_0$ is the unperturbed density of the liquid crystal, $C_p$ and $C_v$ the specific heats, $\lambda_T$ the thermal conductivity, $\eta$ the viscosity, $v$ the speed of sound, $\gamma^e$ the electrostrictive coefficient $[\gamma^e = \rho_0(\rho\varepsilon/\partial\rho)_T]$, $\beta_T$ the coefficient of volume expansion, and $\eta$ a viscosity coefficient.

Equation (7.4) describes the thermal expansion and electrostrictive effects on the density change, whereas (7.5) describes the photoabsorption and the resulting temperature rise and heat diffusion process.

We must emphasize here that in these equations the effective values for most of these parameters are, of course, dependent on the particular phase (isotropic, nematic, smectic, etc.) and temperature (3.4), as well as the laser beam polarization, propagation direction, and nematic axis alignment (i.e., the geometry of the interaction).

## 7.2 DENSITY AND TEMPERATURE CHANGES INDUCED BY SINUSOIDAL OPTICAL INTENSITY

As an example of how one may gain some insight into such a complicated problem, and for other practically useful reasons, we now consider the case where the optical intensity is a spatially periodic function (i.e., an intensity grating). Such an intensity function may be derived from the coherent superposition of two laser fields on the liquid crystal (cf. Figure 7.4).

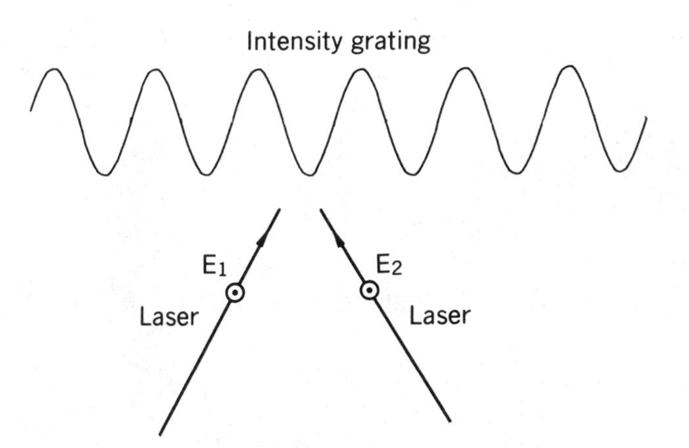


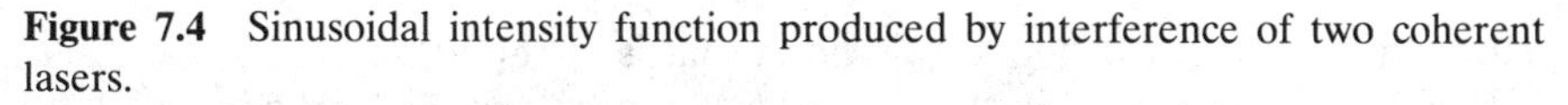
**Figure 7.4** Sinusoidal intensity function produced by interference of two coherent lasers.

As a result of the spatially periodic intensity function, a spatially periodic refractive index change (i.e., an index grating) is induced. If this index grating is probed by a cw laser, side diffractions in the directions $\pm\theta$, $\pm 2\theta$, and so on will be generated. This is shown in Figure 7.5 which depicts a typical so-called dynamic grating setup for studying laser-induced nonlinear diffractions.

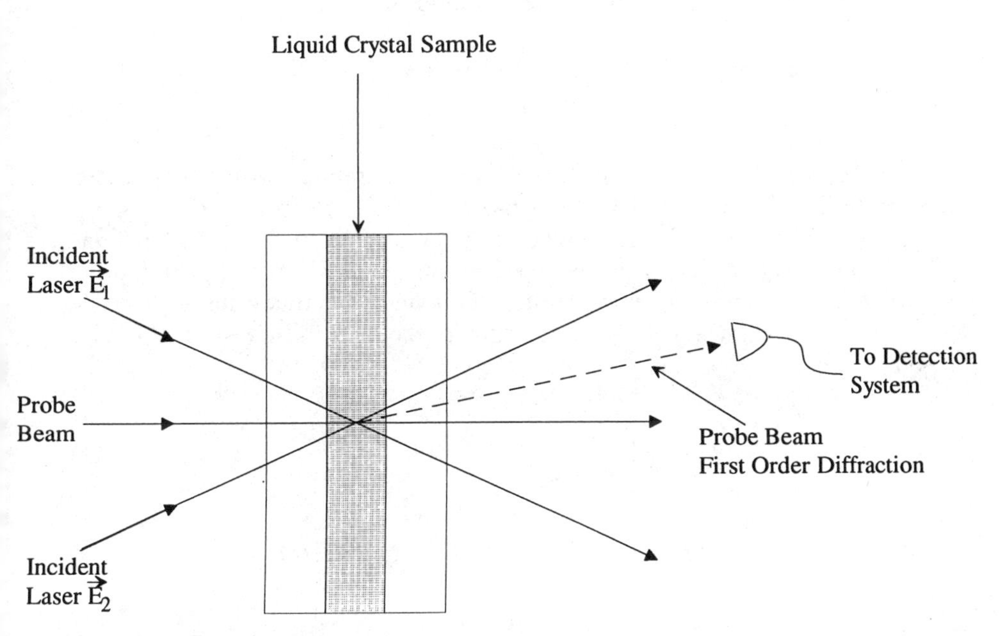


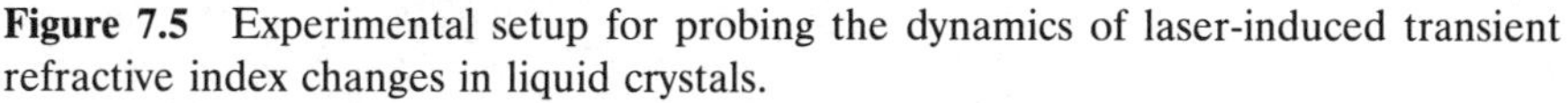
**Figure 7.5** Experimental setup for probing the dynamics of laser-induced transient refractive index changes in liquid crystals.

The diffractions from the probe beam in the $\pm\theta$ directions are termed first-order diffractions. The efficiency of the diffraction $\eta$, defined by the ratio of the intensity of the diffraction to the zero-order (incident) laser intensity, is given by (2):

$$\eta \sim J_1^2\left(\frac{\pi \Delta n d}{\lambda}\right) \tag{7.6}$$

If $\Delta n$, the index grating amplitude, is small [i.e., $(\pi \Delta n d/\lambda) \ll 1$],

$$\eta \sim \left(\frac{\pi \Delta n d}{\lambda}\right)^2 \tag{7.7}$$

The optical intensity inside the liquid crystal is of the form: $E^2 = E_1^2 + E_2^2 + 2|E_1 E_2|\cos((\mathbf{k}_1 - \mathbf{k}_2)\cdot \mathbf{r})$, in the plane wave approximation. The dc part of $E^2$ gives rise to spatially uniform changes in $\rho$ and $T$, and they do not contribute to diffraction of the beam. We may therefore consider only the spatially periodic part and write $E^2 = 2|E_1 E_2|\cos(\mathbf{q}\cdot\mathbf{y})$, where $\mathbf{q} = \mathbf{k}_1 - \mathbf{k}_2$. Furthermore, for simplicity as well as convenience, let $(E_1) = (E_2) = E_0$. This gives

$$E^2 = 2E_0^2(1 + \cos \mathbf{q}\cdot\mathbf{y}) \tag{7.8}$$

Correspondingly, $\Delta\rho$ and $\Delta T$ are of the form

$$\Delta\rho = \rho(t)\cos \mathbf{q}\cdot\mathbf{y} \tag{7.9}$$

$$\Delta T = T(t)\cos \mathbf{q}\cdot\mathbf{y} \tag{7.10}$$

where $\rho(t)$ and $T(t)$ are the density and temperature grating amplitudes. Substituting (7.8) and (7.10) into (7.4) and (7.5), one could solve for the grating amplitudes $\rho(t)$ and $T(t)$. Letting the initial conditions be $t = 0$ and $\rho(0) = T(0) = 0]$ and ignoring the term proportional to $(C_p - C_v)$ in (7.5), a straightforward but cumbersome calculation yields the following results.

For $0 < t < \tau_p$, where $\tau_p$ is the duration of the laser pulse (assumed to be a square pulse),

$$T(t) = \left[\frac{\alpha c n E_0^2}{4\pi\rho_0 C_v \Gamma_R}\right](1 - \exp(-\Gamma_R t)) \tag{7.11}$$

$$\begin{aligned}\rho(t) &= \left[\frac{\gamma^e E_0^2}{4\pi v^2}\right](1 - \exp(-\Gamma_B t)\cos\Omega t) \\ &\quad - \left[\frac{\beta_T \alpha c n E_0^2}{4\pi C_v \Gamma_R}\right](1 - \exp(-\Gamma_R t))\end{aligned} \tag{7.12}$$

In the preceding equations the thermal decay constant $\Gamma_R$ is given by

$$\Gamma_R = \frac{\lambda_T q^2}{\rho C_v} \tag{7.13}$$

the Brillouin decay constant $\Gamma_B$ is given by

$$\Gamma_B = \frac{\eta q^2}{2\rho_0} \tag{7.14}$$

$\Omega$, the sound frequency, is given by

$$\Omega = \sqrt{q^2 v^2 - \Gamma_B^2} \tag{7.15}$$

The corresponding relaxation time constants are, respectively,

$$\tau_R = \Gamma_R^{-1} = \frac{\rho_0 C_v}{\lambda_T q^2} \tag{7.16}$$

and

$$\tau_B = \Gamma_B^{-1} = \frac{2\rho_0}{\eta q^2} \tag{7.17}$$

In liquid crystals the typical values for the various parameters (3, 4) in units are $n = 1.5$, $\eta = 7 \times 10^{-2}$ kg $-$ m$^{-1}$ $-$ s$^{-1}$, $\nu = 1540$ m $-$ s$^{-1}$, $\rho_0 = 10^3$ kg $-$ m$^{-3}$, $\lambda_T/\rho_0 C_v = 0.79 \times 10^{-7}$ m$^2$/s). If a grating constant ($\Lambda = 2\pi|\mathbf{K}_1 - \mathbf{K}_2|^{-1}$) of 20 $\mu$m is used in the experiment, we have, $\tau_R \approx 100$ $\mu$s. For the same set of parameters, we have $\tau_B \approx 200$ ns. These widely different time scales of the thermal and density effects provide a means to distinguish their relative contributions in the nonlinear dynamic grating diffraction experiment.

It is important to note here that the density change $\rho(t)$ given in (7.12) is the sum of two distinct components:

$$\rho(t) = \rho^e(t) + \rho^T(t) \tag{7.18}$$

where

$$\rho^e(t) = \frac{\gamma^e E_0^2}{4\pi v^2}(1 - \exp(-\Gamma_B t)\cos\Omega t) \tag{7.19}$$

and

$$\rho^T(t) = \frac{-\beta_T \alpha c n E_0^2}{4\pi C_v \Gamma_R}(1 - \exp(-\Gamma_R t)) \tag{7.20}$$

The component $\rho^e(t)$ is due to the electrostrictive effect (the movement of molecules under intense electric field); it is proportional to $\gamma^e$ and is characterized by the Brillouin relaxation constant $\Gamma_B$ and frequency $\Omega$. From (7.9) and (7.19) one can see that this $\rho^e$ component gives rise to a propagating wave. On the other hand, the component $\rho^T(t)$ is due to the thermoelastic contribution (proportional to $\beta^T$) and is characterized by the thermal decay constant $\Gamma_R$; it is nonpropagative.

More detailed solutions of the coupled density and temperature equations (7.4) and (7.5) may be found in Batra et al. (1). The preceding simple example, however, will suffice for illustrating the basic processes following photoabsorption and their time evolution characteristics.

## 7.3 REFRACTIVE INDEX CHANGES: TEMPERATURE AND DENSITY EFFECTS

Because of these temperature and density changes, there are corresponding refractive index changes given, respectively, by

$$\Delta n_T = \frac{\partial n}{\partial T} T(t) \tag{7.21}$$

and

$$\Delta n_\rho = \frac{\partial n}{\partial \rho}\rho(t) = \frac{\partial n}{\partial \rho}(\rho^e + \rho^T) \tag{7.22}$$

The thermal index component $\partial n / \partial T$ arises from two effects. One is the spectral shift as a result of the rise in the temperature of the molecule. This effect occurs within the thermalization time $\tau$ (i.e., in the picosecond time scale), and its contribution is usually quite small (5, 6) for the ordered as well as the liquid phases. In the nematic phase $\Delta n_T$ arises from the temperature dependence of the order parameter

$$\frac{\partial n}{\partial T} = \frac{\partial n}{\partial S}\frac{\partial S}{\partial T} \tag{7.23}$$

As we have seen in earlier chapter, this effect is particularly dominant at temperatures in the vicinity of the nematic–isotropic transition. Since the order parameter $S$ exhibits critical slowing down behavior near $T_c$, this

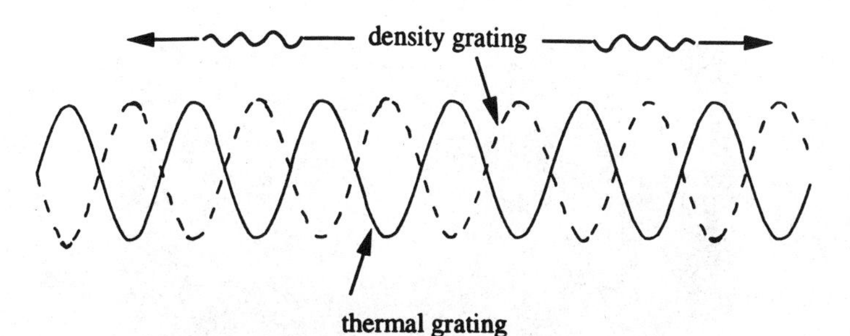


**Figure 7.6** Interference of the propagative index gratings from the electrostrictive density changes with the diffusive thermal components.

component of the refractive index change also exhibits similar behavior. This was discussed in Chapter 2, and more details of it will be discussed in the next sections.

The density contributions, consisting of the electrostrictive component $\rho^e$ and the thermoelastic component $\rho^T$, are effects typically experienced by solids and ordinary liquids; they are not strongly coupled to the order parameter.

The propagation of the electrostrictive component $\rho^e$ will interfere with the nonpropagating thermal component. Consider Figure 7.6 which depicts the presence of both a "static" refractive index grating due to $\Delta n_T$ and $\Delta n_\rho(\rho^T)$ and a propagative one from $\Delta n_\rho(\rho^e)$. The latter is created by two counterpropagating waves right after the laser pulse, with wave vectors $\pm(\mathbf{k}_1 - \mathbf{k}_2)$ and a frequency $\Omega$. Consequently, the maxima and minima of these two index gratings will interfere in time, leading to oscillations (at the frequency $\Omega$ of the sound wave) in the diffractions from the probe beam (cf. Figure 7.7). Since the magnitude of the wave vector $|\mathbf{k}_1 - \mathbf{k}_2|$ is known and the oscillation frequency $\Omega$ can be directly measured, these observed oscillations in the diffraction will provide a means of determining sound velocities in a liquid crystal (3,7). By varying the interaction geometry between the grating wave vectors $\mathbf{q}$ and the director or $c$ axis of the liquid crystal sample, this dynamical diffraction effect could also be used for measuring sound velocity and thermal diffusion anisotropies.

In the earlier experiments reported in Khoo and Normandin (3,7), nanosecond laser pulses from the second harmonic of a $Q$-switched Nd:Yag laser are employed to excite these density and temperature (order parameter) interference effects in nematic and smectic-A liquid crystals.

For the nematic case (3), the liquid crystal used is 5CB. The planar aligned sample is 40 $\mu$m thick. The sample is oriented such that the director axis is either parallel or orthogonal to the pump laser polarization direction. These interaction geometries allow the determination of the thermal diffusion constants $D_\parallel$ and $D_\perp$. Figures 7.8$a$ and $b$ are oscilloscope traces of the relaxation dynamics of the first-order probe beam diffraction from the

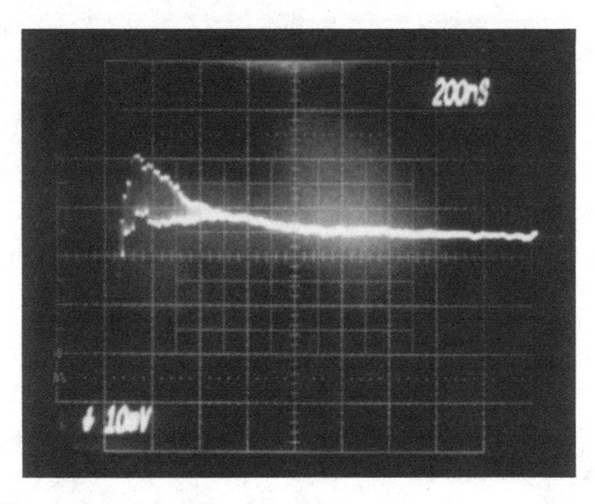


**Figure 7.7** Observed probe diffraction from a room-temperature nematic liquid crystal (E7), showing interference effect between the density and thermal contribution.

thermal gratings for, respectively, thermal diffusions perpendicular and parallel to the director axis. In Figure 7.8*a* the thermal diffusion time constant $\tau_\perp$ is about 100 $\mu$s, while $\tau_\parallel$ from Figure 7.8*b* is about 50 $\mu$s. For these observations the crossing angle in air is 2°, corresponding to a grating constant $\Lambda_\perp = 2\pi/q_\perp$ of 17 $\mu$m and $\Lambda_{11} = 2\pi/q_\parallel$ of 15 $\mu$m.

Using (7.16) for the thermal decay times and the values of the thermal diffusion constants $D_\parallel = 7.9 \times 10^{-4}$ cm$^2$ s$^{-1}$ and $D_\perp = 1.25 \times 10^{-3}$ cm$^2$ s$^{-1}$ (4), the theoretical estimates of $\tau_\parallel$ and $\tau_\perp$ are 110 $\mu$s and 55 $\mu$s, respectively. These are in good agreement with the experimental results. For smaller/larger grating constants, the thermal diffusion times have been observed to be decreasing/increasing roughly in accordance with the $q^{-2}$ dependence.

For the same sample the sound velocity anisotropy was too small to be detectable with the precision of the instruments used. Nevertheless, the crossing-angle dependence of the period of oscillation (caused by acoustic–thermal grating interference, cf. Figure 7.7) in the probe beam diffraction has been measured. The period $T = 2\pi/\Omega$ is related to the grating constant $\Lambda_{\parallel,\perp} = \lambda/[2n_{\parallel,\perp} \sin(\theta/2)]$ by $T = \Lambda/v_s$. (For 5CB, $n_\parallel \sim$ 1.72 and $n_\perp \approx 1.52$.) Figure 7.9 plots the experimentally observed [unpublished data from Khoo and Normandin (3) oscillation period as a function of the crossing angle $\theta$. In general, it follows the theoretical relationship (line) quite closely, with a sound velocity $v_s = 1.53 \times 10^3$ ms$^{-1}$ determined using one of the experimental data points.

In the experiment with the smectic-A liquid crystal (7), the dominant contribution to the grating diffraction seems to come from the density effect.

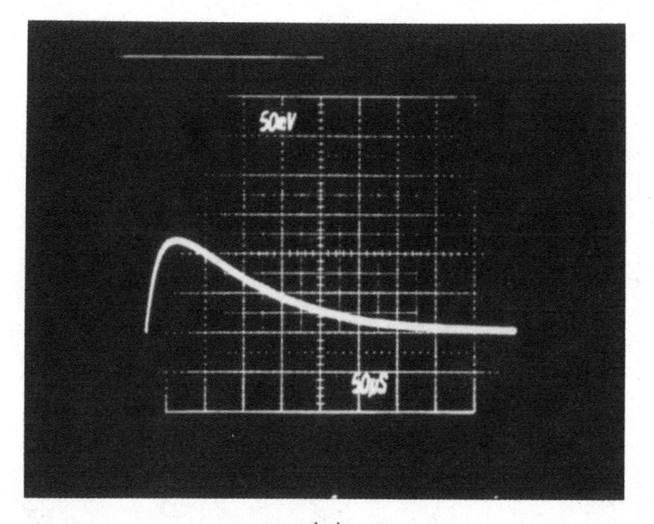

($a$)

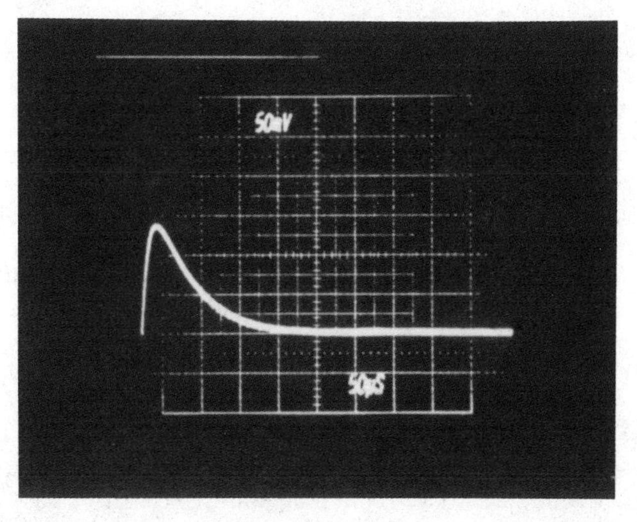

($b$)

**Figure 7.8** Oscilloscope traces of the thermal grating decay dynamics. Time scale, 50 $\mu$s/division. ($a$) Grating wave vector is perpendicular to the director axis. ($b$) Grating wave vector is along the director axis, showing a faster decay.

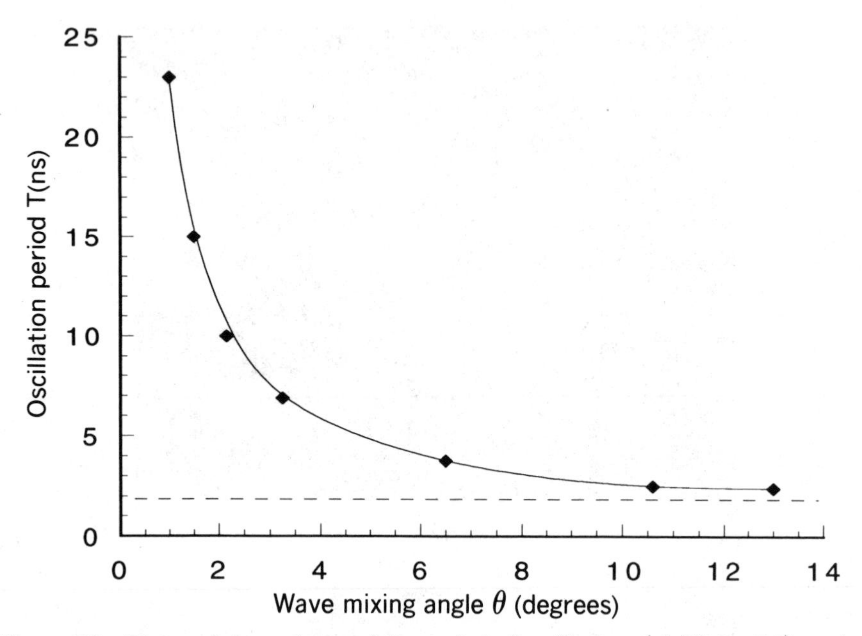


**Figure 7.9** Observed dependence of the period of oscillations (cf. Figure 7.7) on the crossing angle of the excitation laser beams. Solid line is the theoretical dependence. Dotted line shows the detection system dynamic response limit.

The observed acoustic velocity and attenuation time are in accordance with theoretical expectation. An interesting effect reported in Khoo and Normandin (7) is the formation of permanent gratings by intense excitation pulses. These gratings are erasable by warming the sample through the isotropic phase and cooling it back to the aligned smectic-A state.

Since the refractive indices of nematic liquid crystals are very sensitively dependent on the temperature, through its dependence on the order parameter, we expect to see very dramatic dominance of the diffraction by the temperature component when the sample temperature is raised toward $T_c$. Figure 2.3*b* shows an oscilloscope trace of the probe beam diffraction for an E7 sample at about 20°C above room temperature (i.e., at 42°C). Clearly, the temperature component is greatly enhanced compared to the density component (the small "spike" detected at the beginning of the trace is the density component). As noted in Chapter 2, another interesting feature is that the time for the signal to reach its peak has also lengthened to about 20 $\mu$s, compared to about 100 ns at room temperature (cf Figure 7.7). Such lengthening of the thermal component is attributed to a critical slowing down of the order parameter as a phase transition is approached. In both cases the recorded thermal decay time is on the order of 100 $\mu$s.

Recently, these studies of dynamic grating diffraction in nematic liquid crystal (E7) film have also been conducted using short microsecond infrared

($CO_2$) laser pulses (8). Since the wavelength of a $CO_2$ laser is around 10 $\mu$m, the absorption of the laser is via the molecular rovibrational modes of the electronic ground state. Figure 2.3*a* shows the observed diffracted signal. There is an initial spike which is attributed to the "fast" decaying density contributions. On the other hand, the "slower" thermal component associated with the order parameter exhibits a rather long buildup time of about 100 $\mu$s. This implies that, in nematic liquid crystals, the molecular correlation effects leading to ordering among the molecules in the ground (electronic) rovibrational manifold are different from those in the excited (electronic) rovibrational manifold. Such hitherto unexplored dependences of the order parameter dynamics on the electronic and rovibrational excitational states clearly deserve further studies.

## 7.4 THERMAL AND DENSITY OPTICAL NONLINEARITIES OF NEMATIC LIQUID CRYSTALS IN THE VISIBLE–INFRARED SPECTRUM

Since both laser-induced thermal and density changes give rise to intensity-dependent refractive index changes, they may be viewed as optical nonlinearities. From the experimental observations (cf. Figure 7.7), we can say that, in general, the density and thermal contributions to the probe diffraction are comparable for a sample maintained at a temperature far from $T_c$. In the vicinity of $T_c$ the thermal component is much larger. The absolute magnitude

**TABLE 7.1 Typical Values for Some Nematic Liquid Crystal**

| Parameter | Value | Nematic |
|---|---|---|
| Absorption constant | | |
| ($\alpha$ in $cm^{-1}$) | $<1$, visible | E7 |
| | 23, near infrared | E7 |
| | 40–100, infrared | E7 |
| | 69, infrared | 5CB |
| | 44, infrared | E46 (EM Chemicals) |
| Diffusion constant | $D_{\parallel} = 1.95 \times 10^{-3}$ cm$^2$/s | E7 |
| $D = \lambda_T / \rho_0 C_v$ | $D_{\perp} = 1.2 \times 10^{-3}$ cm$^2$/s | E7 |
| Thermal index gradients | | |
| $dn_{\perp}/dT$ | $10^{-3}$ K | 5CB at 25°C |
| | $10^{-2}$ K | 5CB near $T_c$ |
| | (visible–infrared) | |
| $dn_{\parallel}/d7$ | $-2 \times 10^{-3}$ K | 5CB at 25°C |
| | $-10^{-2}$ K | 5CB near $T_c$ |
| | (visible–infrared) | |

Visible → 0.5 $\mu$m; near infrared → CO wavelength ≈ 5 $\mu$m; infrared → 10 $\mu$m.

of the thermal nonlinearity, of course, depends on several factors. From the preceding discussion, some of the obvious factors are: the index gradient $dn/dT$, which depends critically on the order parameter dependence on the temperature; the absorption constant $\alpha$, which varies over several orders of magnitude depending on the laser wavelength (i.e., the spectral regime); the thermal decay constant $\Gamma_R$, which is a function of the laser–nematic interaction geometry (e.g., laser spot size, sample thickness, and diffusion constant); and a collection of liquid crystalline parameters. Some of the parameters for a few typical liquid crystals are listed in Table 7.1.

One of the most striking and important optical properties of liquid crystals is their large birefringence throughout the whole optical spectrum (from near UV to the infrared). The large thermal index gradients noted in Table 7.1 for the visible–infrared spectrum are a consequence of that. In Figures 7.10 and 7.11 the measured refractive indices of E7 and E46 (from EM Chemicals) in the infrared (10.6-$\mu$m $CO_2$ laser line) regime are shown (9).

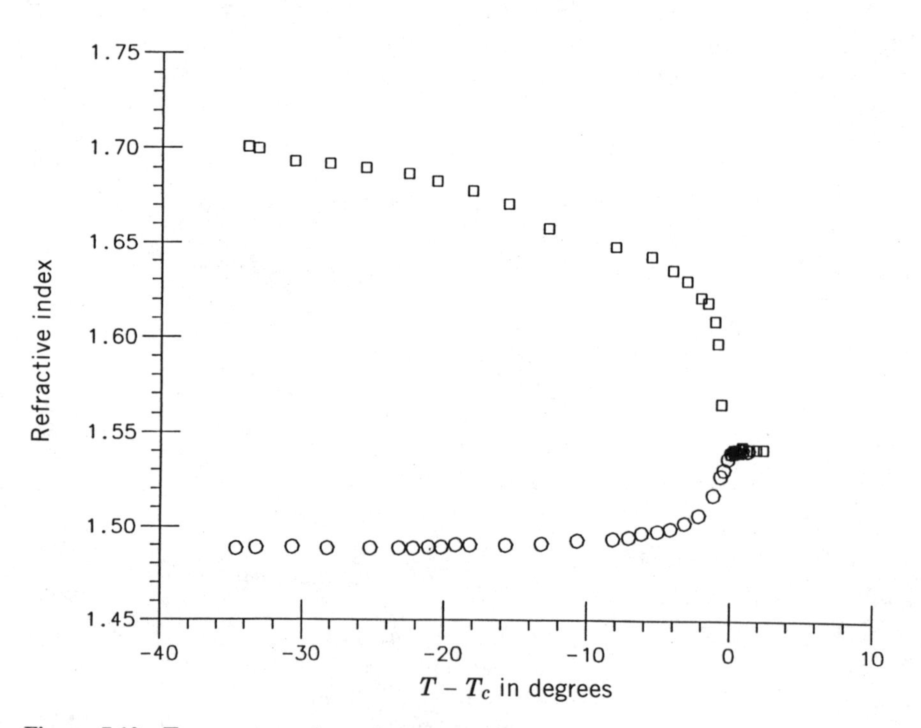


**Figure 7.10** Temperature dependence of refractive indices $n_{\parallel}$ and $n_{\perp}$ of E7 at 10.6 $\mu$m.

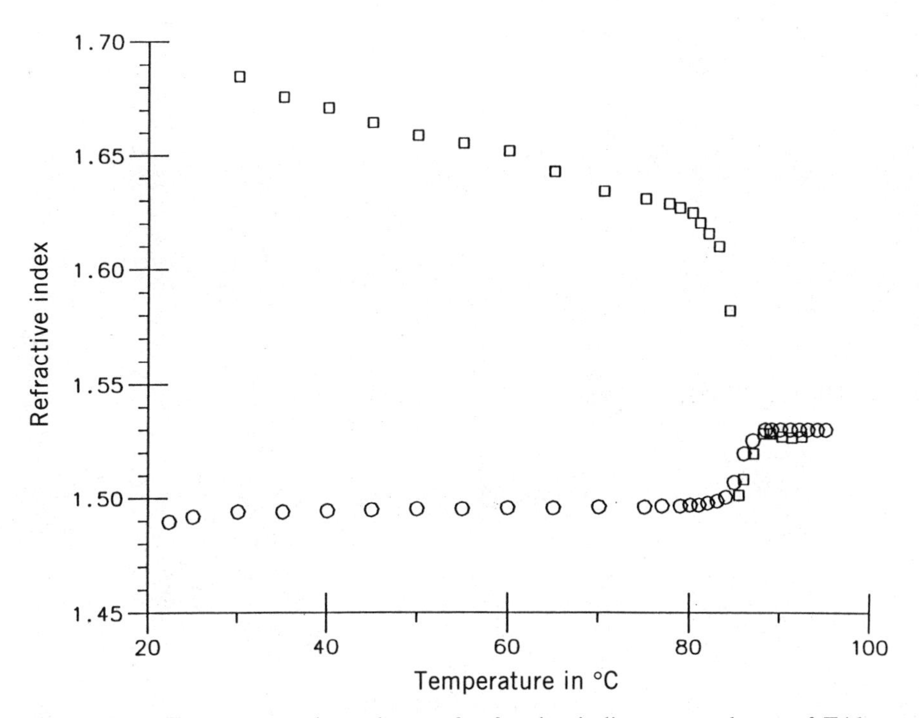


**Figure 7.11** Temperature dependence of refractive indices $n_{\|}$ and $n_{\perp}$ of E46 at 10.6 $\mu$m.

Besides these material parameters, perhaps the single most important factor governing the magnitude of the effective optical nonlinearities is the laser pulse duration relative to the material response time. If the laser pulse is too short, the material response time will be minimal, and the corresponding nonlinearity as "seen" by the laser will be small. This could be cast in more quantitative terms as follows.

We shall consider the following two limiting cases of thermal nonlinearities: (1) steady-state regime, where $\tau_p \gg \tau_R$ and $\tau_S$ ($\tau_S$ is the order parameter response time) and (2) transient regime, corresponding to $\tau_p \ll \tau_R$ and $\tau_S$.

### 7.4.1 Steady-State Thermal Nonlinearity of Nematic Liquid Crystals

From these liquid crystalline parameters, one can estimate the steady-state thermal nonlinear coefficient $\alpha_2(T)$ defined by

$$\Delta n_T = \alpha_2^{\mathrm{SS}}(T) I_{\mathrm{op}} \tag{7.24}$$

where the optical intensity $I_{\mathrm{op}}$ is measured in watts per square centimeter.

From (7.11) and (7.21) we have, in the steady state when $t \gg \Gamma_R^{-1}$,

$$\Delta n_T = \frac{\alpha c n E_0^2}{4\pi \rho_0 C_v \Gamma_R} \frac{\partial n}{\partial T} = \alpha_2^{\rm ss}(T) I_{\rm op} \tag{7.25}$$

that is,

$$\alpha_2^{\rm ss}(T) = \frac{\alpha}{\rho_0 C_v \Gamma_R}\left(\frac{dn}{dT}\right) \tag{7.26}$$

Recall that $\Gamma_R = Dq^2$. Equation (7.26) thus gives

$$\alpha_2^{\rm ss}(T) = \frac{\alpha}{\rho_0 C_v D q^2}\left(\frac{dn}{dT}\right) \tag{7.27}$$

Since $q = 2\pi/\Lambda$, where $\Lambda$ is the grating space of the temperature modulation, we may express $\alpha_2^{\rm ss}(T)$ as

$$\alpha_2^{\rm ss}(T) = \frac{\alpha \Lambda^2}{4\pi^2 \rho_0 C_v D}\left(\frac{dn}{dT}\right) \tag{7.28}$$

This expression is valid for the case where the two interference laser beam sizes are large compared to the liquid crystal cell thickness. If focused laser beams are used, we have to replace the grating constant $\Lambda$ in (7.28) by the characteristic diffusion length. This can be the laser-focused spot size or the thickness of the cell, whichever is associated with the dominant (i.e., shortest time) diffusion process. Therefore, we may write

$$\alpha_2^{\rm ss}(T) = \frac{\alpha \Lambda_{\rm th}^2}{4\pi^2 \rho_0 C_v D}\left(\frac{dn}{dT}\right) \tag{7.29}$$

where $\Lambda_{\rm th}$ is the characteristic thermal diffusion length. Needless to say, this approach has grossly overlooked many of the details involved in laser-induced temperature and refractive index changes in an actual system, where the heat diffusion process is more likely a three-dimensional problem and the laser intensity distribution has a Gaussian envelope, and so on. Nevertheless, (7.28) or (7.29) should provide us with a reasonable means of estimating the thermal nonlinearity.

For example, if the characteristic diffusion length $\Lambda_{\rm th}$ is 20 $\mu$m, then, using typical liquid crystalline parameters, $\rho \sim 1$ g/c.c., $C_p[\approx C_v] \sim 2$ J/g/K, $D \sim 2 \times 10^{-3}$ cm$^2$/s, $\alpha \sim 100$ cm$^{-1}$ (cf. Table 7.1 for $CO_2$ laser wavelength), and $dn/dT \sim 10^{-3}$ K$^{-1}$, we get

$$\alpha_2^{ss}(T) \sim \frac{100}{4\pi^2} \cdot \frac{(20\times 10^{-4})^2}{1\cdot 2\cdot 2\times 10^{-3}} \cdot 10^{-3}$$
$$\sim 2.5\times 10^{-6}\ \text{cm}^2/\text{W} \tag{7.30}$$

In esu units (cf. Chapter 6), the corresponding nonlinear third-order susceptibility is given by

$$\chi_{\text{esu}}^{(3)}(T) \sim 9.5\times n_2(I) \sim 2.4\times 10^{-5}\ \text{esu} \tag{7.31}$$

We remind the reader, again, that these expressions for thermal nonlinearities are rough estimates only. More detailed calculations are clearly needed if we desire more accurate quantitative information; they can be obtained by solving (7.4) and (7.5) for the appropriate interaction geometries and boundary conditions.

### 7.4.2 Short Laser Pulse Induced Thermal Index Change in Nematics and Near-$T_c$ Effect

As in the case of laser-induced molecular reorientation discussed in the preceding chapter, if the duration of the laser pulse is short compared to the thermal decay time $\tau_R^{-1}$, the effective induced optical nonlinearity is diminished.

From (7.11) and (7.12), it is straightforward to show that, for time $t \ll \tau_R^{-1}$, we may write the induced index change $\Delta n(t)$ by

$$\Delta n(t) = \alpha_2(t,T) I_{\text{op}} \tag{7.32}$$

where

$$\alpha_2(t,T) = \alpha_2^{ss}(T)\left(\frac{t}{\tau_R}\right) \tag{7.33}$$

For the same set of parameters used in estimating $\alpha_2^{ss}(T)$ in (7.30), we note that $\tau_R = 0.5\times 10^{-4}$ s. Therefore, for $t = 1\ \mu$s ($10^{-6}$ s), we have

$$\alpha_2(t,T) > (2\times 10^{-2})\alpha_2^{ss}(T) = 5\times 10^{-8}\ \text{cm}^2/\text{W}$$

For $t = 1$ ns ($10^{-9}$ s), we have

$$\alpha_2(t,T) > 5\times 10^{-11}\ \text{cm}^2/\text{W} \tag{7.34}$$

From (7.26) to (7.29), one can see that the diminished thermal nonlinearity for short laser pulses may be improved by optimized choice of molecular and geometrical parameters such as the absorption constant $\alpha$, the grating parameter $q$, and the thermal index. Since the thermal index $dn/dT$ of a nematic liquid crystal is considerably enhanced near $T_c$, one would expect that $\alpha_2^{ss}(t,T)$ will be proportionately increased. This expectation, however, is not borne out in practice because of the critical slowing down (1, 3) in the order parameter $S$ near $T_c$ (cf. Chapter 2 and the discussion in the preceding section). As a result of the long buildup time of the thermal index change, owing to the slower response of the order parameter $S$, the nonlinear coefficient $\alpha_2(t,T)$ for *nanosecond* laser pulses cannot be significantly increased by maintaining the liquid crystal near $T_c$.

To put this in more quantitative terms, let us ascribe the laser-induced order parameter change a time dependence of the form

$$dS = dS_{ss}\left[1 - \exp\left(-\frac{t}{\tau_S}\right)\right] \tag{7.35}$$

where $dS_{ss} = (dS/dT)_{ss}\, dT$ is the steady-state order parameter change associated with a temperature change $dT$ and $\tau_S$ is the order parameter response time. Near $T_c$, $\tau_S$ exhibits a critical slowing down behavior and is therefore of the form

$$\tau_S \sim (T - T_c)^{-\alpha} \tag{7.36}$$

where $\alpha$ is some positive exponent near unity.

For a short laser pulse (i.e., $\tau_p \ll \tau_S 0$, the induced index change due to the order parameter is therefore given by

$$dn \sim \left(\frac{\tau_p}{\tau_S}\right)\left(\frac{dS}{dT}\right)_{ss} \tag{7.37}$$

From (2.20) we have

$$\left(\frac{dS}{dT}\right)_{ss} \sim (T - T_c)^{-0.78} \tag{7.38}$$

Since

$$\left(\frac{\tau_p}{\tau_S}\right) \sim (T - T_c)^{\alpha} \tag{7.39}$$

we therefore have

$$dn \sim (T - T_c)^{\alpha - 0.78} \tag{7.40}$$

In other words, near $T_c$, the short ($\tau_p \ll \tau_S$) laser pulse induced index change will not be enhanced, in spite of the large thermal gradient, owing to the diminishing effect from the critical slowing down of the order parameter response. Since the response times of the order parameter can be as long as microseconds near $T_c$, we could expect enhanced optical nonlinearities to manifest themselves only for relatively long laser pulses (on the order of microseconds or longer). Indeed, experiments which microsecond infrared laser pulses in optical limiting studies (8) and millisecond laser pulses in limiting and wave mixing studies (4, 10) have shown that the efficiency of the processes increases tremendously near $T_c$.

## 7.5 THERMAL AND DENSITY OPTICAL NONLINEARITIES OF ISOTROPIC LIQUID CRYSTALS

A laser-induced change in the temperature of an isotropic liquid crystal can modify its refractive index in two ways, very much as in the nematic phase. One is the change in density $d\rho$ due to thermal expansion. This is the

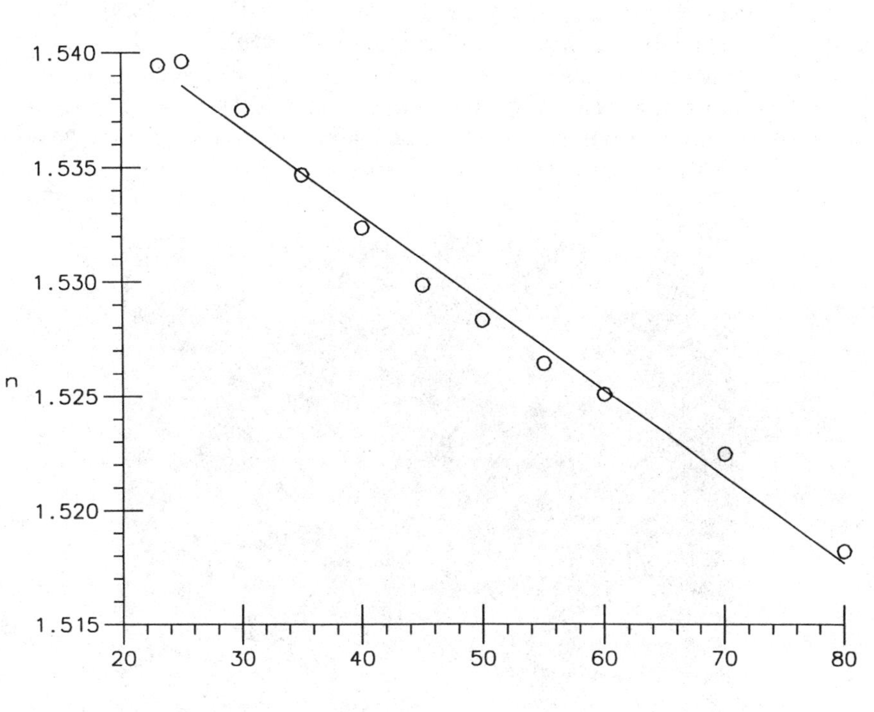


**Figure 7.12** Experimentally measured steady-state refractive index dependence on temperature of an isotropic liquid crystal.

thermal absorptive component discussed before [equation (7.18) for $\rho^T$]; this term may be written as $(\partial n/\partial \rho)\rho^T$. The other is the so-called internal temperature change $dT$ which modifies the spectral dependence of the molecular absorption–emission process; we may express this contribution as $(\partial n/\partial T)_\rho\, dT$. A pure density change effect arises from the electrostrictive component $\rho^e$, which contributes a change in the refractive index by $(\partial n/\partial \rho)\rho^e$. Therefore, the total change in the refractive index $\Delta n$ is given by

$$\Delta n = \left(\frac{\partial n}{\partial T}\right)_\rho dT + \left(\frac{\partial n}{\partial \rho}\right)\rho^T + \left(\frac{\partial n}{\partial \rho}\right)_T \rho^e \tag{7.41}$$

In most liquids the first term is important in the picosecond regime, that is, if the excitation laser pulse duration is in the picosecond regime (5, 6). For longer pulses (e.g., in the nanosecond regime), the second and third terms provide the principal contributions. A good discussion of how the first and the next two terms in (7.41) affect nonlinear light-scattering processes may be found in Mack (5) and Herman and Gray (6), respectively.

The change in the refractive index $dn$ caused by the laser-induced temperature rise $dT$, described by the term $(\partial n/\partial \rho)\rho^T$ in (7.41), is often written in terms of the index gradient $dn/dT$. For most organic liquids, including isotropic liquid crystals, $dn/dT$ is on the order of $10^{-4}$ K. Figure 7.12 shows the temperature dependence of the liquid crystal TM74A (from EM Chemicals) as measured in our laboratory. The liquid crystal is a mixture of four

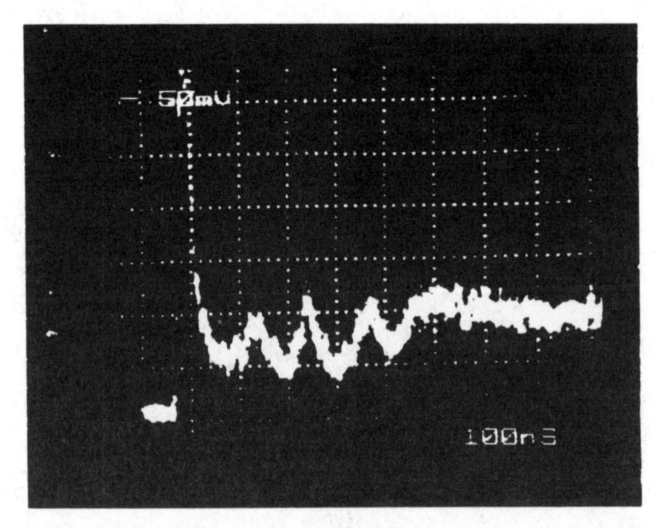


**Figure 7.13** Observed He–Ne probe diffraction from the isotropic liquid crystal TM74A under nanosecond Nd:Yag second-harmonic laser pulse excitation.

chiral nematic materials in the isotropic phase. The measured $dn/dT$ of the material is about $5 \times 10^{-4}\ \mathrm{K}^{-1}$.

The thermal optical nonlinearity of TM74A under short laser pulse excitation has also been studied by a dynamic grating technique (11). For visible laser pulses (the second harmonic of a Nd:Yag laser at 0.53 $\mu$m) and an absorption constant $\alpha > 2\ \mathrm{cm}^{-1}$ (determined by the absorbing dye dopant concentration), the measured nonlinear index coefficient $n_2$ is on the order of about $-2 \times 10^{-11}\ \mathrm{cm}^2/\mathrm{W}$.

In these studies, as in the case of nematic liquid crystals, the dynamic grating diffraction also contains a fast decaying component due to the density contribution, $(\partial n/\partial \rho)\rho^e$ (cf. Figure 7.13). This component decays in about 100 ns. Its peak magnitude is about twice that of the thermal component, which decays in a measured time of about 30 $\mu$s for the grating constants of 11 $\mu$m used in the experiment.

## 7.6 COUPLED NONLINEAR OPTICAL EFFECTS IN NEMATIC LIQUID CRYSTALS

So far, we have singled out and discussed the various nonresonant physical mechanisms that contribute to optical nonlinearities. This approach allows us to understand their individual unique or special properties. However, in reality, these physical parameters are closely coupled to one another; perturbation of one parameter, when it reaches a sufficient magnitude, will inevitably lead to perturbations of the parameters. These coupled responses could give rise to optical nonlinearities of differing signs and dynamical dependences and, consequently, complex behaviors in the nonlinear optical processes under study. In this section we discuss two examples of coupled liquid crystal responses to laser excitation: thermal-orientational coupling and flow-orientational coupling.

### 7.6.1 Thermal-Orientational Coupling in Nematic Liquid Crystals

Consider the interaction of a linearly polarized extraordinary wave laser beam with a homeotropically aligned nematic liquid crystal as shown in Figure 7.14. The extraordinary refractive index as "seen" by a low-intensity laser is given by

$$n_e(\beta, T) = \frac{n_{\perp}(T)n_{\parallel}(T)}{\sqrt{n_{\parallel}^2(T)\cos^2\beta + n_{\perp}^2(T)\sin^2\beta}} \tag{7.42}$$

If the laser induces both molecular reorientation $\Delta\theta$ and a change in

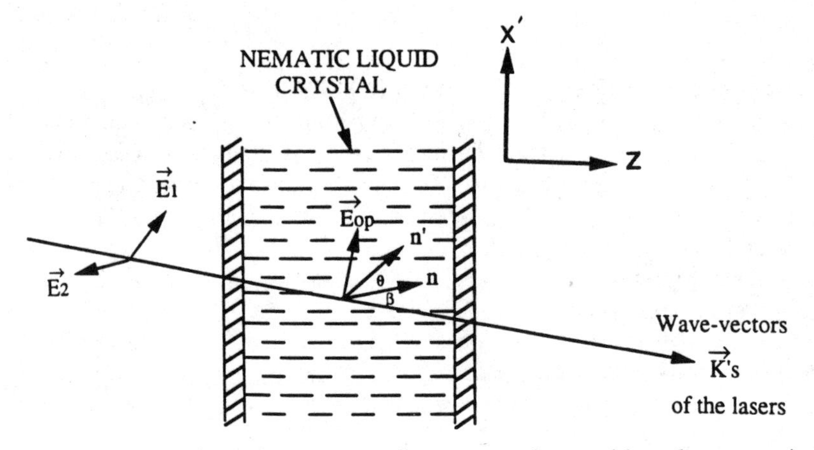


**Figure 7.14** Interaction of an extraordinary wave laser with a homeotropically aligned nematic liquid crystal ($\beta \neq 0$).

temperature, the resulting refractive index becomes

$$n_e(\beta+\Delta\theta, T+\Delta T) = \frac{n_\perp(T+\Delta T)n_\parallel(T+\Delta T)}{\sqrt{n_\parallel^2(T+\Delta T)\cos^2(\beta+\Delta\theta)+n_\perp^2(T+\Delta T)\sin^2(\beta+\Delta\theta)}} \tag{7.43}$$

The resulting change in the extraordinary wave refractive index $\Delta n_e$ is thus given by

$$\Delta n_e = n_e(\beta+\Delta\theta, T+\Delta T) - n_e(\beta, T) \tag{7.44}$$

Its magnitude and sign are determined by the initial value of $\beta$ and $T$.

In general, if both reorientational and thermal effects are present, these processes will act in opposition to each other in terms of their contributions to the net refractive index "seen" by the optical beam.

An example of these competing nonlinear effects may be seen in the CO-laser-induced grating diffraction experiment performed by Khoo, Normandin and Vilks (9) and the nanosecond visible laser experiment of Hsiung et al. (12).

Under nanosecond laser pulse excitation, as we have shown in the preceding sections, the density contribution to the induced index change can be very large. When this occurs, a quantitative analysis of all the contributing effects, their signs and magnitudes, becomes rather complicated. An attempt to analyze such situations was made in Khoo et al. (13).

### 7.6.2 Flow-Orientational Effect

A unique property of nematic liquid crystals is that they flow as conventional liquids. These translational motions are naturally coupled to director axis reorientations and vice versa, as noted in Chapter 3.

Since liquid crystal molecules are highly anisotropic (requiring several viscosity coefficients to describe the viscous forces accompanying various flow processes), it will take a treatise here to formulate a quantitative dynamical theory to describe these flow phenomena which are also coupled to the complex process of density, temperature, and orientational changes. Nevertheless, one could design "simple" experiments to gain some insight into these processes, as was done recently by two groups (14, 15). In these studies using dynamical grating diffraction techniques, picosecond laser pulses are employed to induce density, temperature, and orientational-flow effects in nematic liquid crystals.

In the work by Khoo et al. (14), the pump beams are copolarized and propagate in a plane parallel to the director axis; that is, the optical electric fields are perpendicular to the director axis (cf. Figure 7.14, $\beta = 0$ case). In this case there is no molecular reorientation effect. The principal nonlinear mechanisms are the thermal and density effects and flow. On the other hand, in Eichler and Macdonald (15), the two pump beams are cross-polarized and propagate at an angle $\beta \approx 22°$ with the director axis (cf. Figure 7.14). Instead of having a sinusoidal intensity distribution [as in Khoo et al. (14) with copolarized beams], one has a sinusoidal distribution in terms of the polarization state: the polarization evolves from circular, to elliptical, to linear, as shown in Figure 7.15, while the total intensity of the interfering pump beam

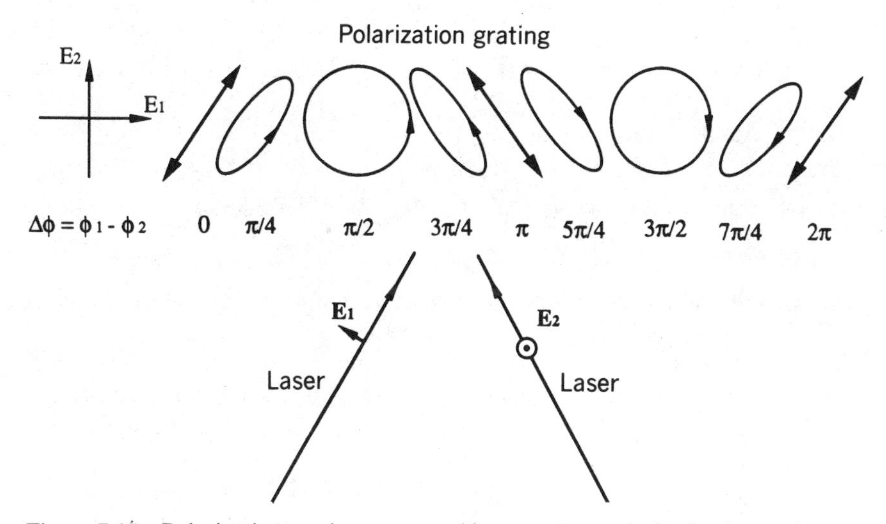


**Figure 7.15** Polarization grating generated by two cross-polarized coherent beams.

is uniform across the distribution. In this case the principal nonlinear mechanism is due to the creation of a reorientation grating.

The flow-orientational coupling can be described by including an extra torque in the equation describing the director axis reorientational angle $\theta$ [cf. Chapter 3, also Eichler and Macdonald (15)]

$$\frac{\mu\,\partial^2\theta}{\partial t^2} + \gamma_1\frac{\partial\theta}{\partial t} + M_{op} + M_{el} + M_{fo} = 0 \tag{7.45}$$

where $M_{op}$ is the optical torque, $M_{el}$ is the elastic torque, and $M_{fo}$ is the torque owing to the flow-orientational coupling. $\mu$ is the moment of inertia of the collection of molecules (as opposed to a single molecule) undergoing the flow-reorientational process, and $\gamma_1$ is a viscosity coefficient.

These torques obviously depend on the interaction geometry. For the geometry employed in Figure 7.16, and assuming that the flow and reorientational direction are principally along the $x$ direction, we have

$$M_{op} = \frac{\varepsilon_\perp}{\varepsilon_\parallel}\frac{\Delta\varepsilon}{16\pi}|E_{op}|^2\sin 2(\beta+\theta) \tag{7.46}$$

$$M_{el} = -K\frac{\partial^2\theta}{\partial z^2} \tag{7.47}$$

$$M_{fo} = -\frac{1}{2}\left(\gamma_1 - \gamma_2\cos^2\theta\right)\frac{\partial V_x}{\partial z} \tag{7.48}$$

Under these approximations (i.e., the flow is along the $x$ direction), the force creating the flow $F_x$ is obtainable from the Maxwell stress tensor:

$$\mathbf{F} = (\mathbf{D}\cdot\nabla)\mathbf{E}^* - \tfrac{1}{2}\nabla(\mathbf{E}\cdot\mathbf{D}^*) \tag{7.49}$$

Under the preceding assumption, this was shown (14) to be

$$F_{x'} = \frac{1}{2}\varepsilon_0\varepsilon_\perp qE^2\cos^2 qy \tag{7.50}$$

where $q$ is the magnitude of the grating wave vector ($q = |\mathbf{k}_1 - \mathbf{k}_2|$, $\mathbf{k}_1$ and $\mathbf{k}_2$ are the wave vectors of the two pump beams).

The time constant characterizing the flow process may be estimated from the dynamical equation for the flow velocity field $\mathbf{V}(\mathbf{r},t)$ discussed in Chapter 3. Assuming that the viscosity coefficient involved is $\eta$, the velocity field $\mathbf{V}(\mathbf{r},t)$ obeys a greatly simplified equation:

$$\rho\mathbf{V}(\mathbf{r},t) - \eta\nabla^2\mathbf{V}(\mathbf{r},t) = \mathbf{F}_{stress} \tag{7.51}$$

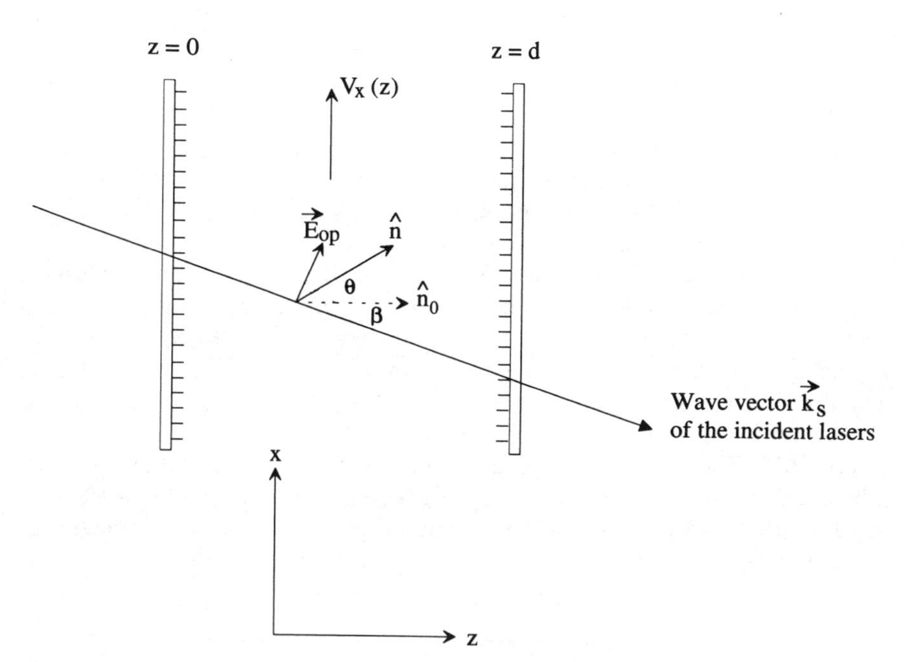


**Figure 7.16** Geometry of interaction for pulsed laser-induced flow-orientational coupling in a nematic liquid crystal film.

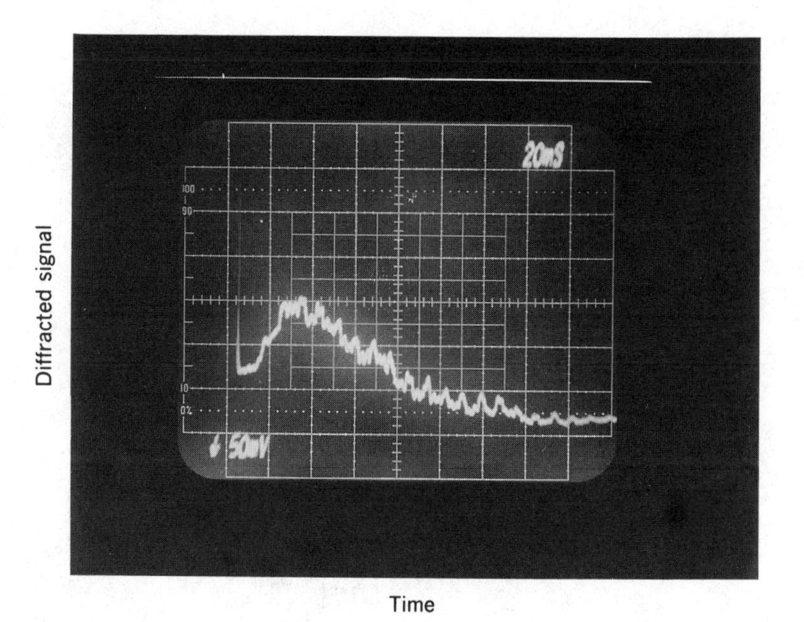


**Figure 7.17** Oscilloscope trace of the probe diffraction showing the observed flow-reorientational effect. The initial spike is the density and thermal effect. The slowly varying component is the flow-reorientational effect.

To estimate the flow *relaxation* time scale, we simply set $\mathbf{F}_{\text{stress}} = 0$ in the preceding equation. This is valid because the picosecond laser pulses, as well as the time scales associated with all the physical processes such as thermal expansion and density change that *initiate* the flow process, are all much shorter than the flow response time.

In Khoo et al. (14), the flow process is assumed to be radial, with a characteristic flow length defined by the width $\omega_0$ of the laser beam; that is, $\nabla^2 V_\beta$ in (7.51) is on the order of $\eta/\omega_0^2 V_\beta$. This gives a flow damping constant $\Gamma_{\text{flow}} \sim \eta\omega_0^{-2}/\rho_0$. Using (mks units) $\rho_0 = 1$ kg/l, $\eta \sim 7\times10^{-2}$, and $\omega_0 = 0.75$ mm $= 750$ $\mu$m, we have a flow relaxation time $\Gamma_{\text{flow}}^{-1} \approx 8$ ms. This is in agreement with the experimental observation (cf. Figure 7.17). The flow is initiated by the large temperature and density changes and is manifested in the initial rise portion of the observed slowly varying component in Figure 7.17. The flow process is terminated in a time on the order of 8 ms, when the accompanying reorientation process also cases. Then the molecules relax to the initial alignment in a time scale characterized by the usual reorientation relaxation time constant $\tau_\theta = Kq^2/\eta \approx 50$ ms (for $K =$

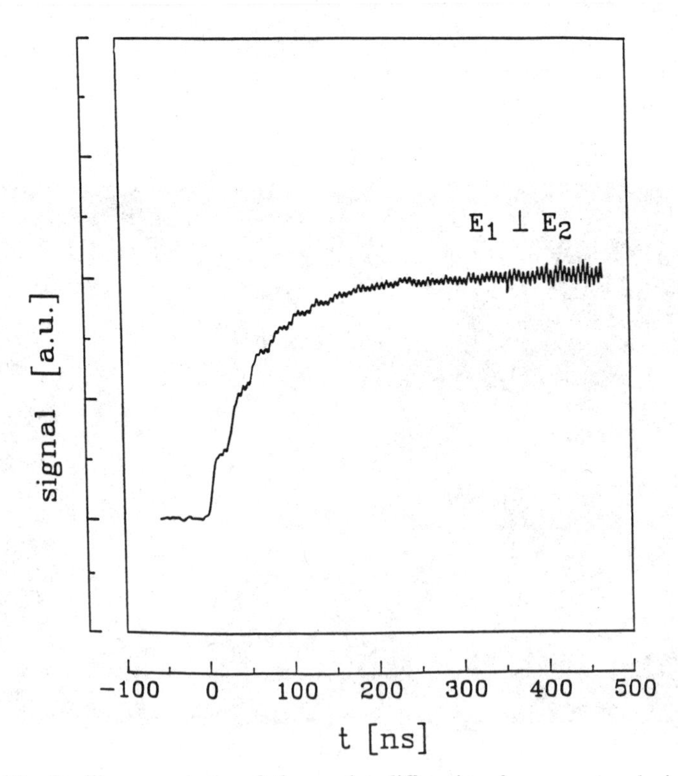


**Figure 7.18** Oscilloscope trace of the probe diffraction from cross-polarized pump beam [after Eichler and MacDonald (15)].

$10^{-11}$ kg·ms$^{-1}$; experimental values of $q = 2\pi/17$ $\mu$m and $\eta = 7\times10^{-2}$, in mks units).

In Eichler and Macdonald (15), the flow is along the $x$ direction and is characterized by two wave vectors: $q_t$ ($\simeq 0.20$ $\mu$m$^{-1}$) along the $y$ direction and $q_1$ ($= 0.25$ $\mu$m$^{-1}$) along the $z$ direction. Accordingly, the flow damping constant $\Gamma_{\text{flow}} \sim \eta(q_1^2 + q_t^2)/\rho$. Note that this damping constant is analogous to the density damping constant [cf. equation (7.14)]; this is because they both originate from a common origin, namely, the translational movement of the molecules. The flow relaxation time $\Gamma_{\text{flow}}^{-1}$ constant may again be estimated as before to give $\Gamma_{\text{flow}}^{-1} \sim 139$ ns, using the values $\eta = 7\times10^{-2}$ and $\rho = 1$ quoted previously. This is on the order of the typical observed time constant characterizing the initial "rise" portion (cf. Figure 7.18) of the flow-reorientational process (15).

As these preceding discussions have demonstrated, laser-induced flow processes and their couplings to other nonlinear mechanisms are very complex problems. These studies have, at best, reviewed only some qualitative aspects of the problems. They do, however, serve as good starting points for more quantitative investigations.

## REFERENCES

1. I. P. Batra, R. H. Enns, and D. Pohl, Stimulated thermal scattering of light. *Phys. Status. Solidi* **48**, 11 (1971).
2. H. J. Eichler, P. Gunter, and D. N. Pohl, "Laser Induced Dynamic Grating." Springer-Verlag, Berlin, 1986.
3. I. C. Khoo and R. Normandin, The mechanism and dynamics of transient thermal grating diffraction in nematic liquid crystal film. *IEEE J. Quantum Electron.* **QE21**, 329 (1985).
4. I. C. Khoo and S. T. Wu, "Optics and Nonlinear Optics of Liquid Crystals." World Scientific, Singapore, 1993; I. C. Khoo, *in* "Progress in Optics" (E. Wolf, ed.), Vol. 26. North-Holland Publ., Amsterdam, 1988.
   C. W. Garland and M. E. Huster, Nematic-smectic-C heat capacity near the nematic-A-smectic-C point. *Phys. Rev. A* **35**, 2365 (1987).
5. M. E. Mack, Stimulated thermal light scattering in the picosecond regime. *Phys. Rev. Lett.* **22**, 13 (1969).
6. R. M. Herman and M. A. Gray, Theoretical prediction of the stimulated thermal Rayleigh scattering in liquids. *Phys. Rev. Lett.* **19**, 824 (1967).
7. I. C. Khoo and R. Normandin, Nanosecond laser-induced transient and permanent gratings and ultrasonic waves in smectic film. *J. Appl. Phys.* **55**, 1416 (1984). see also M. E. Mullen, B. Uthi, and M. J. Stephen, Sound velocity in nematic liquid crystal. *Phys. Rev. Lett.* **28**, 799 (1972); A. E. Lord, Jr., Anisotropic ultrasonic properties of smectic liquid crystal. *ibid.* **29**, 1366 (1972).

8. R. G. Lindquist, P. G. LoPresti, and I. C. Khoo, Infrared and visible laser induced thermal and density optical nonlinearities in nematic and isotropic liquid crystals. *Proc. SPIE—Int. Soc. Opt. Eng.* **1692**, 148 (1992).
9. I. C. Khoo, The infrared nonlinearities of liquid crystals and novel two-wave mixing processes. *J. Mod. Opt.* **37**, 1801 (1990).
10. See, for example, I. C. Khoo, P. Y. Yan, G. M. Finn, T. H. Liu, and R. R. Michael, Low power (10.7 $\mu$m) laser beam amplification via thermal grating mediated degenerate four wave mixings in a nematic liquid crystal film. *J. Opt. Soc. Am. B* **5**, 202 (1988).
11. I. C. Khoo, Sukho Lee, P. G. LoPresti, R. G. Lindquist, and H. Li, Isotropic liquid crystalline film and fiber structures for optical limiting application. *Int. J. Nonlinear Opt. Phys.* **2**, No. 4 (1993).
12. H. Hsiung, L. P. Shi, and Y. R. Shen, Transient laser-induced molecular reorientation and laser heating in a nematic liquid crystal. *Phys. Rev. A* **30**, 1453 (1984).
13. I. C. Khoo, J. Y. Hou, G. L. Din, Y. L. He, and D. F. Shi, Laser induced thermal, orientational and density nonlinear optical effects in nematic liquid crystals. *Phys. Rev. A* **42**, 1001 (1990).
14. I. C. Khoo, R. G. Lindquist, R. R. Michael, R. J. Mansfield, and P. G. LoPresti, Dynamics of picosecond laser-induced density, temperature, and flow-reorientation effects in the mesophases of liquid crystals. *J. Appl. Phys.* **69**, 3853 (1991).
15. H. J. Eichler and R. Macdonald, Flow alignment and inertial effects in picosecond laser-induced reorientation phenomena of nematic liquid crystals. *Phys. Rev. Lett.* **67**, 2666 (1991).

# CHAPTER 8

# ELECTRONIC OPTICAL NONLINEARITIES

## 8.1 INTRODUCTION

In this chapter we treat those nonlinear optical processes in which the electronic wave functions of the liquid crystal molecules are significantly perturbed by the optical field. Unlike the nonelectronic processes discussed in the previous chapters, these electronic processes are very fast; the active electrons of the molecules respond almost instantaneously to the optical field in the form of an induced electronic polarization. Transitions from the initial level to some final excited state could also occur.

Such processes are obviously dependent on the optical frequency and the resonant frequencies of the liquid crystal constituent molecules. They are also understandably extremely complicated, owing to the complex electronic and energy level structure of liquid crystal molecules. Even calculating such basic quantities as the Hamiltonian, the starting point for quantum mechanical calculations (1) of the electronic wave function and energy levels and linear optical properties, require very powerful numerical computational techniques.

In this chapter we present a greatly simplified approach where the liquid crystal molecule is represented as a general multilevel system. Only dipole transitions among the levels are considered. Using this model, we quantitatively illustrate some important basic aspects of the various electronic nonlinear optical processes and their accompanying nonlinearities. Special features pertaining to liquid crystalline materials are then discussed.

## 8.2 DENSITY MATRIX FORMALISM FOR OPTICALLY INDUCED MOLECULAR ELECTRONIC POLARIZABILITIES

Consider the multilevel system depicted in Figure 8.1; the optical transition between any pair of levels ($i$ and $j$) is mediated by an electric dipole.

In the density matrix formalism (2), the expectation value of the dipole moment $\langle \mathbf{d}(t) \rangle$ at any time $t$ following the turn-on of the interaction between the molecule and an optical field is given by

$$\langle \mathbf{d}(t) \rangle = \sum_{i=1}^{N} \sum_{j=1}^{N} d_{ji}\rho_{ij}(t) \tag{8.1}$$

The quantum mechanical equation of motion for the $ij$ component of the density matrix $\rho_{ij}(t)$ is obtainable from the Schrödinger equation:

$$i\hbar \frac{\partial}{\partial t}\rho_{ij} = [H, \rho]_{ij} \tag{8.2}$$

where $H$ is the total Hamiltonian describing the molecule and its interaction with the optical field, that is,

$$H = H_0 - \mathbf{d} \cdot \mathbf{E} \tag{8.3}$$

where $H_0$ is the Hamiltonian describing the unperturbed molecule and $-\mathbf{d} \cdot \mathbf{E}$ is the dipolar interaction between the optical field $\mathbf{E}$ and the molecule.

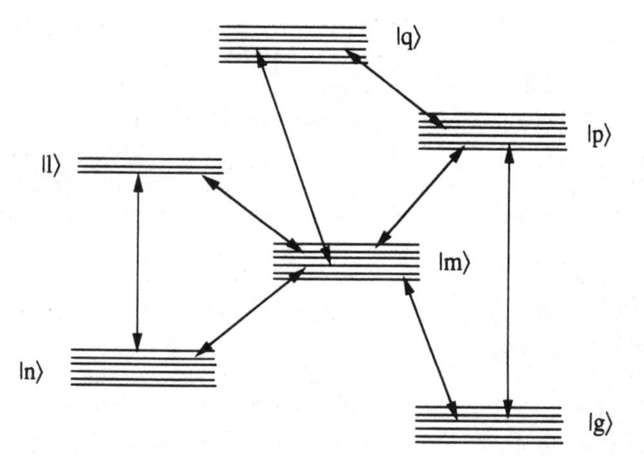

**Figure 8.1** Schematic depiction of the energy level structure of a multilevel molecule.

In general, because of the complexity of the Hamiltonian $H_0$ and the large number of energy levels involved, analytical solution of (8.3) is neither possible nor instructive. On the other hand, we can solve (8.4) in a perturbative manner and obtain a solution for the induced dipole moment, and therefore the electric polarization, in a power series of the interaction $-\mathbf{d}\cdot\mathbf{E}$. This is valid if the dipolar interaction is of a small perturbation magnitude compared to $H_0$. Accordingly, we attach a perturbation parameter $\lambda$ ($\lambda = 1$) to $-\mathbf{d}\cdot\mathbf{E}$ and rewrite (8.3) as

$$H = H_0 - \lambda \mathbf{d}\cdot\mathbf{E} \tag{8.4}$$

Also, $\rho_{ij}(t)$ is expanded in powers of $\lambda$ as follows:

$$\rho_{ij}(t) = \rho_{ij}^{(0)} + \lambda\rho_{ij}^{(1)} + \lambda^2\rho_{ij}^{(2)} + \lambda^3\rho_{ij}^{(3)} \tag{8.5}$$

The reader can easily verify that, upon substituting $\rho_{ij}$ in (8.5) and $H$ in (8.4) into the equation of motion and equating terms containing equal powers of $\lambda$, we have

$$\lambda^0:\quad i\hbar\frac{\partial}{\partial t}\rho_{ij}^{(0)} = \left[H_0, \rho^{(0)}\right]_{ij} \tag{8.6}$$

$$\lambda^1:\quad i\hbar\frac{\partial}{\partial t}\rho_{ij}^{(1)} = \left[H_0, \rho^{(1)}\right]_{ij} - \left[\mathbf{d}\cdot\mathbf{E}, \rho^{(0)}\right]_{ij} \tag{8.7}$$

$$\lambda^2:\quad i\hbar\frac{\partial}{\partial t}\rho_{ij}^{(2)} = \left[H_0, \rho^{(2)}\right]_{ij} - \left[\mathbf{d}\cdot\mathbf{E}, \rho^{(1)}\right]_{ij} \tag{8.8}$$

$$\lambda^3:\quad i\hbar\frac{\partial}{\partial t}\rho_{ij}^{(3)} = \left[H^0, \rho^{(3)}\right]_{ij} - \left[\mathbf{d}\cdot\mathbf{E}, \rho^{(2)}\right]_{ij} \tag{8.9}$$

The first term on the right-hand side of (8.6) to (8.9) may be explicitly written in terms of the energy difference $E_i - E_j$:

$$\begin{aligned}\left[H_0, \rho^{(n)}\right]_{ij} &= \langle i|H_0\rho^{(n)} - \rho^{(n)}H_0|j\rangle \\ &= E_i\langle i|\rho^{(n)}|j\rangle - \langle i|\rho^{(n)}|j\rangle E_j \\ &= (E_i - E_j)\rho_{ij}^{(n)}\end{aligned} \tag{8.10}$$

where we have made use of the fact that $|i\rangle$ and $|j\rangle$ are the eigenfunctions of the Hamiltonian $H_0$, and so $H_0|i\rangle = E_i|i\rangle$ and $H_0|j\rangle = E_j|j\rangle$.

Equations (8.7) to (8.9) become

$$\dot{\rho}_{ij}^{(n)} = -i\omega_{ij}\rho_{ij}^{(n)} - \frac{1}{i\hbar}\left[\mathbf{d}\cdot\mathbf{E}, \rho^{(n-1)}\right] \qquad n=1,2,3,\ldots \tag{8.11}$$

The zeroth-order equation (8.6) gives

$$\dot{\rho}_{ij}^{(0)} = -i\omega_{ij}\rho_{ij}^{(0)} \tag{8.12}$$

If the Hamiltonian $H_0$ of the molecule and therefore its wave functions and eigenvalues are known, as in the case of simple atomic systems, the preceding equations can be solved to any desired order. In dealing with the complex molecules comprising liquid crystals, in fact, for any molecular system, the actual determination of the Hamiltonian $H_0$ is itself quite a feat. In the following discussion we will assume that both $H_0$ and the relevant energy levels involved in the interaction of a molecule with the incident optical field are known (through some quantum mechanical calculations or experiments).

### 8.2.1 Induced Polarizations

An important point regarding the response of a multilevel molecule to the optical field can be immediately deduced from the preceding equation. The driven part of the solution to $\rho_{ij}^{(n)}$ is proportional to the $n$th power of the optical field **E**. From the definition for the dipole moment, equation (8.1), one can see that the induced dipole moment, obtained from the driven part of the solution to $\rho_{ij}^{(n)}$, is of the form of a power series in $E$:

$$\mathbf{d}(t) \sim \alpha:\mathbf{E} + \boldsymbol{\beta}:\mathbf{EE} + \gamma:\mathbf{EEE} \tag{8.13}$$

where the double dots signify tensorial operation between $\alpha$, $\beta$, and $z$ with the vector fields **E**s.

Since **E** is a vector described by three Cartesian components $(E_i, E_j, E_k)$, $\alpha$, $\beta$, and $\gamma$ are tensors of second, third, and fourth rank, respectively; $\alpha = \{\alpha_{ij}\}$, $\beta = \{\beta_{ijk}\}$, and $\gamma = \{\gamma_{ijkh}\}$ are, respectively, linear, second-order, and third-order molecular polarizabilities. The induced electric polarization **P**, defined as the dipole moment per unit volume, is therefore of the form

$$\mathbf{P} = \varepsilon_0\chi^{(1)}:\mathbf{E} + \chi^{(2)}:\mathbf{EE} + \chi^{(3)}:\mathbf{EEE} \tag{8.14}$$

The first, second, and third terms on the right-hand side of (8.14) are, respectively, the linear, second-order, and third-order nonlinear polarizations, with $\chi^{(1)}$, $\chi^{(2)}$, and $\chi^{(3)}$ as the respective susceptibility tensors. The connection between the macroscopic susceptibility tensor $\chi^{(n)}$ and the microscopic polarizabilities $\alpha$, $\beta$, and $\gamma$ is the molecular number density weighted by the local field correction factor (cf. Chapter 3). We will discuss this in more detail in the following sections.

### 8.2.2 Multiphoton Absorptions

The density matrix formalism described previously also shows that, starting from a particular molecular state, the solutions for the population density of the state $\rho_{ij}^{(n)}$ are generally of the following form:

$$\left\langle \rho_{ij}^{(2n)}(t) \right\rangle \neq 0 \qquad n = 1, 2, 3, \ldots \tag{8.15}$$

while the expectation values of all odd-power diagonal elements are vanishing. Since the $n$th-order driven part of the density matrix element is proportional to the $n$th-order power of $E$, we thus have

$$\left\langle \rho_{jj}^{(2)} \right\rangle \sim E^2 \sim I \tag{8.16a}$$

$$\left\langle \rho_{jj}^{(4)} \right\rangle \sim E^4 \sim I^2 \tag{8.16b}$$

$$\left\langle \rho_{jj}^{(6)} \right\rangle \sim E^6 \sim I^3 \tag{8.16c}$$

$$\vdots$$

Equation (8.16$a$) shows that the probability of a molecule's being excited to the $j$th state, starting from the initial state $|i\rangle$, for example, is linearly dependent on the intensity $I$ of the optical field. This is the familiar linear absorption process in which the states $|i\rangle$ and $|j\rangle$ are connected by a single-photon absorption (cf. Figure 8.2). As one may deduce from the detailed calculations given in the following section, $\rho_{jj}^{(2)}$ contains a resonant denominator $(\omega - \omega_{ij})$, the so-called single-photon resonance encountered in traditional absorption spectra.

The next-order nonvanishing diagonal term $\langle \rho_j^{(4)} \rangle$ given in (8.16$b$) is proportional to the square of the optical intensity, $I^2$, and contains resonant denominators of the form $(a_{ij} - 2\omega)$ or $(\omega_{in} - \omega)(\omega_{nj} - \omega)$. This is the so-called two-photon transition (or absorption) process where the initial

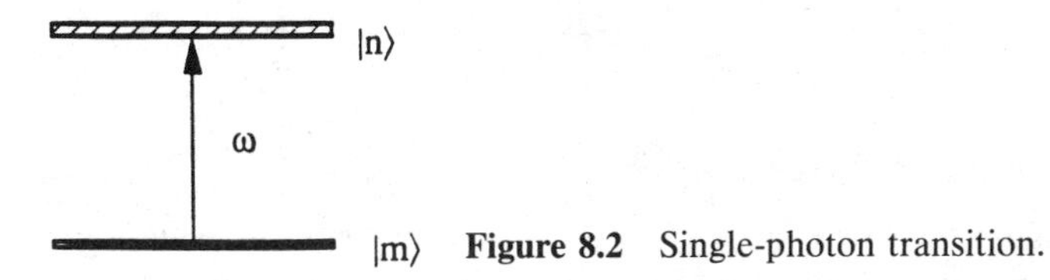

**Figure 8.2** Single-photon transition.

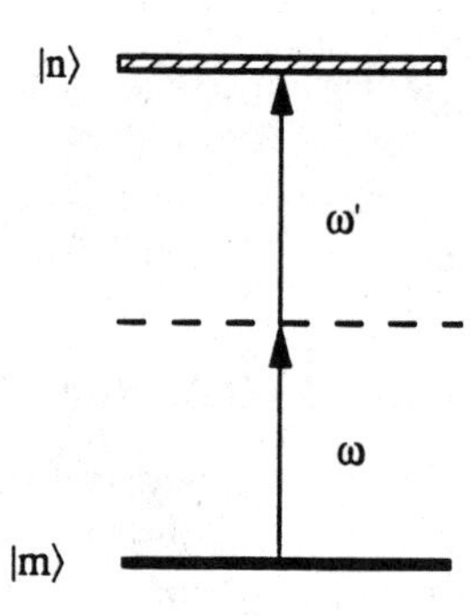


**Figure 8.3** Two-photon absorption process involving simultaneous absorption of two photons at the same frequency.

state $|i\rangle$ is connected to the final state $|j\rangle$ by the simultaneous absorption of two photons (cf. Figure 8.3) such that their sum energy $2\omega$ equals $\omega_{ij}$. Note that by the dipole selection rule, if the $|i\rangle$ and $|j\rangle$ states are connected by a single-photon transition, then the two-photon process is vanishing and vice versa. In the same manner, the $\langle \rho_{jj}^{(6)} \rangle$ term given in (8.16$c$) is proportional to the third-power intensity of the optical field, $I^3$, and is known as a three-photon transition (or absorption) process. The initial state $|i\rangle$ is connected to the final state $|j\rangle$ by a simultaneous absorption of three-photons and is characterized by a three-photon resonant denominator of the form $(\omega_{ij} - 3\omega)$ (cf. Figure 8.4) or $(\omega_{in} - \omega)(\omega_{nm} - \omega)(\omega_{mj} - \omega)$. By the dipole selection rule, if $|i\rangle$ and $|j\rangle$ are connected by a dipole transition, then three-, five-, and higher-order odd-power absorption processes are allowed, whereas all even-power terms are vanishing and vice versa.

Two- and three-photon absorption processes in a liquid crystal (5CB) have been quantitatively studied by Deeg and Fayer (3) and Eichler et al. (4). In Deeg and Fayer (3), two-photon absorptions of the isotropic phase of 5CB were studied with subpicosecond laser pulses. It was determined that the first

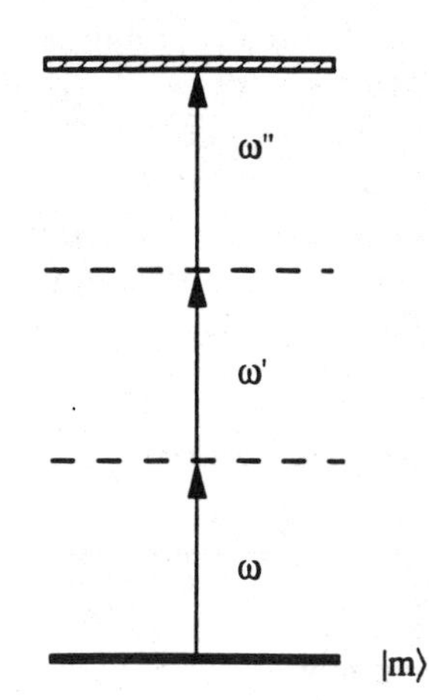


**Figure 8.4** Three-photon transition.

singlet state reached by the two-photon transition relaxes mainly through a radiative mechanism. On the other hand, the study conducted by Eichler et al. (4) with a picosecond laser in the same liquid crystal (in the isotropic as well as the nematic phase) probes the three-photon absorption processes. The excited state reached by such a three-photon transition was found to relax quite substantially through a nonradiative mechanism, resulting in the generation of a large thermal effect. These observations may explain the generation of thermal gratings in the otherwise nonabsorbing (in single-photon language) 5CB material.

As we can see from this simple but general consideration of a multilevel molecule, nonlinear electronic polarizations occur naturally in all materials illuminated by an optical field. The differences among the nonlinear responses of different materials are due to differences in their electronic properties (wave functions, dipole moments, energy levels, etc.) which are determined by their basic Hamiltonian $H_0$. For liquid crystals the extra features we need to take into account are molecular correlations. We will return to specific discussions of liquid crystal nonlinearities following the explicit derivations of their linear and nonlinear polarizations in the following sections.

## 8.3 ELECTRONIC SUSCEPTIBILITIES OF LIQUID CRYSTALS

### 8.3.1 Linear Optical Polarizabilities of a Molecule with No Permanent Dipole

To obtain explicit expressions for the linear and nonlinear polarizabilities $\alpha_{ij}$, $\beta_{ijk}$, and $\gamma_{ijkh}$, consider the solution of $\rho_{ij}(t)$ to the corresponding order. To simplify the calculations, we consider a single molecule illuminated by an optical field $\mathbf{E}(t)$ of the form

$$E(t) = \sum_{\text{all } \omega} \mathbf{E}(\omega) e^{-i\omega t} \tag{8.17}$$

For example, if there are three frequency components $\omega_1$, $\omega_2$, and $\omega_3$ in $E(t)$, equation (8.17) gives

$$E(t) = \left[ E(\omega_1) e^{-i\omega_1 t} + E(\omega_2) e^{-i\omega_2 t} + E(\omega_3) e^{-i\omega_3 t} \right] + \text{c.c.} \tag{8.18}$$

The zeroth-order solution of $\rho_{nm}^{(0)}(t)$ is $\rho_{nm}^{(0)} e^{-i\omega_{nm} t}$. From (8.11) the solution of

$\rho_{ij}^{(1)}(t)$ is given by

$$\rho_{nm}^{(1)}(t) = \frac{i}{\hbar} \int_{-\infty}^{t} e^{-i\omega_{nm}(t-t')} \left[\mathbf{d} \cdot \mathbf{E}(t'), \rho^{(0)}\right]_{nm} dt' \tag{8.19}$$

The commutator in the preceding equation can be evaluated as

$$\left[\mathbf{d} \cdot \mathbf{E}, \rho^{(0)}(t')\right]_{nm} = \sum_{l} \left(\mathbf{d}_{nl} \rho_{lm}^{(0)}(t') - \rho_{nl}^{0}(t') \mathbf{d}_{lm}\right) \cdot \mathbf{E}(t') \tag{8.20}$$

we make a simplifying assumption that there is initially no coherence among the molecular states; that is, all the off-diagonal elements $\rho_{ij}^{(0)} = 0$ for $i \neq j$. (Note that the diagonal elements $\rho_{ii}^{(0)}$ give the population of the $i$th state and some of them are nonvanishing.) In this case the commutator in (8.20) gives

$$\left[\mathbf{d} \cdot \mathbf{E}(t'), \rho^{(0)}\right]_{nm} = \left(\rho_{mm}^{(0)} - \rho_{nn}^{(0)}\right) \mathbf{d}_{nm} \cdot \mathbf{E}(t') \tag{8.21}$$

Substituting (8.21) into (8.19) for $\rho_{nm}^{(1)}(t)$ gives

$$\rho_{nm}^{(1)}(t) = e^{-i\omega_{nm}t} \frac{i}{\hbar} \left(\rho_{mm}^{(0)} - \rho_{nn}^{(0)}\right) \mathbf{d}_{nm} \int_{-\infty}^{t} \mathbf{E}(t') e^{i\omega_{nm}t'} dt' \tag{8.22}$$

$$\rho_{nm}^{(1)}(t) = \frac{i}{\hbar} \left(\rho_{mm}^{(0)} - \rho_{nn}^{(0)}\right) \sum_{\omega} \frac{\mathbf{d}_{nm} \cdot \mathbf{E}(\omega) e^{-i\omega t}}{\omega_{nm} - \omega} \tag{8.23}$$

The expectation value of the induced first-order dipole moment $\langle \mathbf{d}^{(1)}(t) \rangle$ is thus given by

$$\begin{aligned} \langle \mathbf{d}^{(1)}(t) \rangle &= \operatorname{tr}\left(\rho^{(1)} \mathbf{d}\right) = \sum_{nm} \rho_{nm}^{(1)} \mathbf{d}_{mn} \\ &= \sum_{\omega} e^{-i\omega t} \sum_{m,n} \left( \frac{\rho_{mm}^{(0)} - \rho_{nn}^{(0)}}{\hbar} \right) \frac{\mathbf{d}_{mn} \cdot \mathbf{E}(\omega)}{\omega_{nm} - \omega} \end{aligned} \tag{8.24}$$

The induced dipole moment that oscillates at a particular frequency component $\omega$ is thus given by

$$\mathbf{d}(\omega) = \frac{1}{\hbar} \sum_{m,n} \left(\rho_{mm}^{(0)} - \rho_{nn}^{(0)}\right) \frac{\mathbf{d}_{mn}\left(\mathbf{d}_{nm} \cdot \mathbf{E}(\omega)\right)}{\omega_{nm} - \omega} \tag{8.25}$$

If we write **d** and **E** in terms of their Cartesian components [i.e., as $(d^i, d^j, d^k)$ and $(E_i, E_j, E_k)$, respectively], (8.25) gives

$$d_i = \sum_j \alpha_{ij} E_j \tag{8.26}$$

where the (linear) polarizability tensor component $\alpha_{ij}$ is given by

$$\alpha_{ij} \equiv \frac{1}{\hbar} \sum_{m,n} \left(\rho_{mm}^{(0)} - \rho_{nn}^{(0)}\right) \frac{d_{mn}^i d_{nm}^j}{\omega_{nm} - \omega} \tag{8.27}$$

If there are $N$ independent molecules per unit volume, the induced polarization $\mathbf{P} = N\mathbf{d}$. From (8.26), the linear susceptibility $\chi_{ij}^{(1)}$ is thus given by $\varepsilon_0 \chi_{ij}^{(1)} = N\alpha_{ij}$, using the definition $P_i = \varepsilon_0 \chi_{ij}^{(1)} E_j$.

Equation (8.27) shows that the polarizability is dependent on the level populations $\rho_{mm}^{(0)}$ and $\rho_{nn}^{(0)}$. This dependence is more clearly demonstrated if we note that the summation over all values of $m$ and $n$ in (8.27) allows us to rewrite it as [by interchanging the "dummy" indices $m$ and $n$ in the second term of (8.27)]

$$\alpha_{ij} = \frac{1}{\hbar} \sum_{n,m} \rho_{mm}^{(0)} \left[ \frac{d_{mn}^i d_{nm}^j}{\omega_{nm} - \omega} + \frac{d_{nm}^i d_{mn}^j}{\omega_{nm} + \omega} \right] \tag{8.28}$$

Note that $\omega_{nm} = -\omega_{mn}$. If the molecules are in the ground state $|g\rangle$ (i.e., $\rho_{gg}^{(0)} \neq 0$, $\rho_{mm} = 0$ for $m \neq g$) and if we are interested in the contribution from a particular excited state $|n\rangle$ connected to the ground state $|g\rangle$ by the dipole transition, we have

$$\alpha_{ij}(\omega_n) = \frac{1}{\hbar} \left( \frac{d_{gn}^i d_{ng}^j}{\omega_{ng} - \omega} + \frac{d_{ng}^i d_{gn}^j}{\omega_{ng} + \omega} \right) \tag{8.29}$$

In actuality, the ground state $|g\rangle$ may consist of many degenerate or nearly degenerate levels, or a band. In this case it is more appropriate to invoke the so-called oscillator strength (rather than the dipole amount) connecting $n$ and $g$.

The oscillator strength $f_{ng}$ is related to $d_{ng}$ by

$$f_{ng} = \frac{2m\omega_{ng} |\mathbf{d}_{ng}|^2}{3e^2 \hbar} \tag{8.30}$$

If we note that the average value of $d_{gn}^i d_{ng}^j$ in (8.29) is $|\mathbf{d}_{ng}|^2/3$, we may rewrite it as

$$\alpha_{ij}(\omega_n) = \frac{f_{ng}e^2}{2m\omega_{ng}}\left[\frac{1}{\omega_{ng}-\omega} + \frac{1}{\omega_{ng}+\omega}\right] \tag{8.31}$$

The general multilevel linear electric polarizability (8.27) can thus be written as

$$\alpha_{ij} = \sum_n \alpha_{ij}^{(n)}(\omega_n) \tag{8.32}$$

The preceding results are obtained assuming that all the molecular states are sharp. Singularities will occur therefore if the optical interactions with the molecule are at resonances, that is, when the denominators involving frequency differences approach vanishing values (e.g., $\omega_{ng} \approx \omega$). In actuality, the molecular levels are broadened by a variety of homogeneous or inhomogeneous relaxation mechanisms (collisions, natural lifetime broadening, Doppler effects, etc.). These relaxation processes could be phenomenologically accounted for by ascribing a negative imaginary part $-i\gamma_{nm}$ to the frequencies $\omega_{nm}$ (i.e., replacing $\omega_{nm}$ by $\omega_{nm} - i\gamma_{nm}$), where the $\gamma_{nm}$'s are the relaxation rates associated with the $n$ and $m$ states. Note that although $\omega_{nm} = -\omega_{mn}$, $\gamma_{nm} = \gamma_{mn}$. As a result of the inclusion of $-i\gamma_{nm}$, $\alpha_{ij}$ becomes a complex quantity. Consequently, the macroscopic parameters such as the susceptibility $\chi^{(1)}$ and therefore the refractive indices are also complex.

Consider now (8.31) for $\alpha_{ij}(\omega_{ng})$ (i.e., a particular transition involving the ground state). If we replace $\omega_{ng}$ by $\Omega_{ng} = \omega_{ng} - i\gamma_{ng}$ and write $\omega_{ng} = \omega_0$ and $\gamma_{ng} = \gamma_0$, $f_{ng} \equiv f_0$, and $\Delta\omega = \omega_0 - \omega$, we obtain

$$\alpha_{ij}(\omega_0) = \bar{\alpha}\left[\frac{\Delta\omega}{(\Delta\omega)^2 + \gamma_0^2} + \frac{i\gamma_0}{(\Delta\omega)^2 + \gamma_0^2}\right] \tag{8.33}$$

where $\bar{\alpha} \equiv f_0 e^2/2m\omega_0$. If we have $N$ independent molecules per unit volume in the medium, the macroscopic susceptibility $\chi_{ij}$ is given by $\varepsilon_0\chi_{ij} = N\alpha_{ij}$, that is, $\chi_{ij} = (N/\varepsilon_0)\alpha_{ij}(\omega_0)$. The refractive index $n_{ii}$, for example, is thus given by $n_{ii} = \sqrt{1+\chi_{ii}} = n_0 + i\tilde{n}$. The real part $n_0$ and the imaginary part $\tilde{n}$ of $n_{ii}$ are given by

$$n_0(\omega) = 1 + \frac{N}{\varepsilon_0}\bar{\alpha}\left[\frac{\Delta\omega}{(\Delta\omega)^2 + \gamma_0^2}\right] \tag{8.34}$$

and

$$\tilde{n}(\omega) = \frac{N\bar{\alpha}}{\varepsilon_0}\left[\frac{\gamma_0}{(\Delta\omega)^2 + \gamma_0^2}\right] \tag{8.35}$$

The imaginary part $\tilde{n}$ gives a (linear) attenuation factor in the propagation of a plane optical wave $\exp(-\omega/c)\tilde{n}(\omega)l)$, where $l$ is the propagation length into the medium. The attenuation constant $(\omega/c)\tilde{n}$ is usually referred to as the linear absorption constant in units of $(\text{length})^{-1}$. $n_0(\omega)$ is the dispersion

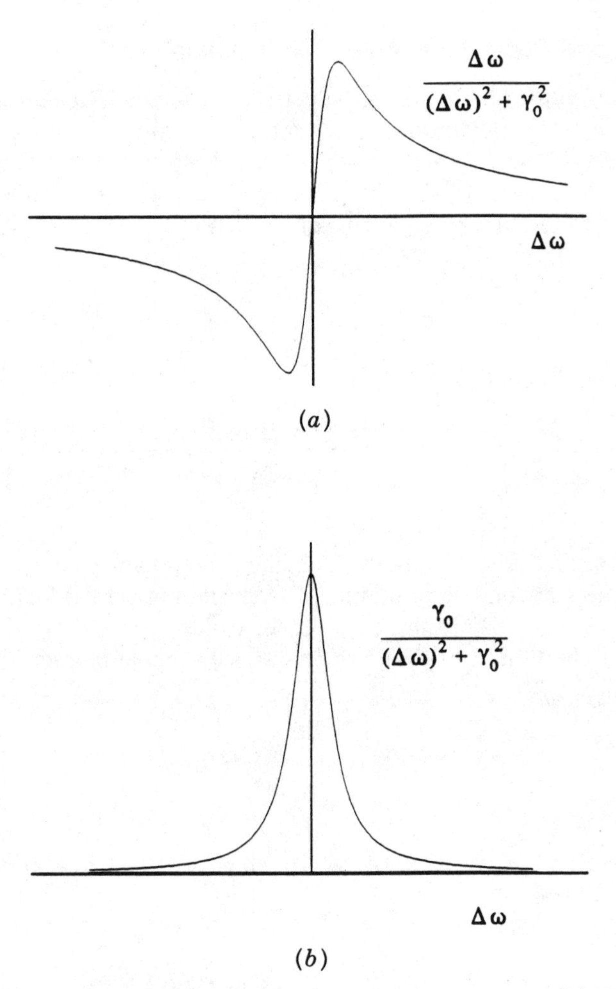


**Figure 8.5** (*a*) Index dispersion $n_0(\omega)$ of a ground-state molecule. (*b*) The imaginary part of the refractive index $\hat{n}(\omega)$.

of the medium (cf. Figure 8.5*a*) $\tilde{n}(\omega)$ gives the linear absorption spectral line shape which is a Lorentzian (cf. Figure 8.5*b*).

It is important to note that, in the preceding derivation of the refractive index $n = n_0 + i\tilde{n}$, we have accounted for only one particular transition $n \to g$; the transition frequency $\omega_{ng} = \omega_0$ is the frequency of interest in the context of resonance with the laser frequency $\omega$. There is, of course, a whole collection (in fact, an infinite number) of transitions that are off resonance with respect to the laser frequency. Collectively, these transactions give rise to a nonresonant background refractive index $n_b$, which is obtainable by performing the sums in (8.32) and (8.31). $n_b$ is, of course, also complex and may be written as $n_b = n_{b(0)} + i\tilde{n}_b$.

### 8.3.2 Second-Order Electronic Polarizabilities

Second- and third-order molecular electronic polarizabilities are obtained by solving for $\rho^{(2)}(t)$. Substituting (8.22) for $\rho^{(1)}(t)$ into the right-hand side of (8.8), evaluating the commutator $[\mathbf{d} \cdot \mathbf{E}(t), \rho^{(1)}(t)]$ as before [now with $\mathbf{E}(t)$ expressed as $\mathbf{E}(t) = \Sigma_{\omega'} \mathbf{E}(\omega') e^{-i\omega' t}$], and performing a simple integration, we obtain, for an exemplary frequency component $(\omega' + \omega)$ of $\rho^{(2)}_{nm}(t)$,

$$\rho^{(2)}_{nm}(t) = e^{-i(\omega'+\omega)t} \left\{ \sum_l \frac{\rho^{(0)}_{mm} - \rho^{(0)}_{ll}}{\hbar^2} \frac{(\mathbf{d}_{nl} \cdot \mathbf{E}(\omega'))(\mathbf{d}_{lm} \cdot \mathbf{E}(\omega))}{[\omega_{nm} - (\omega' + \omega)](\omega_{lm} - \omega)} \right.$$
$$\left. + \sum_l \frac{\rho^{(0)}_{nn} - \rho^{(0)}_{ll}}{\hbar^2} \frac{(\mathbf{d}_{lm} \cdot \mathbf{E}(\omega'))(\mathbf{d}_{nl} \cdot \mathbf{E}(\omega))}{[\omega_{nm} - (\omega' + \omega)](\omega_{nl} - \omega)} \right\} \tag{8.36}$$

There are, of course, as many frequency components in $\rho^{(2)}_{nm}(t)$ as there are possibilities of combining $\omega$ and $\omega'$ from the optical field. Here we focus our attention on a particular example in order to explicitly illustrate the physics. By definition, the second-order induced dipole moment is

$$\langle \mathbf{d} \rangle^{(2)} \equiv \sum_{nm} \rho^{(2)}_{nm} \mathbf{d}_{mn} \tag{8.37}$$

The $i$th Cartesian component of $\langle \mathbf{d}^{(2)} \rangle$, which we shall henceforth write as $d^{(2)}_i$, is of the form

$$d^{(2)}_i = \beta_{ijk}(-(\omega' + \omega); \omega', \omega) E_j(\omega') E_k(\omega) \tag{8.38}$$

with the second-order polarizability tensor $\beta_{ijk}$ given by

$$\beta_{ijk}(-(\omega'+\omega);\omega',\omega) = \sum_{n,m,l}\left(\frac{\rho_{mm}^{(0)}-\rho_{ll}^{(0)}}{\hbar^2}\right)\frac{d_{mn}^i d_{nl}^j d_{lm}^k}{[\Omega_{nm}-(\omega'+\omega)](\Omega_{lm}-\omega)}$$
$$-\sum_{n,m,l}\left(\frac{\rho_{ll}^{(0)}-\rho_{nn}^{(0)}}{\hbar^2}\right)\frac{d_{mn}^i[d_{lm}^j]^j[d_{nl}^k]^k}{[\Omega_{nm}-(\omega'+\omega)](\Omega_{nl}-\omega)} \quad (8.39)$$

where we have used $\Omega_{nm}=\omega_{nm}-i\gamma_{nm}$ instead of $\omega_{nm}$ in order to account for the $n-m$th-level relaxation mechanism.

The expression for $\beta_{ijk}$ may be rewritten in many equivalent ways by appropriate relabeling of the indices $n$, $m$, and $l$. In particular, one can show in a manner analogous to the preceding section that, for a molecule in a ground state, the second-order polarizability tensor component $\beta_{ijk}$ becomes

$$\beta_{ijk}^{\text{ground}}(-(\omega'+\omega);\omega,\omega) = \sum_{nl}\left[\frac{d_{gn}^i d_{nl}^j d_{lg}^k \hbar^{-2}}{[\Omega_{ng}-(\omega'+\omega)](\Omega_{lg}-\omega)} + \frac{d_{nl}^i d_{gn}^j d_{lg}^k \hbar^{-2}}{[\Omega_{ln}-(\omega'+\omega)](\Omega_{lg}-\omega)}\right]$$
$$-\sum_{nm}\left[\frac{d_{mn}^i d_{ng}^j d_{gm}^k \hbar^{-2}}{[\Omega_{nm}-(\omega'+\omega)](\Omega_{gm}-\omega)} + \frac{d_{ng}^i d_{mn}^j d_{gm}^k \hbar^{-2}}{[\Omega_{gn}-(\omega'+\omega)](\Omega_{gm}-\omega)}\right] \quad (8.40)$$

Notice that, in general, both single-photon resonances (e.g., $\Omega_{gm}-\omega$) and two-photon resonances [e.g., $\Omega_{ln}-(\omega'+\omega)$] are involved.

### 8.3.3 Third-Order Electronic Polarizabilities

Using the same straightforward though cumbersome algebra, the third-order density matrix element $\rho_{mm}^{(3)}(t)$ and the corresponding nonlinear polarizability tensor component $\gamma_{ijkh}$ can be obtained. It suffices to note that $\rho_{mn}^{(3)}(t)$ will involve a triple product of $\mathbf{d}\cdot\mathbf{E}$ and oscillate as $e^{-i(\omega+\omega'+\omega'')t}$, where $\omega$, $\omega'$, and $\omega''$ are the frequency components contained in the optical field $\mathbf{E}(t)$. The interested reader, with the practice gained from the preceding section, can easily show that if $\mathbf{E}(t)$ is now expressed as $\mathbf{E}(t)=\sum_{\omega'}\mathbf{E}(\omega'')e^{-i\omega'' t}$ in the commutator $[\mathbf{d}\cdot\mathbf{E}(t),\rho^{(2)}(t)]$, the component of $\rho^{(3)}_{nm}$ that oscillates at a particular frequency $(\omega+\omega'+\omega'')$ is given by

$$\begin{aligned}
\rho^{(3)}_{nm}(t) &= e^{-i(\omega''+\omega'+\omega)t} \\
&\times \sum_{pq}\Bigg\{\frac{1}{\hbar}\frac{[\mathbf{d}_{np}\cdot\mathbf{E}(\omega'')]}{[\omega_{nm}-(\omega''+\omega'+\omega)]} \\
&\quad\times\Bigg[\frac{(\rho^{(0)}_{mm}-\rho^{(0)}_{qq})}{\hbar^2}\frac{[\mathbf{d}_{pq}\cdot\mathbf{E}(\omega')][\mathbf{d}_{qm}\cdot\mathbf{E}(\omega)]}{[\omega_{pm}-(\omega'+\omega)](\omega_{qm}-\omega)} \\
&\quad-\frac{(\rho^{(0)}_{qq}-\rho^{(0)}_{pp})}{\hbar^2}\frac{[\mathbf{d}_{qm}\cdot\mathbf{E}(\omega')][\mathbf{d}_{pq}\cdot\mathbf{E}(\omega)]}{[\omega_{pm}-(\omega'+\omega)](\omega_{pq}-\omega)}\Bigg] \\
&\quad-\frac{1}{\hbar}\frac{[\mathbf{d}_{pm}\cdot\mathbf{E}(\omega'')]}{[\omega_{nm}-(\omega''+\omega'+\omega)]} \\
&\quad\times\Bigg[\frac{(\rho^{(0)}_{pp}-\rho^{(0)}_{qq})}{\hbar^2}\frac{[\mathbf{d}_{nq}\cdot\mathbf{E}(\omega')][\mathbf{d}_{qp}\cdot\mathbf{E}(\omega)]}{[\omega_{np}-(\omega'+\omega)](\omega_{qp}-\omega)} \\
&\quad-\frac{(\rho^{(0)}_{qq}-\rho^{(0)}_{nn})}{\hbar^2}\frac{[\mathbf{d}_{qp}\cdot\mathbf{E}(\omega'')][\mathbf{d}_{nq}\cdot\mathbf{E}(\omega)]}{[\omega_{np}-(\omega'+\omega)](\omega_{nq}-\omega)}\Bigg]\Bigg\}
\end{aligned} \tag{8.41}$$

From this, the third-order polarizability tensor component $\gamma_{ijkh}(-(\omega''+\omega'+\omega);\omega'',\omega',\omega)$ can be identified from

$$d_i(\omega''+\omega'+\omega)=\gamma_{ijkh}E_j(\omega'')E_k(\omega')E_h(\omega) \tag{8.42}$$

where $d_i$ is the $i$th Cartesian component of the expectation value of the dipole moment

$$d_i=\hat{i}\cdot\langle\mathbf{d}\rangle=\sum_{n,m}\rho^{(3)}_{nm}d^i_{mn} \tag{8.43}$$

This gives

$$\gamma_{ijkh}(-(\omega''+\omega'+\omega);\omega'',\omega',\omega)$$

$$= \sum_{nmpq} \frac{\hbar^{-3} d^i_{mn}}{[\omega_{nm}-(\omega''+\omega'+\omega)]}$$

$$\left\{ \frac{\left(\rho^{(0)}_{mm}-\rho^{(0)}_{qq}\right) d^j_{np} d^k_{pq} d^h_{qm}}{[\omega_{pm}-(\omega'+\omega)](\omega_{qm}-\omega)} - \frac{\left(\rho^{(0)}_{qq}-\rho^{(0)}_{pp}\right) d^j_{np} d^k_{qm} d^h_{pq}}{[\omega_{pm}-(\omega'+\omega)](\omega_{pq}-\omega)} \right.$$

$$\left. - \frac{\left(\rho^{(0)}_{pp}-\rho^{(0)}_{qq}\right) d^j_{pm} d^k_{nq} d^h_{qp}}{[\omega_{np}-(\omega'+\omega)](\omega_{qp}-\omega)} + \frac{\left(\rho^{(0)}_{qq}-\rho^{(0)}_{nn}\right) d^j_{pm} d^k_{qp} d^h_{nq}}{[\omega_{np}-(\omega'+\omega)](\omega_{nq}-\omega)} \right\} \quad (8.44)$$

If we account for relaxation mechanisms, again, all the $\omega$'s will be replaced by the corresponding $\Omega = \omega - i\gamma$. From (8.44) we can see that besides one- and two-photon resonances, $\gamma_{ijkh}$ also contain three-photon resonances, characterized by denominators of the form $\Omega_{nm} - (\omega + \omega' + \omega'')$.

From these expressions for the second- and third-order nonlinear polarizabilities, several observations can be made:

- The magnitudes and signs of the nonlinear responses of a molecule are highly dependent on the state (e.g., $\rho^{(0)}_{nn}$) it is in; an excited-state molecule obviously has a nonlinear response very different from a ground-state molecule.
- The magnitudes of the nonlinear polarizabilities depend on the dipole matrix moments involved, which are basically governed by the electronic structure of the molecules.
- The contribution from any pair of levels (e.g., $p$ and $q$) to the nonlinear response will be greatly enhanced as the resonance condition $\omega_{pq} - \omega = 0$ is approached.
- Away from resonances, the nonlinear response depends mainly on the oscillator strength, which is governed by the electronic structure of the molecule.

## 8.4 ELECTRONIC NONLINEAR POLARIZATIONS OF LIQUID CRYSTALS

The expressions for the molecular polarizabilities derived in the last two sections can be applied to a liquid crystal molecule if the Hamiltonian (and wave functions) of the liquid crystal molecule is known. To obtain the right Hamiltonian function and solve for the wave functions and the dipole matrix elements required for evaluating $\alpha_{ij}$, $\beta_{ijk}$, and $\gamma_{ijkh}$ of the actual liquid crystal molecule is a quantum chemistry problem that requires a great deal of numerical computation.

Several methods (5, 6) have been developed for calculating these molecular nonlinear polarizabilities. These methods have proven to be fairly reliable for predicting the second- and third-order molecular polarizabilities of organic molecules and polymers in general and liquid crystalline molecules in particular. For treating the ordered phases of liquid crystals, one has to account for molecular correlations and ordering by introducing an orientational distribution function (7, 8).

Other important factors that should be accounted for are the local field corrections and the anisotropy of the molecules. In general, liquid crystalline molecules are complex and possess permanent dipole moments. Furthermore, quadrupole moments could also contribute significantly in second-order nonlinear polarizations (9), and thus the treatment presented in the preceding sections, which ignores these points, has to be appropriately modified.

In the rest of this chapter, we address these and other macroscopic symmetry properties of liquid crystalline electronic optical nonlinearities.

### 8.4.1 Local Field Effects and Symmetry

The molecular polarizabilities obtained in the last two sections are microscopic parameters that characterize an *individual* molecule. To relate them to the bulk or macroscopic parameters, namely, the susceptibilities $\chi^{(1)}$, $\chi^{(2)}$, $\chi^{(3)}$, and so forth, one needs to account for the intermolecular fields as well as the symmetry properties of the bulk materials. These intermolecular fields will result in a collective response that is different from the one obtained by treating the bulk materials as being composed of noninteracting molecules. In other words, one has to account for the so-called local field correction factors.

The local field correction factors for linear susceptibility were discussed in Chapter 3. In the usual studies (2) of local field correction factors for nonlinear susceptibilities, molecular correlations that characterize liquid crystals are not taken into account. With this gross approximation, the second- and third-order nonlinear susceptibilities of an isotropic medium or a medium with cubic symmetry containing $N$ molecules per unit volume are given by (2):

$$\chi_{ijk}^{(2)}(-(\omega' + \omega); \omega') = N\mathscr{L}^{(2)}(\omega' + \omega, \omega', \omega)\beta_{ijk}(-(\omega + \omega'); \omega', \omega) \tag{8.45a}$$

where

$$\mathscr{L}^{(2)}(\omega' + \omega, \omega', \omega) = \left[\frac{\varepsilon^{(1)}(\omega' + \omega) + 2}{3}\right]\left[\frac{\varepsilon^{(1)}(\omega') + 2}{3}\right]\left[\frac{\varepsilon^{(1)}(\omega) + 2}{3}\right] \tag{8.45b}$$

Similarly,

$$\chi^{(3)}_{ijkh}(-(\omega'' + \omega' + \omega); \omega'', \omega', \omega)$$
$$= N\mathscr{L}^{(3)}(\omega'' + \omega' + \omega, \omega'', \omega', \omega)\gamma_{ijkh}(-(\omega'' + \omega' + \omega); \omega'', \omega', \omega) \tag{8.46a}$$

where

$$\mathscr{L}^{(3)} = \left[\frac{\varepsilon^{(1)}(\omega'' + \omega' + \omega) + 2}{3}\right]\left[\frac{\varepsilon^{(1)}(\omega'') + 2}{3}\right]\left[\frac{\varepsilon^{(1)}(\omega') + 2}{3}\right] \tag{8.46b}$$

Because of their relatively recent "discovery," local field correction factors for the nonlinear susceptibilities of liquid crystals, taking into account the noncubic symmetries of the various liquid crystalline phases, have hitherto not been quantitatively performed. Most studies of macroscopic nonlinear polarization that do account for the molecular orderings proceed by first adopting the macroscopic relationships given previously.

### 8.4.2 Symmetry Considerations

Real liquid crystal molecules are much more complex than the multilevel molecule studied in the preceding section. In general, they are anisotropic and possess permanent dipoles (i.e., the molecules themselves are polar). Nevertheless, in bulk form, these liquid crystalline molecules tend to align themselves such that their collective dipole moment is of vanishing value.

Put another way, most phases of liquid crystals are characterized by a centrosymmetry, due to the equivalence of the $-\mathbf{n}$ and $\mathbf{n}$ directions. This holds in the nematic phase (which belongs to the $D_{\infty h}$ symmetry group), the smectic-A phase ($D_{\infty}$ symmetry), and the smectic-C phase ($C_{2h}$ symmetry). As a result of such centrosymmetry, the macroscopic second-order polarizability $\chi_{ijk} \equiv 0$.

The most well-known and extensively studied non-centrosymmetric liquid crystalline phase is the smectic-C* phase. The helically modulated smectic-C* system, with a spatially varying spontaneous polarization, is locally characterized by a $C_2$ symmetry; the average polarization of the bulk is still vanishing. By unwinding such a helical structure, the system behaves as a crystal with $C_2$ point symmetry. Such unwound SmC* phases possess sizeable second-order nonlinear polarizabilities, with $\chi^{(2)}$ ranging from $8 \times 10^{-16}$ m/V for DOBAMBC (10) to about $0.2 \times 10^{-12}$ m/V for *o*-nitroalkoxyphenyl-biphenyl-carboxylate (11).

### 8.4.3 Permanent Dipole and Molecular Ordering

The fact that liquid crystal molecules themselves could possess a permanent dipole moment (even though the bulk dipole moment has vanished) has been taken into account in Saha and Wong's (7) treatment of the molecular correlation effects in third-order nonlinear polarizabilities. Their calculations show that, in general, the experimentally measurable third-order nonlinear polarizabilities $\bar{\gamma}_{xxxx}$, $\bar{\gamma}_{yyyy}$, and $\bar{\gamma}_{zzzz}$, for example, are related to their microscopic counterparts $\beta_{ijk}$ and $\gamma_{ijkl}$ by the following relationships:

$$\bar{\gamma}_{zzzz} = \gamma_{\text{iso}} + \frac{2}{7}\alpha\langle P_2\rangle + \frac{8}{7}\beta\langle P_4\rangle \tag{8.47}$$

$$\bar{\gamma}_{xxxx} = \bar{\gamma}_{yyyy} = \gamma_{\text{iso}} - \frac{1}{7}\alpha\langle P_2\rangle + \frac{3}{7}\beta\langle P_4\rangle \tag{8.48}$$

$$\begin{aligned}\gamma_{\text{iso}} = {} & \tfrac{1}{5}(\gamma_{1111} + \gamma_{2222} + \gamma_{3333} + 2\gamma_{1122} + 2\gamma_{3322} + 2\gamma_{3311}) \\ & + (5kT)^{-1}(\mu_3\beta_{333} + \mu_3\beta_{322} + \mu_3\beta_{311} + \mu_2\beta_{233} + \mu_2\beta_{222} \\ & + \mu_2\beta_{211} + \mu_1\beta_{133} + \mu_1\beta_{122} + \mu_1\beta_{111})\end{aligned} \tag{8.49}$$

$$\begin{aligned}\alpha = {} & (2\gamma_{3333} - \gamma_{1111} - \gamma_{2222} - 2\gamma_{1122} + 2\gamma_{3322} + \gamma_{3311}) \\ & + (kT)^{-1}(2\mu_3\beta_{333} + \tfrac{1}{2}\mu_3\beta_{322} + \tfrac{1}{2}\mu_3\beta_{311} + \tfrac{1}{2}\mu_2\beta_{233} - \mu_2\beta_{222} \\ & - \mu_2\beta_{211} + \tfrac{1}{2}\mu_1\beta_{133} - \mu_1\beta_{122} - \mu_1\beta_{111})\end{aligned} \tag{8.50}$$

where $\mu_i$ $(i = 1, 2, 3)$ are the components of the permanent dipole moment in the molecule's major axes frame, $\langle P_l\rangle$ are the moments of the molecular orientational distribution function $f(\theta)$ defined by

$$\langle P_l \rangle = \int_{-1}^{1} f(\theta) P_l(\cos\theta)\, d\cos\theta \tag{8.51}$$

and $P_l$ is the $l$th-order Legendre polynomial.

This follows from the definition for the orientational distribution (7,8) function $f(\theta)$:

$$f(\theta) = \sum_{l=0}^{\infty} \left( \frac{2l+1}{2} \right) a_l P_l(\cos\theta) \tag{8.52}$$

for uniaxial molecules (most liquid crystals are in this category).

### 8.4.4 Quadrupole Contribution and Field-Induced Symmetry Breaking

It is important to note at this juncture that so far our discussion of the nonlinear polarization is based on the dipole interaction [cf. equation (8.3)]. We have neglected quadrupole and higher-pole contributions as they are generally small in comparison. However, in situations where the dipolar contribution is vanishing, either because the molecules are symmetric (12) or the bulk is centrosymmetric, the quadrupole contribution is not necessarily vanishing and may give rise to sizeable second-order nonlinear polarization (13). Nematic liquid crystal second-order nonlinear susceptibilities associated with quadrupole moment contributions were studied in detail by Ou-Yang and Xie (9). The starting point of their analysis is the general expression for second-order nonlinear polarization:

$$P_i^{\mathrm{NL}} = \chi_{ijk} E_j E_k + \gamma_{ijkl} E_j \frac{\partial E_l}{\partial n_k} + \cdots \qquad i,j,k,l = 1,2,3 \tag{8.53}$$

The first term on the right-hand side of (8.53) vanishes in media with centrosymmetry, such as nematic liquid crystals. It is, however, nonvanishing if the nematic director **n** is distorted [e.g., curvature deformation that produces flexoelectric effects (14)]. The second term is the quadrupole moment contribution, which is, in general, nonzero.

By considering the symmetry and coordinate transformation properties of a nematic liquid crystal under the action of an applied electric field, these authors were able to obtain explicit expressions for the susceptibility tensors $\chi_{ijk}$ and $\chi_{ijkl}$ and explain the experimental observations of second-harmonic generations by several groups (7, 15).

### 8.4.5 Molecular Structural Dependence of Nonlinear Susceptibilities

The magnitudes of the various tensorial components of the nonlinear susceptibilities of liquid crystals are determined fundamentally by their electronic structure. Since liquid crystal molecules are large and complex, quantitative numerical ab initio computations of these susceptibilities are rather involved.

Most studies of liquid crystal electronic nonlinear susceptibilities, as in other organic or polymeric molecular systems, are therefore phenomenological in nature. Early studies included the systematic examination of about 100 different organic compounds by Davydov et al. (16). They found the conjugated $\pi$ electrons in the benzene ring to be responsible for large second-harmonic signal generation (cf. Figure 8.6). This observation was also confirmed by other studies (17) which showed that the delocalized electronics in the conjugated molecules could produce large second- and third-order nonlinearities.

In the past two decades, an extensive volume of research has been performed in the field of molecular nonlinear optics (for details see 1, 5, and 18). In spite of the tremendous variations in the organic materials' molecular structures, the following conclusions seem to be well documented.

- The nonlinear optical response of thermotropic liquid crystals and similar organic materials is due mainly to the delocalization of the $\pi$-electron wave functions of the so-called polarizable core of a liquid crystal. The polarizable core usually consists of more than one benzene ring connected by a variety of linkages or bonds.
- The nonlinear polarizations of a molecule, as well as its linear optical properties (e.g., absorption spectrum, etc.), can be drastically modified by substituents. In general, the interactions of substituents with the $\sigma$ bonds are rather mild, whereas their effects on the $\pi$ electron are much more pronounced and extend over the entire delocalization range.

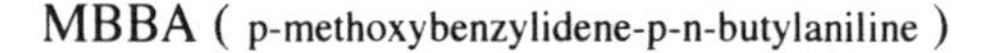


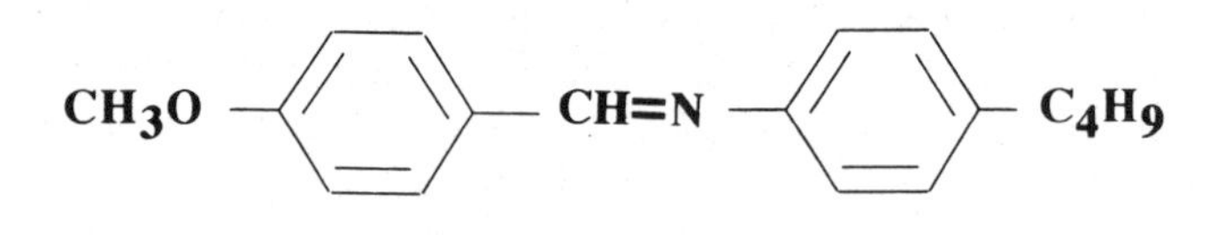


Solid-21 °C-Nematic-48 °C-Isotropic

**Figure 8.6** Molecular structure of MBBA showing the presence of two benzene rings. The six $\pi$ electrons in each ring may be considered as free electrons confined to the conjugated bond.

There are two types of substituents: acceptors and donors, in analogy to semiconductor physics. If a substituent group has vacant low-lying $\pi$ orbitals, it will attract electrons from the host conjugated molecule; such a substituent is called an acceptor. Examples include $NO_2$, $C{\equiv}N$, and the $CONH_2$ group. On the other hand, substituents that have occupied high-lying $\pi$ orbitals will tend to share their electronic charges with the conjugated molecule; they are thus classified as donors. Examples include $N(CH_3)_2$, $NH_2$, and OH.

- In the engineering of new molecules, it has been observed that multiple substituents by weakly interacting substituents tend to produce an additive effect. On the other hand, when strong interacting substituents are used, for example, a strong donor and a strong acceptor linked to the same conjugated molecule as follows:

$$\underset{\text{acceptor}}{NO_2} - C_6H_4 - \underset{\text{donor}}{NH_2}$$

the result could be a tremendous enhancement in the oscillator strength and susceptibilities to external (optical) fields.

- It is important, however, to bear in mind that although the nonlinear polarizabilities (and susceptibilities) of a particular class of molecules may be improved by molecular engineering, other physical properties (e.g., viscosity, size, elastic constant, absorption, etc.) are also likely to be affected. These physical properties play an equally (in some cases even more) important role in actual nonlinear optical processes and applications. A simple example of this necessity to consider the overall picture, rather than an optimization procedure based solely on a particular molecular parameter, is discussed in Chapter 6.
- In some nematic liquid crystals (e.g., MBBA), theories and experiments (19) have shown that, when photoexcited, the molecules will undergo some structural or conformational changes and will exhibit usually large optical nonlinearities near the phase transition temperature. In some absorbing (at the 5145-Å line of an $Ar^+$ laser) liquid crystals [4-4′-bis(Heptyloxy)azoxybenzene], a study (20) has shown peculiar polarization dependences on the optical fields. These effects are undoubtedly related to the electronic structures of these liquid crystals and their changes following photoabsorption, but the exact mechanisms remain to be ascertained.

## REFERENCES

1. See, for example, chapters dealing with Hamiltonian and quantum chemistry in D. S. Chemla and J. Zyss, "Nonlinear Optical Properties of Organic Molecules and Crystals." Academic Press, Orlando, FL, 1987.

2. R. W. Boyd, "Nonlinear Optics." Wiley (Interscience), New York, 1993.
3. F. W. Deeg and M. D. Fayer, *J. Chem. Phys.* **91**, 2269 (1989).
4. H. J. Eichler, R. Macdonald, and B. Trösken, *Mol. Cryst. Liq. Cryst.* **231**, 1–10 (1993).
5. See, for example, Chemla and Zyss (1) also P. N. Prasad and D. J. Williams, "Introduction to Nonlinear Optical Effects in Molecules and Polymers." Wiley (Interscience), New York 1990.
6. J. D. LeGrange, M. G. Kuzyk, and K. D. Singer, *Mol. Cryst. Liq. Cryst.* **150B**, 567 (1987).
7. S. K. Saha and G. K. Wong, *Appl. Phys. Lett.* **34**, 423 (1979).
8. S. Jen, N. A. Clark, P. S. Pershan, and E. B. Priestley, *J. Chem. Phys.* **66**, 4635 (1977).
9. Z.-C. Ou-Yang and Y.-Z. Xie, *Phys. Rev. A* **32**, 1189 (1985).
10. N. M. Shtykov, M. I. Barnik, L. A. Beresnev, and L. M. Blinov, *Mol. Cryst. Liq. Cryst.* **124**, 379 (1985).
11. J. Y. Lin, M. G. Robinson, K. M. Johnson, D. M. Wabba, M. B. Ros, N. A. Clark, R. Shao, and D. Doroski, *J. Appl. Phys.* **70**, 3426 (1991).
12. J. Prost and J. P. Marcerou, *J. Phys. (Paris)* **38**, 315 (1977).
13. J. E. Bjorkholm and A. E. Siegman, *Phys. Rev.* **154**, 851 (1967).
14. R. B. Meyer, *Phys. Rev. Lett.* **22**, 918 (1969); S.-J. Gu, S. K. Saha, and G. K. Wong, *Mol. Cryst. Liq. Cryst.* **69**, 287 (1981).
15. S. M. Arakelyan, G. L. Grigoryan, S. Ts. Nersisyan, M. A. Nshanyan, and Yu. S. Chilingaryan, *Pis'ma Zh. Eksp. Teor. Fiz.* **28**, 202 (1978); *JETP Lett. (Engl. Transl.)* **28**, 186 (1978).
16. B. L. Davydov, L. D. Derkacheva, V. V. Duna, M. E. Zhabotinskii, V. F. Zolin, L. G. Kereneva, and M. A. Samokhina, *Pisma Zh. Eksp. Teor. Fiz.* **12**, 24 (1970); *JETP Lett. (Engl. Transl.)* **12** 16 (1970).
17. B. F. Levine and C. G. Bethen, *J. Chem. Phys.* **63**, 2666 (1975); J. L. Oudar, *ibid.* **67**, 446 (1977); see also Chemla and Zyss (1).
18. A detailed review of recent work is I. Ledoux and J. Zyss, *Int. J. Nonlinear Opt. Phys.* **3**, No. 2 (1994).
19. I. P. Pinkevich, Y. A. Reznikov, V. Y. Reshetnyak, and O. V. Yaroshchuk, *Int. J. Nonlinear Opt. Phys.* **1**, No. 3, (1992); see also I. Janossy, L. Csillag, and A. D. Lloyd, *Phys. Rev. A* **44**, 8410 (1992).
20. I. C. Khoo, *Mol. Cryst. Liq. Cryst.* **207**, 317 (1991).

# CHAPTER 9

# INTRODUCTION TO NONLINEAR OPTICS

## 9.1 NONLINEAR SUSCEPTIBILITY AND INTENSITY-DEPENDENT REFRACTIVE INDEX

In Chapter 2 we discussed the refractive indices of liquid crystals in terms of the induced polarization **P** and the optical electric field **E**. There **P** is linearly related to **E**. Generally speaking, a material is said to be optically nonlinear when the induced polarization **P** is not linearly dependent on **E**. This could happen if the optical field is very intense. It could also happen if the material's physical properties are easily perturbed by the optical field. Both types of phenomena, that is, those associated with an intense field and those associated with the material's "easy" susceptibility to perturbation, are found in liquid crystals. In this chapter we describe the theoretical framework for studying these nonlinear optical phenomena. Nonlinear optical phenomena in liquid crystals will be presented in Chapter 10.

There are two basic approaches. In many systems (e.g., atoms, molecules, and semiconductors), the primary processes responsible for nonlinear polarizations are associated with electronic transitions. To describe such processes and obtain the correct polarization, it is necessary to employ quantum mechanical theories. On the other hand, many processes are essentially classical in nature. In liquid crystals, for example, processes such as thermal and density effects, molecular orientations, flows, and electrostrictive effects require only classical mechanics and electromagnetic theories. In this chapter the fundamentals of nonlinear optics are described within the framework of classical electromagnetic theories. Some of the quantum mechanical aspects of electronic nonlinearities were given in Chapter 8.

### 9.1.1 Nonlinear Polarization and Refractive Index

All optical phenomena occurring in a material arise from the optical field-induced polarization **P**. In general, the total polarization **P** may be written in the form

$$\mathbf{P} = \mathbf{P}_L + \mathbf{P}_{NL} \tag{9.1}$$

where the subscripts L and NL denote linear and nonlinear responses, respectively.

We discussed the linear part of the polarization $\mathbf{P}_L$ in Chapter 3. Following the same procedure for treating the linear polarization, the wave equation for the case when the nonlinear part of the polarization **P**, $\mathbf{P}_{NL}$ in (9.1), is also included can be simply derived to yield

$$\nabla^2 E_i - \mu_i \varepsilon_i \frac{\partial^2 E_i}{\partial t^2} = \mu_0 \frac{\partial^2 P_{NL}^{(i)}}{\partial t} \tag{9.2}$$

Since $P_{NL}$ is not linear in $E_i$, its effects on the propagation go far beyond the simple velocity change caused by the linear polarization term $P_L$ discussed before. Nevertheless, let us first consider here the simple but very important case where the effect of the nonlinear polarization is manifested in the form of a change in the refractive index. This is in analogy to the case of linear optics, where the refractive index (unity) of the vacuum is modified to a value ($n_i$) owing to the linear polarization. To see this more clearly, consider, for example, a commonly occurring $\mathbf{P}_{NL}$ that is proportional to the third power of $\mathbf{E}_i$ in the form

$$\mathbf{P}_{NL} = \chi_{NL} \langle E_i^2 \rangle \mathbf{E}_i \tag{9.3}$$

Substituting (9.3) into (9.2) and assuming a plane wave form for $E_i(\mathbf{r}, t)$ and $\mathbf{P}_{NL}$, that is,

$$E_i(\mathbf{r}, t) = E_i \exp i(\mathbf{k} \cdot \mathbf{r} - \omega t) \tag{9.4a}$$

$$P_{NL}(\mathbf{r}, t) = P_{NL} \exp i(\mathbf{k} \cdot \mathbf{r} - \omega t) \tag{9.4b}$$

we have

$$\nabla^2 E_i(\mathbf{r}, t) - \mu_i' \varepsilon_i' \frac{\partial^2 E_i}{\partial t^2}(\mathbf{r}, t) = 0 \tag{9.5}$$

where

$$\mu_i' \varepsilon_i' = \mu_i \varepsilon_i + \mu_0 \chi_{NL} \langle E_i^2 \rangle \tag{9.6}$$

Equation (9.6) allows us to define an effective optical dielectric constant $\varepsilon'$ given by

$$\varepsilon' = \varepsilon_i + \chi_{\mathrm{NL}} \langle E_i^2 \rangle \tag{9.7}$$

where we have let $\mu'_i = \mu_i = \mu_0$ (for nonmagnetic materials). The effective refractive index $n$ is therefore given by

$$n^2 = \frac{\mu'_i \varepsilon'_i}{\mu_0 \varepsilon_0} = n_0^2 + \frac{\chi_{\mathrm{NL}} \langle E_i^2 \rangle}{\varepsilon_0} \tag{9.8}$$

For the typical case where $\chi_{\mathrm{NL}} \langle E \rangle^2 / \varepsilon_0 \ll n_0^2$, we can approximate $n$ in (9.8) by

$$n \simeq n_0 + n_2(E) \langle E_i^2 \rangle \tag{9.9}$$

where

$$n_2(E) = \frac{\chi_{\mathrm{NL}}}{2\varepsilon_0 n_0} \tag{9.10}$$

In the current literature $n_2(E)$ is often referred to as the nonlinear coefficient. Since $|E|^2$ is related to the optical intensity $I_{\mathrm{op}}$:

$$I_{\mathrm{op}} = \frac{\varepsilon_0 n c}{2} \langle E^2 \rangle \tag{9.11}$$

equation (9.9) may also be written in the form

$$n = n_0 + n_2(I) I_{\mathrm{op}} \tag{9.12}$$

with

$$n_2(E) = n_2(I) \left( \frac{n_0 \varepsilon_0 c}{2} \right) \tag{9.13}$$

In SI units the factor $\varepsilon_0 c/2$ amounts to $1/753$. It is important to note here that the special form of the nonlinear polarization [cf. equation (9.3)], as well as the associated nonlinear coefficient $n_2(I)$ or $n_2(E)$, is appropriate for describing only a specific class of nonlinear optical phenomena. In the next section we will describe other more general forms of nonlinear polarization. For the present, however, we will continue with this form of polarization in discussing some conventional usage and systems of units.

### 9.1.2 Nonlinear Coefficient and Units

In electrostatic units, the nonlinear polarization is often written as

$$P_i = \chi_{\text{esu}}^{(3)} E_i E_i^* E_i \tag{9.14}$$

where the superscript (3) on $\chi_{\text{esu}}$ signifies that this is a third-order nonlinear polarization; that is, the resulting polarization involves a triple product of the field $E$.

The corresponding polarization in SI units involves a third-order susceptibility $\chi_{\text{SI}}^{(3)}$, that is,

$$P_i = \chi_{\text{SI}}^{(3)} E_i E_i^* E_i \tag{9.15}$$

Since $P_i(\text{esu})/P_i(\text{SI}) = 3 \times 10^5$ and $E_i(\text{esu}) = (3 \times 10^4)^{-1} E_i(\text{SI})$, we have

$$\chi_{\text{esu}}^{(3)} = 3^4 \times 10^{17} \chi_{\text{SI}}^{(3)} \tag{9.16}$$

Also, from (9.10) and noting that $\langle E^2 \rangle = |E_i|^2/2$ (using $\langle \cos^2 \omega t \rangle = \frac{1}{2}$), we have the following relationship:

$$\chi_{\text{SI}}^{(3)} = \varepsilon_0 n_0 n_2(E) \tag{9.17}$$

The difference between $\chi_{\text{SI}}^{(3)}$ here and $\chi_{\text{NL}}$ in (9.10) is due to the different form in which the nonlinear polarization $P_{\text{NL}}$ is written [cf. equations (9.15) and (9.3)] and the fact that $\langle E_i^2 \rangle = (E_i E_i^*)/2$.

Another useful relationship to know is the one between the nonlinear coefficient $n_2(I)$ (usually expressed in standard units) and the nonlinear susceptibility $\chi^{(3)}$ (usually quoted in the literature in electrostatic units). Using (9.16), (9.17), and (9.13), we have

$$\begin{aligned} \chi_{\text{esu}}^{(3)} &= 3^4 \times 10^{17} \chi_{\text{SI}}^{(3)} = 3^4 \times 10^{17} \times \varepsilon_0 n_0^2 n_2(I) \cdot \tfrac{1}{753} \\ &= \left[9.54 \times 10^4 n_0^2\right] n_2(I) \end{aligned} \tag{9.18}$$

where $n_2(I)$ is in square meters per watt.

If one uses the usual "mixed" unit of square centimeters per watt for $n_2(I)$ (owing to the prevalent usage of the "mixed" unit for the optical intensity $I$ in watts per square centimeter), then $\chi_{\text{esu}}^{(3)}$ is given by, from (9.18),

$$\chi_{\text{esu}}^{(3)} = 9.54 n_0^2 n_2(I)_{\text{mixed}} \tag{9.19}$$

where the unit for $n_2(I)_{\text{mixed}}$ is square centimeters per watt.

In liquid crystalline research mixed systems of units have been employed by various workers as some of the numerical examples in the preceding chapter have demonstrated. A mastery of these unit conversion techniques is almost a prerequisite in the study of liquid crystals and/or nonlinear optics.

## 9.2 GENERAL NONLINEAR POLARIZATION AND SUSCEPTIBILITY

As we mentioned in the beginning of this chapter, in the electromagnetic approach to nonlinear optics, all quantum mechanical aspects (e.g., electronic transitions and relaxations, energy level population changes, etc.) are not explicitly considered; the nonlinear response of a medium to an applied electromagnetic field is described by a functional dependence of the form

$$P(\mathbf{r};t) = f(\mathbf{E}(\mathbf{r},t),\mathbf{E}(\mathbf{r}',t')) \tag{9.20}$$

The nonlinear response at the space–time point $(\mathbf{r},t)$ is due not only to the action of the applied fields at the same space–time point, but also of fields at a different space–time point $(\mathbf{r}',t')$. In other words, the response is a *nonlocal* function of time and space.

A more explicit form for $\mathbf{P}(\mathbf{r},t)$ can be obtained if we make some simplifying assumptions.

An approximation that is often made is the expression of the resulting polarization as a power series in the fields. We have

$$\mathbf{P}(\mathbf{r},t) = \mathbf{P}^{(1)}(\mathbf{r},t) + \mathbf{P}^{(2)}(\mathbf{r},t) + \mathbf{P}^{(3)}(\mathbf{r},t) + \cdots \tag{9.21}$$

The first term is the linear polarization discussed before and given by

$$\mathbf{P}^{(1)}(\mathbf{r},t) = \varepsilon_0 \int_{-\infty}^{\infty} \overline{\overline{\chi}}^{(1)}(\mathbf{r}-\mathbf{r}',t-t') : \mathbf{E}(\mathbf{r}',t')\, d^3\mathbf{r}'\, dt' \tag{9.22}$$

where the double dots between $\overline{\overline{\chi}}^{(1)}$ and $\mathbf{E}$ signify a tensorial operation. Note that $\chi$ is a second-rank tensor, and $E$, a first-rank tensor.

Upon Fourier transformation, this gives

$$\mathbf{P}(\mathbf{k};\omega) = \varepsilon_0 \chi^{(1)}(\mathbf{k};\omega) \cdot \mathbf{E}(\mathbf{k};\omega) \tag{9.23}$$

which is the generalized nonlocal version of $(3.30a)$.

In the same token, the second-order polarization term $\mathbf{P}^{(2)}(\mathbf{r},t)$ can be written as

$$\mathbf{P}^{(2)}(\mathbf{r},t) = \int_{-\infty}^{\infty} \chi^{(2)}(\mathbf{r}-\mathbf{r}',t-t';\mathbf{r}-\mathbf{r}'',t-t'') : \mathbf{E}(\mathbf{r}',t')$$
$$\times \mathbf{E}(\mathbf{r}'',t'')\, d\mathbf{r}'\, d\mathbf{r}''\, dt'\, dt'' \tag{9.24}$$

where $\chi^{(2)}$ is a third-rank tensor. Upon Fourier transformation, this gives

$$P^{(2)}(\mathbf{k},\omega) = \chi^{(2)}(\mathbf{k},\mathbf{k}',\mathbf{k}'';\omega,\omega',\omega'') : \mathbf{E}(\mathbf{k}'_x,\omega')\mathbf{E}(\mathbf{k}'',\omega'') \tag{9.25}$$

Similarly, we have

$$\mathbf{P}^{(3)}(\mathbf{r},t)=\int_{-\infty}^{\infty}\chi^{(3)}(\mathbf{r}-\mathbf{r}',t-t';\mathbf{r}-\mathbf{r}'',t-t'';\mathbf{r}-\mathbf{r}''',t-t'''):$$
$$\mathbf{E}(\mathbf{r}',t')\mathbf{E}(\mathbf{r}'',t'')\mathbf{E}(\mathbf{r}''',t''')\,d\mathbf{r}',d\mathbf{r}''\,d\mathbf{r}'''\,dt'\,dt''\,dt''' \quad (9.26)$$

where $\chi^{(3)}$ is a fourth-rank tensor, and

$$\mathbf{P}^{(3)}(\mathbf{k},\omega)=\chi^{(2)}(\mathbf{k},\mathbf{k}',\mathbf{k}'',\mathbf{k}''';\omega,\omega',\omega'',\omega'''):\mathbf{E}(\mathbf{k}',\omega')\mathbf{E}(\mathbf{k}'',\omega'')\mathbf{E}(\mathbf{k}''',\omega''') \quad (9.27)$$

and so on for $\mathbf{P}^{(n)}, \chi^{(n)}$.

If the $n$th-order nonlinear susceptibility $\chi^{(n)}(\mathbf{r},t)$ is independent of $\mathbf{r}$, the Fourier transform $\chi^{(n)}(\mathbf{k},\omega)$ will be independent of $k$. In this sense the nonlinear response is termed local. Also, if the response of the medium is *instantaneous*, the expressions for the nonlinear polarizations $P^{(2)}$, $P^{(3)}$ and so on become simple tensorial products of $E$:

$$P^{(2)}(t)=\chi^{(2)}:E(t)E(t) \quad (9.28)$$

and

$$P^{(3)}(t)=\chi^{(3)}:E(t)E(t)E(t) \quad (9.29)$$

## 9.3 CONVENTION AND SYMMETRY

Since the nonlinear polarization $\mathbf{P}_{\mathrm{NL}}$ involves various powers of the optical electric field (in the form **EE**, **EEE**, etc.) and if the total electric field **E** comprises many frequency components [e.g., $E=E(\omega_1)+E(\omega_2)+\cdots E(\omega_n)$], there are several possible combinations of the frequency components that will contribute to a particular frequency component for $P_{\mathrm{NL}}$.

As an example consider the second-order polarization $P^{(2)}$. Also, let us ignore all tensorial relationships and focus our attention on copolarized fields for simplicity. One can see that the frequency component $\omega_3=\omega_1+\omega_2$ in $P_{\mathrm{NL}}$ can arise from two possible combinations: $E(\omega_1)E(\omega_2)$ and $E(\omega_2)E(\omega_1)$.

In actual calculation, it is easier to adopt the convention similar to that mentioned in Shen (1) and Boyd (2), where these various combinations are correctly distinguished by distinctive labeling of the nonlinear susceptibility $\chi$ involved. For example, the combinations $E(\omega_1)E(\omega_2)$ and $E(\omega_2)E(\omega_1)$ are associated with $\chi(-\omega_3,\omega_1,\omega_2)$ and $\chi(-\omega_3,\omega_2,\omega_1)$, respectively. In this case we need not invoke the degeneracy factor $g$ in the definition of the nonlinear polarization. Also, it will account for possible differences between $\chi(-\omega_3,\omega_1,\omega_2)$ and $\chi(-\omega_3,\omega_2,\omega_1)$.

To express the relationships between $P_{\text{NL}}$ and $E$ in a more detailed form, we now write the electric field in terms of their spatial coordinate components:

$$\mathbf{E} = \hat{\imath}E_i + \hat{\jmath}E_j + \hat{k}E_k \tag{9.30$a$}$$

$$\mathbf{P}_{\text{NL}} = \hat{\imath}P_i + \hat{\jmath}P_j + \hat{k}P_k \tag{9.30$b$}$$

Also, the $E_l$'s and $P_l$'s $(l = i, j, k)$ are expressed in the plane wave form

$$E_l = \sum_{\omega_n} \frac{1}{2}\left[A_l^{\omega_n}(\mathbf{r}, t)e^{i(\mathbf{k}\cdot\mathbf{r} - \omega_n t)} + \text{c.c.}\right] \tag{9.31$a$}$$

$$P_l = \sum_{\omega_m} \frac{1}{2}\left[P_l^{\omega_m}(\mathbf{r}, t)e^{i(\mathbf{k}\cdot\mathbf{r} - \omega_m t)} + \text{c.c.}\right] \tag{9.31$b$}$$

The nonlinear polarization $P_i^{\omega_3}$, for example, involving the mixing of two frequency components $\omega_1$ and $\omega_2$ to produce a third wave of frequency $\omega_3$ (see Figure 9.1$a$), will thus be given by

$$\begin{aligned} \tfrac{1}{2}P_i^{\omega_3} = 2^{-2}\big[&\chi_{ijk}(-\omega_3, \omega_1, \omega_2)A_j^{\omega_1}A_k^{\omega_2} \\ &+ \chi_{ijk}(-\omega_3, \omega_2, \omega_1)A_j^{\omega_2}A_k^{\omega_1}\big] \end{aligned} \tag{9.32}$$

On the right-hand side of (9.32), the two terms in the square brackets come from the two possible ways of ordering $\omega_1$ and $\omega_2$ (i.e., we can have $A_j^{\omega_1}A_k^{\omega_2}$ and $A_j^{\omega_2}A_k^{\omega_1}$). The denominator $2^2$ comes from the factor $\frac{1}{2}$ involved in the definition of $A$ [equations (9.31$a$) and (9.31$b$)]. The subscripts $i$, $j$, $k$, and so forth signify the corresponding Cartesian components of the field. The frequencies $\omega_1$, $\omega_2$, and $\omega_3$ appearing in the parentheses following $\chi_{ijk}$ represent a conventional way of expressing the fact that $\omega_3 = \omega_1 + \omega_2$ (i.e., $-\omega_3 + \omega_1 + \omega_2 = 0$). On the left-hand side, the factor $\frac{1}{2}$ comes from the definition of the amplitude $P_l^{\omega_m}$ in (9.31$b$). With these factors properly

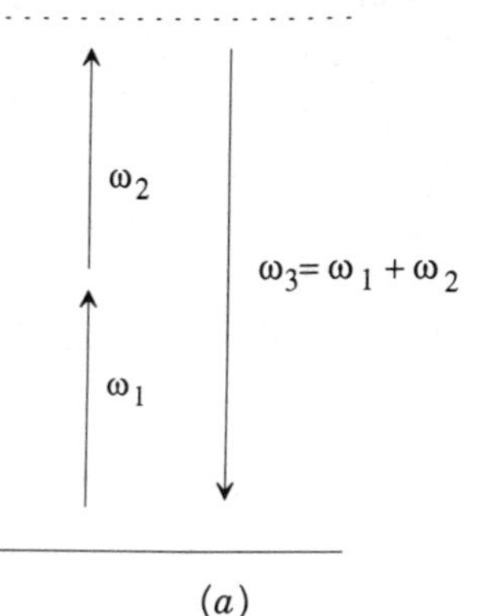


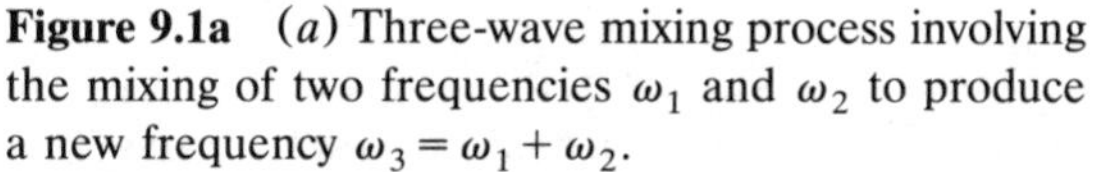
**Figure 9.1a** ($a$) Three-wave mixing process involving the mixing of two frequencies $\omega_1$ and $\omega_2$ to produce a new frequency $\omega_3 = \omega_1 + \omega_2$.

accounted for, (9.32) reads

$$P_i^{\omega_3} = \tfrac{1}{2}\big[\chi_{ijk}^{(2)}(-\omega_3, \omega_1, \omega_2)\, A_j^{\omega_1} A_k^{\omega_2} + \chi_{ijk}^{(2)}(-\omega_3, \omega_2, \omega_1)\, A_j^{\omega_2} A_k^{\omega_1}\big] \tag{9.33}$$

where $\chi_{ijk}^{(2)}$ is a third-rank tensor.

If $\omega_1 = \omega_2 = \omega$ (i.e., $\omega_3 = \omega_1 + \omega_2 = 2\omega$, second-harmonic generation), then the number of distinct permutations of $\omega_1$ and $\omega_2$ is reduced to one. We therefore have

$$P_i^{2\omega} = \tfrac{1}{2}\chi_{ijk}^{(2)}(-2\omega; \omega, \omega)\, A_j^{\omega} A_k^{\omega} \tag{9.34}$$

It is important to notice here that a totally different process is involved if we have $\omega_1 = \omega$ and $\omega_2 = -\omega$. This process is called optical rectification, since it results in a dc response with $\omega_3 = \omega_1 + \omega_2 = 0$. In this case the number of distinct permutations of $\omega_1$ and $\omega_2$ is two. It is also important here to remind ourselves that $i$, $j$, and $k$ refer to the coordinate axes, and that in (9.34) and other equations involving tensorial relationships, a sum over nonrepeated indices is implied [i.e., on the right-hand side of (9.34), for example, one needs to sum over $j = 1,2,3$ and $k = 1,2,3$.]

In a similar manner, the nonlinear polarization component $\omega_4$ generated by the mixing of three waves $\omega_1$, $\omega_2$, and $\omega_3$ (i.e., $\omega_4 = \omega_1 + \omega_2 + \omega_3$) is given by

$$P_i^{\omega_4} = \frac{1}{2^2}\big[\chi_{ijkl}^{(3)}(-\omega_4, \omega_1, \omega_2, \omega_3)\, A_j^{\omega_1} A_k^{\omega_2} A_l^{\omega_3} + \cdots \tag{9.35}$$

where $\chi_{ijkl}^{(3)}$ is a fourth-rank tensor (see Figure 9.1*b*). With a little practice, the reader can readily write down the corresponding nonlinear polarization

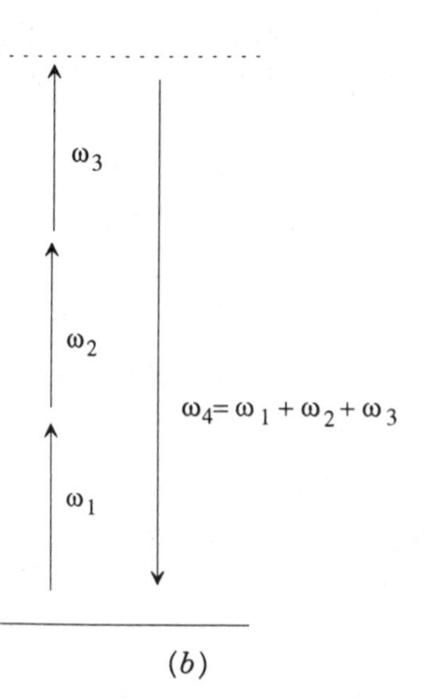


**Figure 9.1b** (*b*) Four-wave mixing process involving $\omega_1$, $\omega_2$, and $\omega_3$ to produce $w_4 = w_1 + w_2 + w_3$.

components for $n$-wave mixing processes. It is important to note here that $\chi^{(3)}_{ijkl}$ is related to $\chi^{(3)}_{SI}$ [cf. equation (9.15)] by

$$\frac{1}{2^2}\chi^{(3)}_{ijkl} = \chi^{(3)}_{SI} \tag{9.36}$$

This relationship, generated by the different plane wave forms used for expressing the field and polarization [cf. equations (9.4) and (9.31)], *should be borne in mind* as one compares values of the nonlinear coefficients quoted or reported in the literature.

In accordance with current convention, $n$ stands for the total number of interacting waves (counting the incident and the generated waves) within the nonlinear media (Figure 9.2). Therefore, interactions involving nonlinear polarization of the form given in (9.33) and (9.34), where two incident waves combine to give a generated wave, are called three-wave mixing processes. Processes associated with the combination of three waves to yield a fourth one [cf. equation (9.35)] are called four-wave mixing processes. If all the frequencies involved are the same, the processes are called degenerate. We have, for example, degenerate four-wave mixing; otherwise they are called nondegenerate wave mixings.

In Cartesian coordinates obviously there are altogether $3^4$ elements in the third-order susceptibility $\chi^{(3)}_{ijkl}$, a fourth-rank tensor, since $(i, j, k, l)$ each has three components 1,2,3. In an isotropic medium with inversion symmetry, however, it can be shown (3) that there are only four different components, three of which are independent:

$$\begin{aligned}
&\chi_{1111} = \chi_{2222} = \chi_{3333} \\
&\chi_{1122} = \chi_{1133} = \chi_{2211} = \chi_{2233} = \chi_{3311} = \chi_{3322} \\
&\chi_{1212} = \chi_{1313} = \chi_{2121} = \chi_{2323} = \chi_{3232} = \chi_{3131} \\
&\chi_{1221} = \chi_{1331} = \chi_{2112} = \chi_{2312} = \chi_{3113} = \chi_{3223} \\
&\chi_{1111} = \chi_{1122} + \chi_{1212} + \chi_{1221}
\end{aligned} \tag{9.37}$$

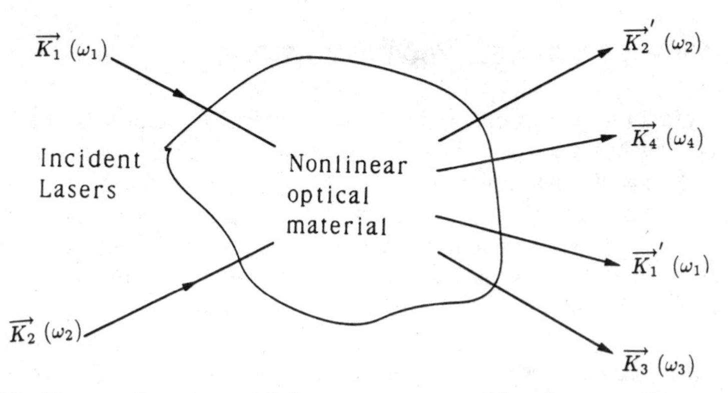


**Figure 9.2** Temporal and spatial frequency wave mixing in a nonlinear optical material.

In an anisotropic medium such as a liquid crystal, these symmetry considerations can be very complex (2).

These microscopic susceptibilities are derived from explicit calculations of the dipolar interaction between a particular atom or molecule with the incident optical fields (cf. Chapter 8). In the presence of other atoms or molecules, and their polarizations, an atom or molecule will experience a net depolarization field in addition to the incident optical field (4). In condensed matter or inhomogeneous media, these depolarization fields are quite complex and, in general, very difficult to estimate. On the other hand, in simple liquids or gases (i.e., isotropic, homogeneous media), the depolarization field may be accounted for by multiplying each of the field components by a so-called Lorentz correction factor (2, 5):

$$L = \frac{\varepsilon(\omega_i) + 2}{3} \tag{9.38}$$

The macroscopic susceptibilities $\chi^{(3)}$, $\chi^{(2)}$, and $\chi^{(1)}$ discussed previously are therefore related to the microscopic susceptibilities by

$$\chi^{(1)} = \left(\frac{\varepsilon(\omega) + 2}{3}\right)^2 \chi_{\text{mic}}^{(1)}(\omega) \tag{9.39}$$

$$\chi_{\text{mac}}^{(3)} = \left[\frac{\varepsilon(\omega_4) + 2}{3}\right]\left[\frac{\varepsilon(\omega_1) + 2}{3}\right]\left[\frac{\varepsilon(\omega_2) + 2}{3}\right]\left[\frac{\varepsilon(\omega_3) + 2}{3}\right] \times \chi_{\text{mic}}^{(3)}(-\omega_4, \omega_1, \omega_2, \omega_3) \tag{9.40}$$

In liquid crystalline media, owing to large molecular correlation effects, these local field effects are considerably more complex (6). Several formalisms have been developed to treat the various phases; some of these studies are pointed out in Chapter 8 and in Khoo and Wu (6).

## 9.4 COUPLED MAXWELL WAVE EQUATIONS

The interaction of various optical fields inside a nonlinear medium are described by Maxwell equations. For $n$-wave mixing processes, there are $n$ coupled Maxwell wave equations.

The solutions for the optical electric fields at any arbitrary penetration depth into the nonlinear medium depend on the initial boundary conditions on the incident field intensity and their phases and boundary conditions. Usually, in order to solve the equations in a meaningful way, some approximations are made.

An approximation that is generally valid is the so-called slowly varying envelope approximation. In effect, a particular frequency component $E_j$ of the total electric field can be represented by a product of a fast (temporally and spatially) oscillatory term $\exp[i(\mathbf{k}_j \cdot \mathbf{r} - \omega_j t)]$ and an amplitude function $A_j(\mathbf{r}, t)$ whose temporal and spatial variations are on a much slower scale. Therefore, we have

$$E_j = \tfrac{1}{2}\left[A_j(\mathbf{r};t)\exp\left[i(\mathbf{k}_j \cdot \mathbf{r} - \omega_j t)\right] + \text{c.c.}\right] \qquad j = 1, \ldots n \tag{9.41}$$

where $n$ is the number of frequency components present. Although these $n$ interacting fields, in general, propagate in $n$ different directions, one can define a reference direction, usually denoted as the $z$ direction, to describe their interactions and propagation. Accordingly, the fast oscillatory term $\exp[i(\mathbf{k}_j \cdot \mathbf{r} - \omega t)]$ may be written as $\exp[i(k_j z - \omega t)]$, where $k_j$ is now understood as the $z$ component of the wave vector $\mathbf{k}_j$.

In more quantitative terms, then, the slowly varying envelope approximation (SVEA) translates into the following inequalities:

$$\left|\frac{\partial^2 A_j}{\partial z^2}\right| \ll \left|k_j \frac{\partial A_j}{\partial z}\right| \tag{9.42$a$}$$

$$\left|\frac{\partial^2 A_j}{\partial t^2}\right| \ll \left|\omega_j \frac{\partial A_j}{\partial t}\right| \tag{9.42$b$}$$

In the slowly varying approximation, the corresponding nonlinear polarization vector components are expressed as

$$P_{\text{NL}}^{(j)} = \tfrac{1}{2}\left[P_j(\mathbf{r};t)e^{i(k_j^p z - \omega_j t)} + \text{c.c.}\right] \tag{9.43}$$

that is, the total nonlinear polarization, $P_{\text{NL}} = \Sigma_j P_{\text{NL}}^{(j)}$, contains $n$ frequency components. The superscript $p$ on $k_j^p$ in (9.43) denotes that it is the wave vector associated with the $j$th component of the *polarization*; in general, it is *different* from the wave vector of the $j$th component of the electric field, as we will see presently.

Substituting (9.41) to (9.43) into the wave equation (9.2) which we rewrite here

$$\nabla^2 E - \mu\varepsilon \frac{\partial^2 E}{\partial t^2} = \mu_0 \frac{\partial^2 P}{\partial t^2} \tag{9.44}$$

we obtain

$$\nabla_\perp^2 A_j + 2ik_j \frac{\partial A_j}{\partial z} + 2i\mu\varepsilon\omega \frac{\partial A_j}{\partial t} = -\mu_0 \omega^2 P_j^{\text{NL}} \exp i\,\Delta k_j z \tag{9.45}$$

where $\Delta k_j = k_j^p - k_j$ is the wave vector mismatch for the $j$th wave. Also, we have separated the three-dimensional $\nabla^2$ into a transverse term $(\nabla_\perp^2)$ and a longitudinal term $(\partial^2/\partial z^2)$. Since $k_j = \omega_j / v_j$, (9.45) becomes

$$\nabla_\perp^2 A_j + 2ik_j\left(\frac{\partial A_j}{\partial z} + \frac{1}{v_j}\frac{\partial A_j}{\partial t}\right) = -\mu_0\omega^2 P_j^{\mathrm{NL}} \exp i\,\Delta k_j z \qquad (9.46)$$

If we transform this $(z,t)$ coordinate system to one that moves at a velocity $v_j$ along $z$, that is, $(z',t')$ given by

$$z' = z \qquad t' = t - \frac{z}{v_j} \qquad (9.47)$$

(9.46) becomes

$$\nabla_\perp^2 A_j + 2ik_j\frac{\partial A_j}{\partial z} = -\mu_0\omega^2 P_j^{\mathrm{NL}} \exp i\,\Delta k_j z \qquad (9.48)$$

for $j = 1, 2\ldots$ (the number of interacting waves), where we have written $z'$ as $z$ as they are the same.

In deriving (9.48) we have also assumed that the process under study is in a steady state; the time dependence of the process is "removed" by the substitution of $t' = t - z/v_j$ [cf. equations (9.46) and (9.47)]. In many actual experiments, especially in those involving pulsed lasers, this is definitely not a correct assumption, and one has to revert to using the full Maxwell equation.

## 9.5 NONLINEAR OPTICAL PHENOMENA

Numerous distinctly different nonlinear optical phenomena (1, 2) can be created via the nonlinear polarizations discussed previously. Here we discuss some widely studied ones.

### 9.5.1 Stationary Degenerate Four-Wave Mixings

Figure 9.3 shows a typical interaction geometry. Two equal-frequency laser beams $\mathbf{E}_1$ and $\mathbf{E}_2$ are incident on a nonlinear optical material of thickness $d$; they intersect at an angle $\theta$ (within the nonlinear material). On the exit side, there are new, generated waves $E_3$, $E_4$, and so forth of the same frequency on the side of the beams $E_1$ and $E_2$. These generated waves are due to spatial mixings of the input waves.

For example, the wave $E_3$ propagating in the $\mathbf{k}_3$ direction is due to a nonlinear polarization term $P_3$ of the form $P_3 \sim E_1 E_2^* E_1$, with an associated wave vector mismatch $\Delta\mathbf{k}_3 = \mathbf{k}_3 - (\mathbf{k}_1 + \mathbf{k}_1 - \mathbf{k}_2)$ (cf. Figure 9.4).

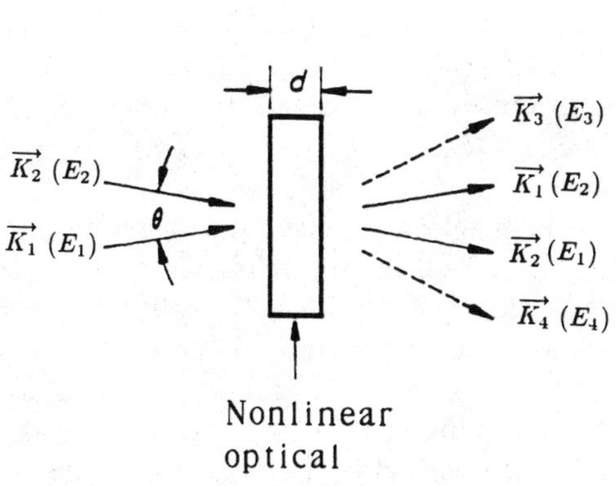


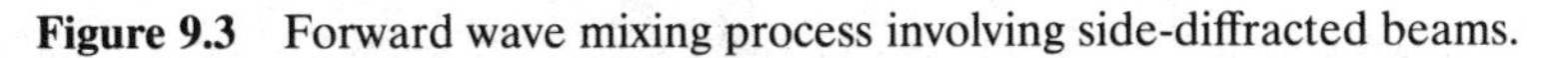
**Figure 9.3** Forward wave mixing process involving side-diffracted beams.

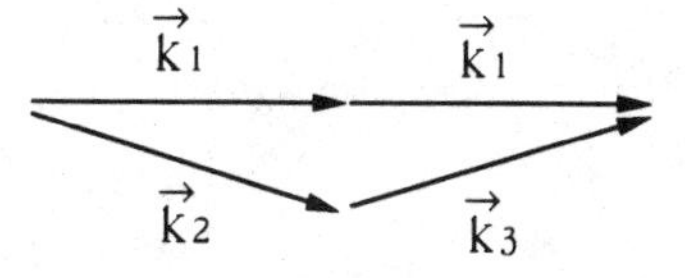


**Figure 9.4** Wave vector addition diagram for the process $\mathbf{k}_3 = 2\mathbf{k}_1 - \mathbf{k}_2$.

Similarly, the wave $E_4$ is due to a nonlinear polarization $P_4$ of the form $P_4 \sim E_2 E_2^* E_1$, with an associated wave vector mismatch $\Delta \mathbf{k}_4 = \mathbf{k}_4 - (2\mathbf{k}_2 - \mathbf{k}_1)$ (cf. Figure 9.5).

To illustrate the physics, we assume that all waves are polarized in the $\hat{x}$ direction. In plane wave form the total electric field **E** and the nonlinear polarization **P** are given by

$$\mathbf{E} = \sum_{j=1}^{4} \mathbf{E}_j = \hat{x} \sum_{j=1}^{4} \frac{1}{2}\left[A_j \exp i(k_j \cdot z - \omega t) + \text{c.c.}\right] \tag{9.49}$$

and

$$\mathbf{P} = \sum_{n=1}^{4} \mathbf{P}_j = \hat{x} \sum_{j=1}^{4} \frac{1}{2}\left[P_j \exp i(k_j \cdot z - \omega t) + \text{c.c.}\right] \tag{9.50}$$

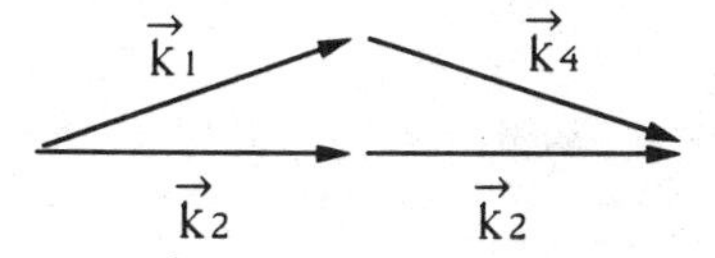


**Figure 9.5** Wave vector addition diagram for the process $\mathbf{k}_4 = 2\mathbf{k}_2 - \mathbf{k}_1$.

Furthermore, we impose a simplifying condition, namely, that the intensity of $E_1$ is much higher than the intensity of $E_2$; in this case the generated wave $E_3$ is much more intense than the wave $E_4$. Under this simplifying assumption, the terms contributing to the generation of waves in the $E_3$ direction, for example, are $(E_3E_3^*)E_3$, $(E_2E_2^*)E_3$, $(E_1E_1^*)E_3$, $(E_1E_2^*)E_3$, $(E_3E_1^*)E_1$, and $(E_3E_2^*)E_2$.

In writing down these terms, we have put in parentheses products of the electric field amplitude which represent interference among the waves. The terms $E_3E_3^*$, $E_2E_2^*$, and $E_1E_1^*$ are the so-called static or dc components (i.e., the intensity is spatially and temporally nonvarying). On the other hand, the terms $E_3E_1^*$ and $E_3E_2^*$ are the so-called (intensity) grating components, which are spatially periodic with spatial grating wave vectors given by $(\mathbf{k}_3-\mathbf{k}_1)$ and $(\mathbf{k}_3-\mathbf{k}_2)$, respectively. From this point of view, the term $(E_3E_1^*)E_1$, for example, is associated with the scattering of wave $E_1$ from the refractive index grating generated by the intensity grating $E_3E_1^*$ into the direction of $E_3$.

The effects associated with these static and grating terms become clearer when we write down the corresponding coupled Maxwell wave equations. From (9.48), ignoring the transverse Laplacian term, the equations for the amplitudes $A$ of these three waves can be derived to give

$$\frac{dA_1}{dz}=ig\left(|A_1|^2+2|A_2|^2+2|A_3|^2\right)A_1 + ig\left[(A_2A_1^*)A_3+(A_3A_1^*)A_2\right]\exp i\,\Delta k_3\,z \tag{9.51}$$

$$\frac{dA_2}{dz}=ig\left(|A_2|^2+2|A_1|^2+2|A_3|^2\right)A_2 + ig\left[(A_1A_3^*)\exp -i\,\Delta k_3\,z\right]A_1 \tag{9.52}$$

$$\frac{dA_3}{dz}=ig\left(|A_3|^2+2|A_1|^2+2|A_2|^2\right)A_3 + ig\left[(A_1A_2^*)\exp -i\,\Delta k z\right]A_1 \tag{9.53}$$

where

$$g=\frac{\mu_0\omega^2\chi_{xxxx}}{8} \tag{9.54}$$

$$\Delta k=|2\mathbf{k}_1-\mathbf{k}_2-\mathbf{k}_3| \tag{9.55}$$

For simplicity, we have let all the nonlinear coefficients associated with the various contributing terms ($E_3E_3^*E_3$, $E_3E_1^*E_1$, etc.) in these equations be equal to $\chi^{(3)}_{xxxx}$. A typical nonlinear polarization term (for $P_3^{NL}$) associated with $E_1 E_2^* E_1$ is given by, in terms of the amplitudes $A$,

$$P_3^{NL} = \frac{1}{4}\chi_{xxxx} A_1 A_2^* A_1 \tag{9.56}$$

Because of nonlinear coupling among the waves, (9.51) to (9.53), in general, cannot be solved analytically. However, if one assumes that the pump beam is so strong that its depletion is negligible (i.e., $|A_1|^2 = \text{const}$) and that $|A_1|^2 \gg |A_2|^2, |A_3|^2$ so that only the phase modulation caused by the pump beam (the term $-ig|A_1|^2$ on the right-hand side of these equations) is retained, (9.51) to (9.53) may be readily solved to yield

$$A_1(d) = A_1(0)\exp\left[ig|A_1(0)|^2 d\right] \tag{9.57}$$

$$A_2(d) = A_2(0)\exp\left[ig|A_1(0)|^2 d\right]\left[\cosh(qd) - i\frac{P}{q}\sinh(qd)\right] \tag{9.58}$$

and

$$A_3(d) = iA_2^*(0)\exp\left[ig|A_1(0)|^2 d\right]\left[\frac{p^{*2} + q^{*2}}{q^*}\sinh(q^* d)\right] \tag{9.59}$$

where

$$p = \frac{\Delta k}{2} - g|A_1(0)|^2 \tag{9.60}$$

$$q = \left[g|A_1(0)|^2 - \frac{\Delta k}{4}\right]^{1/2} \tag{9.61}$$

From these solutions one can also see that the intensity of the transmitted probe beam acquires a gain factor $G(d)$ given by

$$G(d) = \frac{I_2(d)}{I_2(0)} = \left|\cosh(qd) - \frac{iP}{q}\sinh(qd)\right|^2 \tag{9.62}$$

Notice that the generated wave $A_3(d)$ in (9.59) is proportional to the complex conjugate of the wave $A_2$; for this reason, it is sometimes called the forward phase conjugate of $A_2$. In the following section we will discuss another phase conjugation process.

### 9.5.2 Optical Phase Conjugation

Figure 9.6 shows a particular degenerate four-wave mixing process involving three input waves $E_1$, $E_2$, and $E_3$ and the generated wave $E_4$ which is counterpropagating to $E_3$. The wave $E_4$ can arise from various combinations of these three input waves.

For example, it could be due to the scattering of $E_1$ from the (refractive index) grating formed by the interference of $E_2$ and $E_3$ [i.e., a nonlinear polarization term of the form $(E_2E_3^*)E_1$]. Another possibility is the scattering of $E_2$ from the grating formed by the interference of $E_1$ and $E_3$ [i.e., a nonlinear polarization term of the form $(E_1E_3^*)E_2$].

Both processes carry a wave vector mismatch $\Delta\mathbf{k}_4 = \mathbf{k}_4 - (\mathbf{k}_1 + \mathbf{k}_2 - \mathbf{k}_3)$. Since $\mathbf{k}_2 = -\mathbf{k}_1$ and $\mathbf{k}_4 = -\mathbf{k}_3$, we have $\Delta\mathbf{k}_4 = 0$ (i.e., a perfectly phase-matched four-wave mixing process). However, the grating constant associated with $(E_2E_3^*)$ is much smaller than that associated with $(E_1E_3^*)$.

If the physical mechanism responsible for the formation of the refractive index grating is not dependent on the grating period, then these two terms will contribute equally to the generation of the wave $E_4$. Usually, owing to

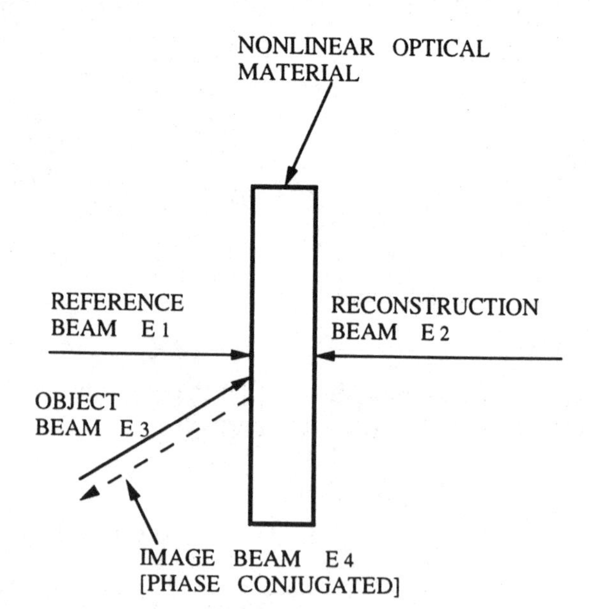


**Figure 9.6** Schematic depiction of optical wave front conjugation by degenerate four-wave mixing process.

diffusion or other intermolecular physical processes (e.g., in liquid crystals the torque exerted between molecules situated at the intensity minima on molecules at the intensity maxima or heat diffusion in the thermal grating), the larger is the grating period, the higher will be the wave mixing efficiency. In this case $(E_1 E_3^*)$ is the main contributing term.

Taking into account only the grating term $(E_1 E_3^*)$, the coupled equations for the four interacting waves within the nonlinear medium are given by

$$\frac{dA_1}{dz} = ig'\left[|A_1|^2 + 2|A_2|^2 + 2|A_3|^2 + 2|A_4|^2\right] + ig'\{\text{wave mixing terms}\} \tag{9.63}$$

$$\frac{dA_2}{dz} = -ig'\left[|A_2|^2 + 2|A_1|^2 + 2|A_3|^2 + 2|A_4|^2\right] - ig'\{\text{wave mixing terms}\} \tag{9.64}$$

$$\frac{dA_3}{dz} = ig\left[|A_3|^2 + 2|A_1|^2 + 2|A_2|^2 + 2|A_4|^2\right] + ig\{A_1 A_2 A_4^*\} \tag{9.65}$$

$$\frac{dA_4}{dz} = -ig\left[|A_4|^2 + 2|A_1|^2 + 2|A_2|^2 + 2|A_3|^2\right] - ig\{A_1 A_2 A_3^*\} \tag{9.66}$$

where $g = \mu_0 \omega^2 \chi / 8$, $g' = g/\cos\theta = \mu_0 \omega^2 \chi / 8\cos\theta$, and $\chi$ is the nonlinear susceptibility involved in this process.

As pointed out in the preceding section, the first terms (in square brackets) on the right-hand sides of these equations are phase modulation effects caused by the spatially static terms in the optical intensity function. Notice that, because of the presence of these phase modulations, the initially perfect phase-matching condition ($\Delta k = 0$) will be degraded as these waves interact and propagate deeper into the nonlinear medium.

To qualitatively examine the results and implications of optical phase conjugation, we ignore here the phase modulation terms (7). We also assume that both waves $E_1$ and $E_2$ (usually referred to as the pump waves) are very strong compared to the probe wave $E_3$ and the generated wave $E_4$ (i.e., $|A_1|$ and $|A_2|$ are constant). We thus have

$$\frac{dA_3}{dz} = igA_1 A_2 A_4^* \tag{9.67}$$

$$\frac{dA_4}{dz} = -igA_1 A_2 A_3^* \tag{9.68}$$

For the configuration depicted in Figure 9.6, the initial conditions are $A_3 = A_3(0)$ at $z = 0$ and $A_4 = 0$ at $z = L$, where $L$ is the thickness of the nonlinear material. The solutions for $A_3$ and $A_4$ are

$$A_3(z) = A_3(0)\frac{\cos[g|A_1A_2|(z-L)]}{\cos[g|A_1A_2|L]} \tag{9.69}$$

$$A_4(z) = -iA_3^*(0)\frac{\sin[g|A_1A_2|(z-L)]}{\cos[g|A_1A_2|L]} \tag{9.70}$$

These equations show the following:

- The transmitted probe beam $E_3$ is amplified by an amplitude gain factor

$$G_a = \frac{A_3(L)}{A_{3(0)}} = \sec[g|A_1A_2|L] \tag{9.71}$$

  just as in the previous four-wave mixing case.
- The generated wave $E_4$ is proportional to the complex conjugate of the probe beam $E_3$. This feature of the generated wave bears a resemblance to the result obtained in the previous section [cf. equation (9.59)]. In fact, they are both generated by the same kind of wave mixing process except that a different set of propagation wave vectors is involved. For the same reason, the wave $E_4$ is termed the optical phase conjugate of $E_3$. The generated wave $E_4$ carries a *phase* which is the conjugate of that carried by the input $E_3$. A consequence of this, which has been well demonstrated in the literature, is that if $E_3$ passes through a phase-aberrating medium, picking up a phase distortion factor $\exp i\phi_D$, the generated wave $E_4$ (if allowed to traverse the phase-aberrating medium backward) will pick up a phase factor $\exp - i\phi_D$. The net result is therefore the exact cancelation of these phase distortion factors in the signal beam $E_4$.

Another interesting feature of (9.69) and (9.70) is that if $g|A_1A_2|L = \pi/2$, both $A_3(z)$ and $A_4(z)$ will assume finite values for a vanishing input $A_3(0)$ (i.e., "oscillations" will occur). To put it another way, imagine that a cavity is formed with its optical propagation axis along $\mathbf{k}_3$ (or $\mathbf{k}_4$). Then any coherent noise generated in this direction may be set into oscillation.

An example of *coherent* noise is the scattering noise from the pump beam, which is coherent with respect to, and can therefore interfere with, the pump beam. Numerous reports in the current literature of self-pumped phase conjugators, oscillators, and so forth are based on this process or its variants (8). In Chapter 10, we describe two recent observations of self-starting phase conjugation using nematic liquid films.

### 9.5.3 Nearly Degenerate and Transient Wave Mixing

As pointed out before, laser-induced nonlinear polarizations amount to changes in the dielectric constant. If these dielectric constant changes are spatially coincident with the imparted intensity grating, the materials are said to possess local nonlinearity. In some situations, however, the induced dielectric constant change is spatially shifted from the optical intensity; that is, there is a phase shift $\phi$ between the dielectric constant and the intensity grating function (cf. Figure 9.7).

A phase shift between the intensity and index grating function can arise, for instance, if the frequencies of the two interfering waves $E_1$ and $E_2$ are not equal (i.e., $\omega_2 = \omega_1 + \Omega$) (cf. Figure 9.8). In this case the intensity grating imparted by $E_1$ and $E_2$ on the nonlinear medium is moving with time. Because of the finite response time $\tau$ of the medium, the induced index grating will be delayed in the steady state (i.e., a phase shift $\phi$ will occur between the intensity and the index grating).

The magnitude and sign of the phase shift depend on various material, optical, and geometrical parameters. For the purpose of our present

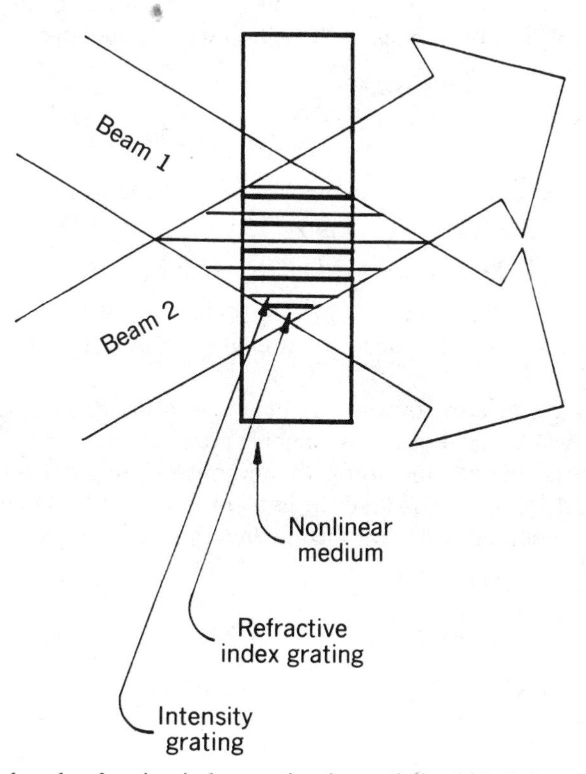


**Figure 9.7** Induced refractive index grating is spatially shifted from the imparted optical intensity grating (e.g., in photorefractive material).

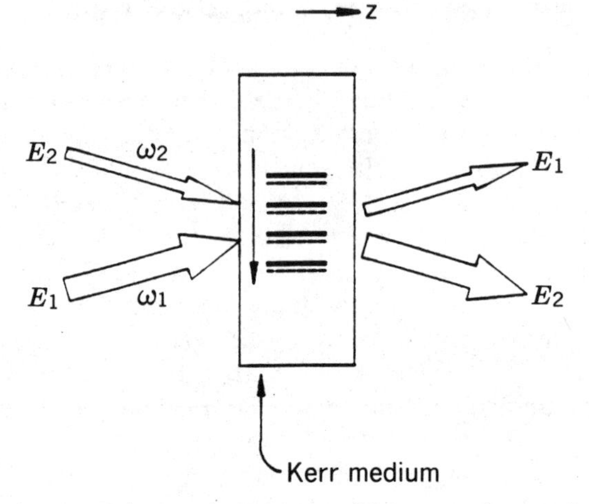


**Figure 9.8** Moving intensity grating generated by two coherent incident lasers of different frequency.

discussion on the consequence of having $\phi$ in the wave mixing process, we will represent it generally as $\phi$. If the total intensity function is given by

$$I_{\text{total}} \propto (E_1 + E_2) \cdot (E_1 + E_2)^* = |E_1|^2 + |E_2|^2 + E_1 E_2^* + E_2 E_1^* \quad (9.72)$$

then the induced index grating function will be of the form

$$\Delta n \sim \Delta n_0 + n_2 \left( E_1 E_2^* e^{i\phi} + E_2 E_1^* e^{-i\phi} \right) \quad (9.73)$$

where $\Delta n_0$ is the (spatially) static index change caused by the static intensity $(|E_1|^2 + |E_2|^2)$.

Consider the scattering of wave $E_1$ or $E_2$ from the induced index grating; one can associate third-order nonlinear polarization terms of the form $E_1^* E_2 E_1 e^{i\phi}$, $E_1 E_2^* E_2 e^{i\phi}$, and so on. If we limit our discussion to only these two waves and follow the procedure used in deriving (9.64) and (9.65), the corresponding coupled two-wave mixing equations are

$$\frac{dA_1}{dz} = ig\left[|A_1|^2 + |A_2|^2\right]A_1 + ig(A_2^* A_1)e^{i\phi}A_2 \quad (9.74)$$

$$\frac{dA_2}{dz} = ig\left[|A_1|^2 + |A_2|^2\right]A_2 + ig(A_1^* A_2)e^{-i\phi}A_1 \quad (9.75)$$

The effects of the phase shift $\phi$ on these two-wave couplings are more transparent if these equations are expressed in terms of the intensities $I_1$ and $I_2$ and individual phases $\phi_1$ and $\phi_2$ of the waves. Then we have

$$I_{1,2} = \frac{\varepsilon_0 nc}{2}\langle E^2 \rangle = \frac{\varepsilon_0 nc}{2}\frac{1}{2}A_1 A_1^* = \frac{\varepsilon_0 nc}{4}|A_1|^2 \tag{9.76}$$

and

$$A_1 = |A_1| e^{i\phi_1} \tag{9.77}$$

$$A_2 = |A_2| e^{i\phi_2} \tag{9.78}$$

The complex coupled wave equations become two sets of real variable equations. For the intensities we have

$$\frac{dI_1}{dz} = -4g' \sin\phi\, I_1 I_2 \tag{9.79}$$

$$\frac{dI_2}{dz} = +4g' \sin\phi\, I_1 I_2 \tag{9.80}$$

For the phases we have

$$\frac{d\phi_1}{dz} = g'[(I_1 + I_2) + I_2] \tag{9.81}$$

$$\frac{d\phi_2}{dz} = g'[(I_2 + I_1) + I_1] \tag{9.82}$$

where $g' = 4g/\varepsilon_0 nc$.

From these equations one can readily see that if $\sin\phi$ is nonvanishing, there is a flow of energy between the waves $E_1$ and $E_2$. In materials having local nonlinearity (i.e., phase shift $\phi = 0$), side diffractions provide a means for energy exchanges between the two incident beams.

Temporally nonlocal nonlinearities naturally arise if the response of the medium is not instantaneous and generally will be manifested if short laser pulses are used. Because of the delayed reaction of the induced refractive index changes, a *time-dependent* phase shift between the intensity and the index grating function will occur (cf. Figure 9.9). There is therefore a time-dependent energy exchange process between the incident beams.

Usually, the process is such that energy flow is from the stronger incident beam to the weaker one. Detailed formalisms, with quantitative numerical solutions, for treating such time-dependent wave mixing processes are given in Khoo and Zhou (8).

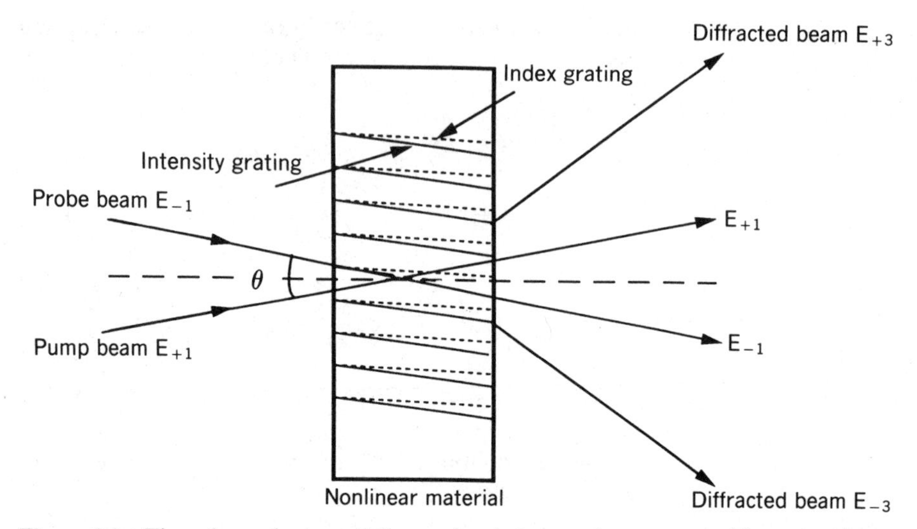


**Figure 9.9** Time-dependent spatially moving index grating generated by two coherent equal-frequency short laser pulses.

Temporally and spatially shifted grating effects are naturally present in stimulated scattering processes. A frequently studied process is stimulated scattering, where an incident optical field $E_1$ (with a frequency $\omega_1$) generates a coherent backward propagating wave $E_2$ (with a frequency $\omega_2 = \omega_1 - \omega_s$, where $\omega_s$ is the second frequency of the medium). The physical reason for the transfer of energy from $E_1$ to $E_2$ is analogous to what we discussed previously, although the actual dynamics and mechanisms are a lot more complicated. This is discussed in detail in Section 9.5.6.

### 9.5.4 Nondegenerate Optical Wave Mixings; Harmonic Generations

Nondegenerate optical wave mixings, where the frequencies of the incident and the generated waves are different, are due to nonlinear polarization of the form given by (9.35). For example, the incident frequencies $\omega_1$, $\omega_2$, and $\omega_3$ can be combined to create new frequencies $\omega_4 = \omega_1 \pm \omega_2 \pm \omega_3$ involving sums or differences. These wave mixing processes are sometimes termed sum–difference frequency generations.

In this section we discuss two basically different sum frequency generation processes, namely, second- and third-harmonic generations. These processes are associated with nonlinear polarizations of the following form.

Second-harmonic generation:

$$P_i^{2\omega} = \tfrac{1}{2}\left[\chi_{ijk}^{(2)}(-2\omega, \omega, \omega)\, A_j^{\omega} A_k^{\omega}\right] \tag{9.83}$$

and third-harmonic generation:

$$P_i^{3\omega} = \tfrac{1}{4}\left[\chi_{ijkl}^{(3)}(-3\omega, \omega, \omega, \omega) A_j^{\omega} A_k^{\omega} A_l^{\omega}\right] \tag{9.84}$$

where $i$, $j$, and $k$ refer to the crystalline axes and a sum over $j = 1,2,3$ and $k = 1,2,3$ in the preceding equations is implicit.

The main fundamental difference between these two processes is that second-harmonic generations involve a polarization that is *even order* in the electric field (i.e., $P \sim EE$), whereas third-harmonic generations involve an *odd-order* polarization ($P \sim EEE$). Consider a *centrosymmetric* nonlinear medium. If we invert the coordinate system, then (9.83) becomes

$$\tilde{P}_i^{2\omega} = \tfrac{1}{2}\tilde{\chi}_{ijk}^{(2)} \tilde{A}_j^{\omega} \tilde{A}_k^{\omega} \tag{9.85}$$

Since the polarization and the electric field are vectors, they must change sign (i.e., $\tilde{P}_i^{2\omega} = -P_i^{2\omega}$, $\tilde{A}_i = -A_i$, $\tilde{A}_j \sim -A_j$], whereas $\tilde{\chi}_{ijk}^{(2)} = \chi_{ijk}$ since it is a measure of the material response. Equation (9.85) therefore gives

$$P_i^{2\omega} = -\tfrac{1}{2}\chi_{ijk}^{(2)} A_j^{\omega} A_k^{\omega} \tag{9.86}$$

Comparing (9.85) and (9.86), the only logical conclusion is that $\chi_{ijk}^{(2)} = 0$ in a medium possessing centrosymmetry. Using similar symmetry considerations, one can show that in centrosymmetric media, all even-power nonlinear susceptibilities are vanishing. Second-, fourth-, and so on harmonic generations are possible therefore only in nonlinear media that are non-centrosymmetric. Khoo and Wang (7) present a detailed discussion of the symmetry relationships involved in second-harmonic generation in liquid crystals.

The other fundamentally important issue is phase matching, owing to the much larger phase mismatch between the incident fundamental wave at frequency $\omega$ and the generated second or third harmonic at $2\omega$ or $3\omega$, respectively. In the case of second-harmonic generation, the phase mismatch $\Delta\mathbf{k}$ in the coupled wave equation [cf. equation (9.60)] is given by

$$\Delta\mathbf{k} = \mathbf{k}_{2\omega} - 2\mathbf{k}_{\omega}$$

with a magnitude

$$\Delta k = \frac{2\pi n(2\omega)}{\lambda/2} - 2\left(\frac{2\pi}{\lambda}\right) n(\omega) = \frac{4\pi}{\lambda}\left(n(2\omega) - n(\omega)\right) \tag{9.87}$$

One possibility for achieving phase matching (i.e., $\Delta k = 0$) is to have the fundamental and the harmonic waves propagate as different types of waves (extraordinary and ordinary), that is, utilize the birefringence of the medium. For example, one can have the fundamental wave as an ordinary wave with a refractive index $n_o^{\omega}$ that is independent of the direction of propagation, while the harmonic wave is an extraordinary wave propagating at an angle $\theta_m$ with respect to the optical axis; that is, its refractive index is $n_e^{2\omega}(\theta_m)$ such that

$$n_e^{2\omega}(\theta_m) = n_o^{\omega} \tag{9.88}$$

This may be seen in Figure 9.10, which is drawn for a crystal where $n_e^{2\omega} < n_o^{2\omega}$ (i.e., a negative uniaxial crystal where $n_e < n_o$). From (9.87) and (9.88) we have

$$\frac{1}{n_o^{\omega}} = \frac{1}{n_e^{2\omega}(\theta_m)} = \frac{\cos^2\theta_m}{\left(n_o^{2\omega}\right)^2} + \frac{\sin^2\theta_m}{\left(n_e^{2\omega}\right)^2} \tag{9.89}$$

which gives, for $n_e < n_o$,

$$\sin^2\theta_m = \frac{\left(n_o^{\omega}\right)^{-2} - \left(n_o^{2\omega}\right)^{-2}}{\left(n_e^{2\omega}\right)^{-2} - \left(n_o^{2\omega}\right)^{-2}} \tag{9.90}$$

By drawing a figure similar to Figure 9.10 and going through the preceding analysis for a *positive* ($n_e > n_o$) uniaxial crystal, the reader can readily show that the phase-matching angle for this case is given by

$$\sin^2\theta_n = \frac{\left(n_o^{\omega}\right)^{-2} - \left(n_o^{2\omega}\right)^{-2}}{\left(n_o^{\omega}\right)^{-2} - \left(n_e^{\omega}\right)^{-2}} \tag{9.91}$$

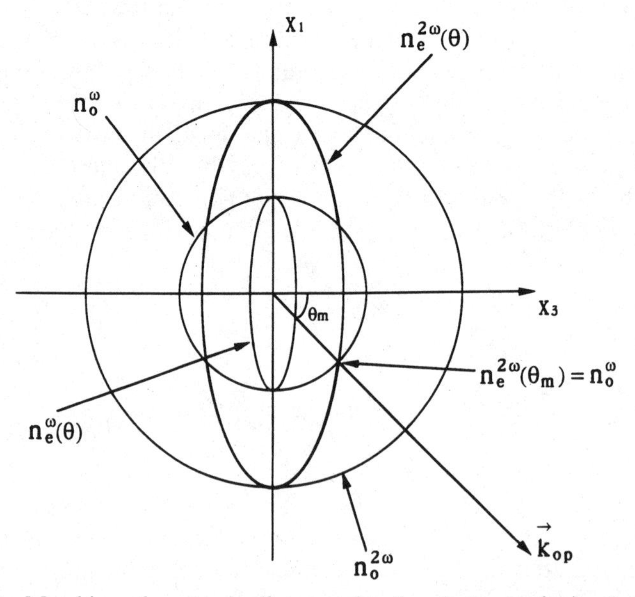


**Figure 9.10** Matching the extraordinary refractive index $n_3(\theta_m)$ of the second harmonic (at $2\omega$) with the ordinary refractive index $n_o$ of the fundamental (at $\omega$).

### 9.5.5 Self-Focusing and Self-Phase Modulation

The passage of a laser beam through a nonlinear optical material is inevitably accompanied by an intensity-dependent phase shift on the wave front of the laser, as a result of the intensity-dependent refractive index. Since the intensity of a laser beam is a spatially varying function, the phase shift is also spatially varying. This, together with large amplitude changes, leads to severe distortions on the laser in the form of self-focusing, defocusing, trapping, beam breakups, filamentations, spatial ring formations, and others.

For thick media the problem of calculating the beam profile, by solving (9.46) or (9.48) within the medium, is extremely complicated. Numerical solutions are almost always the rule (1).

If the material is thin (e.g., in nematic liquid crystal film), the problem of optical propagation is greatly simplified. In its passage through such a nonlinear film, the laser beam is only phase modulated and suffers negligible amplitude change. In other words, the transmitted far-field intensity pattern is basically a diffraction pattern of the incident field, with the nonlinear film playing the role of an intensity-dependent phase screen.

Consider now a *steady-state self-phase modulation* effect (e.g., that induced by a cw laser). If the response of the medium to the laser field is local, the spatial profile of the phase shift follows that of the laser; that is, given that the laser transverse profile is $I(r)$ (assuming cylindrical symmetry), the nonlinear part of the phase shift is simply $\phi(r) = d\,\Delta n(r) 2\pi/\lambda$, where $d$ is the thickness of the film.

If the incident laser is a Gaussian, that is,

$$I(\text{laser}) = I_0 \exp\left(-\frac{2r^2}{\omega^2}\right) \tag{9.92}$$

the nonlinear (intensity-dependent) phase shift $\phi_{\text{NL}}$ is given by

$$\phi_{\text{NL}} = kn_2 dI_0 \exp\left(-\frac{2r^2}{\omega^2}\right) \tag{9.93}$$

where $k = 2\pi/\lambda$ and $\omega$ is the laser beam waist. This is the phase shift imparted on the laser upon traversing the nonlinear film.

On the exit side (cf. Figure 9.11) at a distance $z$ from the film, the intensity distribution on the observation plane $P$ is given by the Kirchhoff diffraction integral (9, 10):

$$I(r_1, Z) = \left(\frac{2\pi}{\lambda Z}\right)^2 I_0 \left| \int_0^\infty r\,dr J_0\left(\frac{2\pi r r_1}{\lambda Z}\right) \times \exp\left(-\frac{2r^2}{\omega^2}\right) \exp[-i(\phi_D + \phi_{\text{NL}})] \right|^2 \tag{9.94}$$

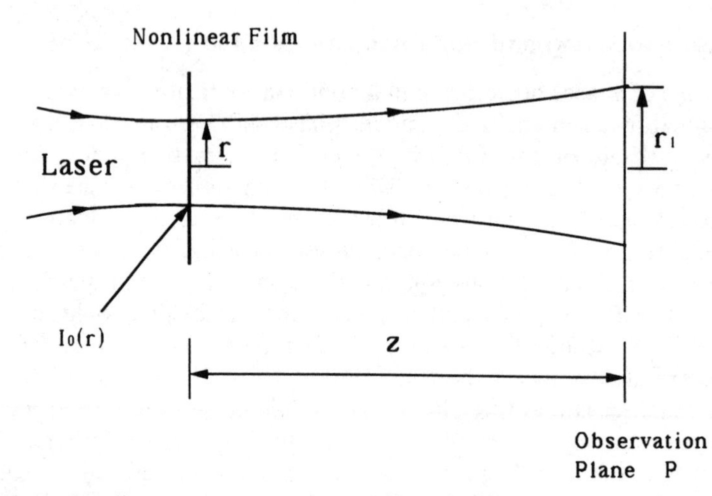


**Figure 9.11** Nonlinear transverse phase shift and diffraction of a laser beam by a thin nonlinear film.

where the diffractive phase $\phi_D$ is given by

$$\phi_D = k\left(\frac{r^2}{2Z} + \frac{r^2}{2R}\right) \tag{9.95}$$

Using the following definitions:

$$y = \frac{r}{\omega} \tag{9.96a}$$

$$C_1 = \frac{4\pi}{(\lambda z)^2}\omega^4 \tag{9.96b}$$

$$C_2 = \frac{2\pi}{\lambda}\omega \tan\alpha_0 \qquad \alpha = \frac{\lambda}{\pi\omega_0} \tag{9.96c}$$

$$C_a = \frac{2\pi}{\lambda}\bar{n}_2 d I_0 \tag{9.96d}$$

$$C_b = \frac{\pi}{\lambda}\omega^2\left(\frac{1}{Z} + \frac{1}{R}\right) \tag{9.96e}$$

and

$$\theta = \frac{r_1}{Z\tan\alpha_0} \tag{9.96f}$$

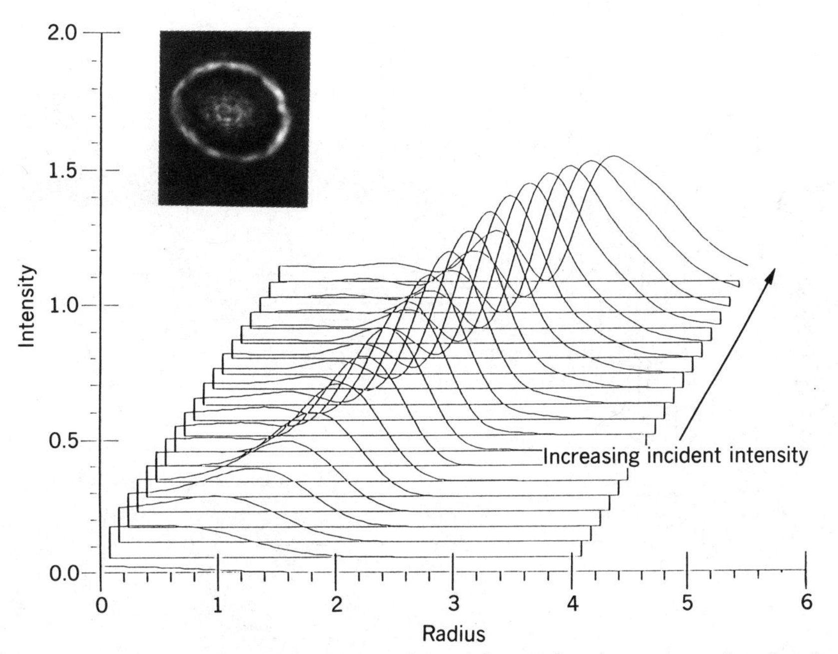


**Figure 9.12** Radial intensity as a function of increasing input intensity showing self-phase-modulation-induced redistribution of the axial intensity to the radial portion. Insert is a photograph of the observed pattern on the observation plane $P$ as a result of the passage of a laser beam through a nematic film.

we can rewrite (9.94) as

$$I(r_1, Z) = C_1 I_0 \left| \int_0^{\infty} \exp(-2y^2) y \times \exp\{-i[C_a \exp(-2y^2) + C_b y^2]\} J_0(C_2 \theta y)\, dy \right|^2 \quad (9.97)$$

Khoo et al. (9) discuss in detail the various types of intensity distributions and how they evolve as a function of the input intensity. Essentially, the principal deciding factor is the sign of $C_b$ (i.e., the sign of $1/Z + 1/R$).

If $n_2 > 0$, we have the following cases:

1 $(1/Z + 1/R) < 0$ case. The intensity distribution evolves from a Gaussian beam to one where the central area tends to be dark (low intensity) surrounded by bright rings (cf. Figure 9.12).

2 $(1/Z + 1/R) > 0$ case. The intensity distribution evolves as the incident intensity $I_0$ is increased, from a Gaussian shape to one with a bright central spot and concentric bright and dark rings (cf. Figure 9.13).

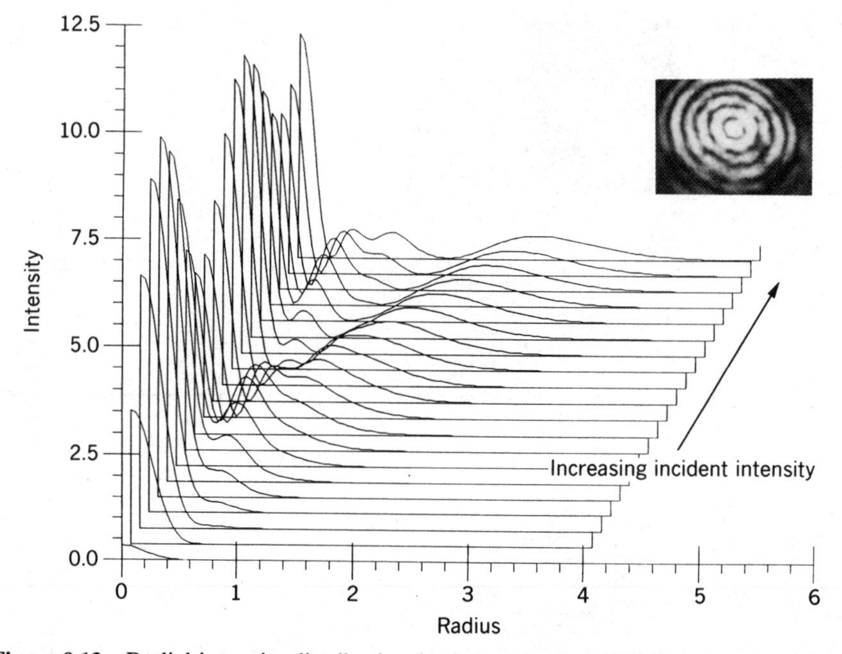


**Figure 9.13** Radial intensity distribution for increasing input intensity for a different geometry showing the intensification of the axial laser intensity. Insert is a photograph of the actual pattern using a nematic film.

3 $(1/Z + 1/R) \approx 0$ case. This is the so-called intermediate case, where the intensity distribution at the central area assumes forms that are intermediate between those of cases 1 and 2.

If $n_2 < 0$, the intensity distribution is almost a "mirror" image of the $n_2 > 0$ case; that is, the distributions for cases 1 and 2 are interchanged.

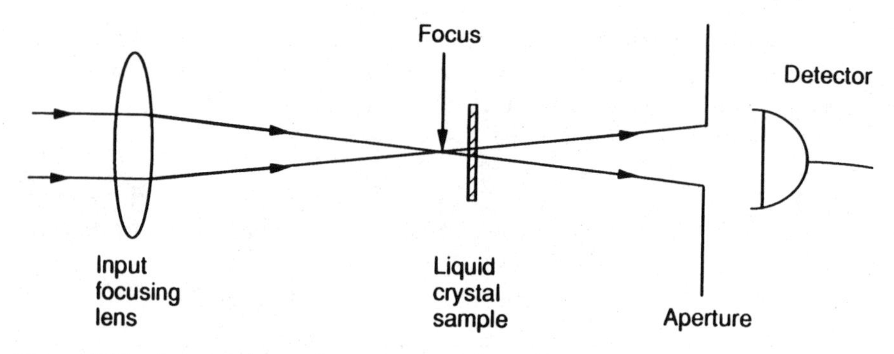


**Figure 9.14** An experimental arrangement for achieving the optical limiting effect using the self-phase modulation or self-focusing/self-defocusing effect.

Such intensity redistribution effects, in which the transmitted on-axis laser intensity is reduced owing to the nonlinear self-defocusing effect caused by the laser-induced index change, may be employed for optical limiting applications. A simple setup is shown in Figure 9.14. The aperture situated at the observation plane $P$ will transmit totally a low-power laser, as well as the signal/image. If the laser is intense, the defocusing effect will reduce the transmission through the aperture, thereby protecting the detector or sensor placed behind the aperture.

For practical self-limiting applications (11) dealing with a pulsed laser, a dynamical theory is needed. While a dynamical theory is extremely complicated (almost insoluble) for self-focusing effects in long media, such a theory is relatively easy to construct for *thin* nonlinear media.

Consider a Gaussian–Gaussian incident laser pulse:

$$I(\mathbf{r},t) = I_0 \exp\left(-\frac{2r^2}{\omega^2}\right)\exp\left(-\frac{t^2}{\tau_p^2}\right) \tag{9.98}$$

(cf. the input pulse shape of Figure 9.15). The intensity distribution at the

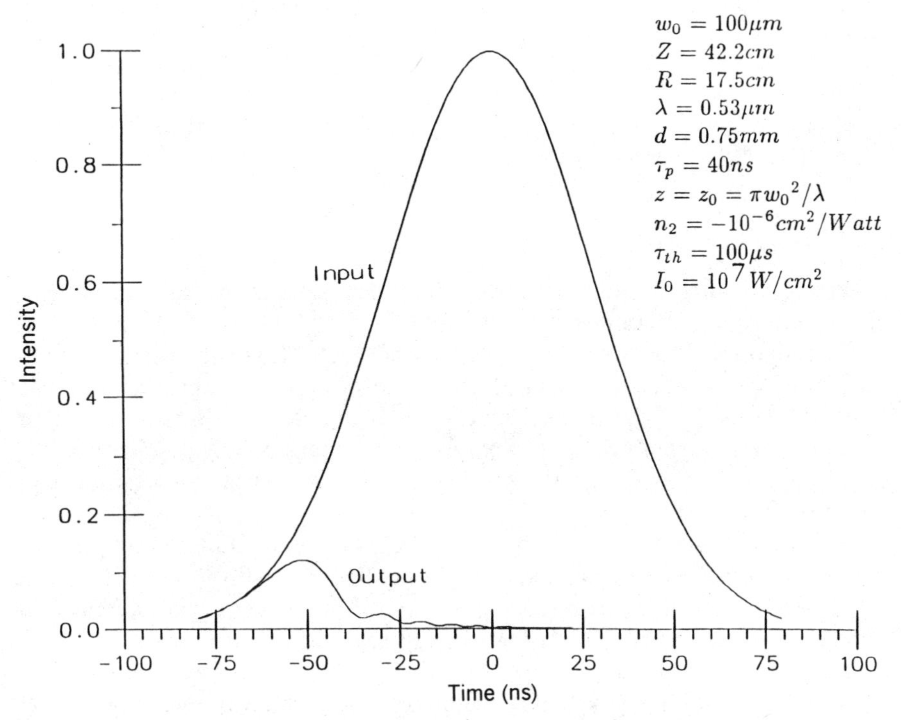


**Figure 9.15** Theoretical plot of the output (energy limited) and input laser pulses.

observation plane can be deduced from the diffraction theory discussed previously:

$$I(\mathbf{r}_p, z, t) = \left(\frac{2\pi}{\lambda z}\right)^2 I_0 \exp\left(-\frac{t^2}{\tau_p^2}\right) \times \left| \int_0^\infty r\,dr J_0\left(\frac{2\pi r r_p}{\lambda z}\right) \exp\left(-\frac{2r^2}{\omega^2}\right) \times \exp[-i(\Delta\phi_D + \Delta\phi_{\mathrm{NL}})] \right|^2 \tag{9.99}$$

where the diffractive phase shift is

$$\Delta\phi_D = k\left(\frac{r^2}{2z} + \frac{r^2}{2R}\right) \tag{9.100a}$$

and the nonlinear phase shift is

$$\begin{aligned} \Delta\phi_{\mathrm{NL}}(\mathbf{r}_p, z, t) &= kd\,\Delta n(t) \\ &= kd \sum_i \left[\frac{1}{\tau_i}\int_{-\infty}^{t} n_{2i}(t')\, I(\mathbf{r}_p, z, t')\, dt'\right] \\ &= kdI_0 \exp\left(-\frac{2r^2}{\omega^2}\right) \sum_i \left[\frac{1}{\tau_i}\int_{-\infty}^{t} n_{2i}(t') \exp\left(-\frac{t'^2}{\tau_p^2}\right) dt'\right] \end{aligned} \tag{9.100b}$$

where $n_{2i}(t')$ are the time-dependent nonlinear coefficients associated with all the contributing nonlinear mechanisms.

The final form for $\Delta\phi_{\mathrm{NL}}(t)$, as well as the output intensity distribution on the observation plane $P$, depends on the form of $n_{2i}(t')$ and the response time $\tau_i$ versus the laser pulse duration $\tau_p$. If the response time $\tau_i$ is longer than $\tau_p$, the self-limiting effect is basically an integrated effect, and thus it is more appropriate to discuss the process in terms of energy/area rather than intensity (power/area).

Figure 9.15 shows a theoretical plot of the input and output on-axis intensity profile for the case of a single nonlinearity with $\tau_i \gg \tau_p$. It shows that as the intensity-dependent phase modulation effect is accumulated in time, the output intensity is progressively reduced, finally to highly diminished value. The energy contained in the transmitted pulse, when plotted against the input energy, will also exhibit a limiting behavior. More details on this are given in Section 10.3.

## 9.6 STIMULATED SCATTERINGS

Stimulated scatterings occupy a special place in nonlinear optics. They have been investigated ever since the dawn of nonlinear optics and lasers. Devoting only a section to the discussion of this subject is not likely to do justice to the voluminous work that has been done in this area. The main objective of this section, however, is to outline the fundamental principles involved in these scattering processes in the context of the wave mixing processes discussed previously.

There are four main types of stimulated scattering processes (1, 2):

1 Stimulated Raman scattering (SRS)
2 Stimulated Brillouin scattering (SBS)
3 Stimulated thermal Brillouin and thermal Rayleigh scattering
4 Stimulated Rayleigh wing (orientational) scattering

We will discuss processes 1 and 2 in detail, as they represent two distinctive types of material excitations. In Raman processes, as in orientational and thermal processes, the excitations are not propagative. On the other hand, Brillouin processes involve propagating sound waves. These differences will be reflected in the coupling and propagation of the incident and generated optical waves in the material.

### 9.6.1 Stimulated Raman Scatterings

Stimulated Raman scatterings involve the nonlinear interaction of the incident and generated lasers with the vibrations or rotations (or other Raman transitions) of the material. These Raman transitions involve two energy levels of the material connected by a two-photon process, as discussed in Chapter 5.

From (5.51) the induced polarization associated with Raman scattering is of the form

$$P_{\text{ind}} \sim q_v E \cdot \varepsilon_0 \left( \frac{\partial \alpha}{\partial q} \right)_{q_0} \left( \frac{N}{V} \right) \tag{9.101}$$

where $(N/V)$ is the density of the molecules. For simplicity, we also neglect the vector nature of the field **E** and the polarization **P**. The displacement of the normal coordinate $q_v$ obeys an equation of motion for a forced harmonic oscillator, which is of the form

$$\dot{q}_v^0 + \omega_R q = F \tag{9.102}$$

where the force $F$ is related to the optical field molecular interaction energy

$u$ (owing to the induced polarization) given by

$$F = \frac{\partial u}{\partial q_v} \tag{9.103}$$

where

$$u = \frac{1}{2} \Delta\varepsilon\, E^2 = \frac{1}{2} q_v \mathbf{E} \cdot \mathbf{E} \varepsilon_0 \left( \frac{\partial \alpha}{\partial q} \right)_{q_0} \tag{9.104}$$

that is,

$$F \propto E^2 \tag{9.105}$$

From (9.102),

$$q_v \sim E^2 \tag{9.106}$$

Therefore, the induced polarization, from (9.101), is of the form

$$P_{\text{ind}} \sim E^3 \tag{9.107}$$

In other words, the nonlinear polarization associated with stimulated Raman scattering is a third-order one.

To delve further into this process, we note from (9.102) and (9.106) that, because of the vibrational frequency $\omega_R = \omega_2 - \omega_1$, only terms in $F \sim E^2$ containing the frequency component $\omega_2 - \omega_1$ (i.e., $E_2 E_1^*$ or $E_1 E_2^*$) contribute to $q_v$. In other words,

$$q_v \sim \{E_2 E_1^* + \text{c.c.}\} \tag{9.108}$$

The induced third-order nonlinear polarization $P_{\text{NL}}$ is therefore of the form

$$P_{\text{NL}} \sim (E_2 E_1^* + \text{c.c.})(E_1 + E_2 + \text{c.c.}) \tag{9.109}$$

The nonlinear polarization therefore produces scattering at various frequencies $\omega_1$, $\omega_2$, $2\omega_1 - \omega_2$, and $2\omega_2 - \omega_1$.

Focusing our attention on the Stokes waves at frequency $\omega_2$, the nonlinear polarization is given by

$$P \sim (E_2 E_1^*) E_1 \tag{9.110}$$

Using the formalism given in Section 9.2, we can thus write

$$P^{\omega_2} = \frac{1}{2^2}\chi_s(-\omega_2, \omega_1, -\omega_1, \omega_1) A^{\omega_2} A^{*-\omega_1} A^{\omega_1}$$
$$= \frac{1}{2^2}\chi_s(-\omega_2, \omega_2, -\omega_1, \omega_1) A^{\omega_2} |A^{\omega_1}|^2 \tag{9.111}$$

where the subscript $s$ on $\chi_s$ signifies that it is for the Stokes wave. $\chi_s$ can be deduced from (9.101) to (9.108). If the incident laser beam suffers negligible depletion, we can let $|A^{\omega_1}|^2$ be a constant. In this case $\chi_s|A^{\omega_1}|^2$ is equivalent to the induced refractive index change experienced by the Stokes wave.

The equation describing the amplitude $A_2$ of the Stokes wave, from Section 9.5, becomes

$$\frac{\partial A_2}{\partial z} = ig|A_1|^2 A_2 \tag{9.112}$$

where

$$g = \frac{\mu_0 \omega_2^2}{2k_2}\left(\frac{\chi_s}{4}\right) \tag{9.113}$$

Accounting for the loss $\alpha$ (due to random scatterings, absorption, etc.) experienced by $E_2$ in traversing the medium, (9.112) becomes

$$\frac{\partial A_2}{\partial z} = \left[ig|A_1|^2 - \alpha\right] A_2 \tag{9.114}$$

Since the nonlinear susceptibility $\chi_s$ is complex, we may write it as

$$\chi_s = \chi_s' - i\chi_s'' \tag{9.115}$$

Substituting (9.115) into (9.112), we can readily show that the amplitude $A_2$ will grow exponentially with $z$, with a *net gain* constant given by

$$\bar{g}_s = \left(\frac{\mu_0 \omega_2^2}{2k_2}\frac{\chi_s''}{4} - \alpha\right) \tag{9.116}$$

In Boyd (2), the Stokes susceptibility $\chi_s$ (sometimes called the Raman susceptibility) is explicitly derived for a multilevel system. However, in practice, it is easier and quantitatively correct to relate the gain constant to the scattering cross section as indicated in Chapter 5. The reader is reminded of the role played by the population difference $N_1 - N_2$ in $\chi_s$, which could reverse sign depending on $N_1$ and $N_2$.

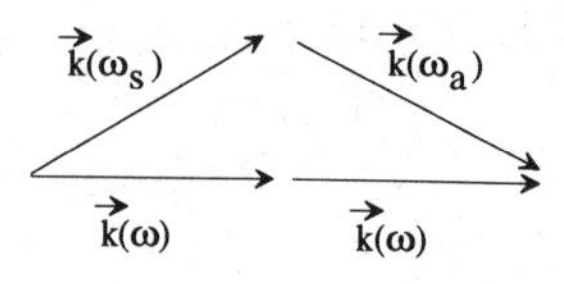


**Figure 9.16** Coherent anti-Stokes Raman scattering (CARS) involving the pump beam at $\omega$ and its Stokes wave $\omega_s$.

From (9.109), as we mentioned earlier, other frequency components besides the Stokes waves can also be generated. An example is the so-called coherent anti-Stokes Raman scattering (CARS), which involves the mixing of the incident laser with its (phase coherent) Stokes waves (cf. Figure 9.16). The nonlinear polarization is of the form

$$P_{\text{NL}} \sim \tfrac{1}{4}\chi(-\omega_a, \omega_1, \omega_1, -\omega_2)\bar{A}_1\bar{A}_1\bar{A}_2^* \tag{9.117}$$

with a phase mismatch

$$\Delta\mathbf{k} = (\mathbf{k}_a - (2\mathbf{k}_1 - \mathbf{k}_2)) \tag{9.118}$$

Notice that this is the analog of the degenerate four-wave mixing process discussed in Section 9.5; the anti-Stokes beam assumes the role of the diffracted beam $E_3$ (cf. Figure 9.4).

### 9.6.2 Stimulated Brillouin Scatterings

Brillouin scatterings are caused by acoustic waves in the material. In analogy to Raman scattering, the nonlinear polarizations responsible for the coupled-wave equations are obtainable from consideration of the appropriate material excitations. There are, in fact, many distinctly different physical mechanisms that will alter the density $\rho$ of a material, such as pressure $p$, entropy $S$, or temperature $T$. These parameters are in most cases strongly coupled to one another. Accordingly, there are quite a variety of stimulated Brillouin scatterings depending on the underlying physical mechanisms.

In this section we focus our attention on *nonabsorptive materials*, where the density change is due to the spatially and temporally varying *strain* caused by a corresponding optical intensity. An example of such an intensity pattern is the one obtained by the interference of two lasers of different frequencies $\omega_1$ and $\omega_2$ (cf. Figure 9.8). Analogous to Raman scattering, the frequency difference $\omega_1 - \omega_2$ should match the acoustic frequency.

There is one very important difference between Raman scattering and Brillouin scattering. Raman scattering is associated with *nonpropagating* material oscillations; Brillouin scattering, however, involves the creation of propagating sound waves. Since the wave vector $\mathbf{k}_s$ of the sound wave carries a momentum $\mathbf{k}_s$, the resulting directions of the generated acoustic wave (at $\mathbf{k}_s$ and $\omega_s$) and the frequency downshifted optical wave ($\mathbf{k}_2$, $\omega_2 = \omega_1 - \omega_s$) must obey the wave vector matching conditions. The equation governing the induced density change $\Delta\rho$ is given by (2, 12), in SI units,

$$\left(\frac{\partial^2}{\partial t^2}+2\Gamma_B\frac{\partial}{\partial t}-v_s^2\nabla^2\right)\Delta\rho=\nabla\cdot\left(-\frac{\gamma_e}{2}\nabla(E\cdot E)\right) \tag{9.119}$$

If we write $\Delta\rho$ in terms of a slowly varying envelope $\bar{\rho}$ as

$$\Delta\rho=\frac{1}{2}\left[\bar{\rho}(\mathbf{r}_s)\exp[i(\mathbf{k}_s\cdot\mathbf{r}-\omega_s t)]+\text{c.c.}\right] \tag{9.120}$$

then $\nabla$ in (9.119) is given by $\nabla\equiv\partial/\partial r_s$. In (9.119) $v_s^2$ is the sound velocity and $\Gamma_B$ is the damping constant.

Writing now the optical fields in terms of their amplitudes $A_1$ and $A_2$ in plane wave form

$$E_1=\tfrac{1}{2}\hat{x}_1\left[A_1\exp[i(\mathbf{k}_1\cdot\mathbf{r}-\omega_1 t)]+\text{c.c.}\right] \tag{9.121}$$

$$E_2=\tfrac{1}{2}\hat{x}_2\left[A_2\exp[i(\mathbf{k}_2\cdot\mathbf{r}-\omega_2 t)]+\text{c.c.}\right] \tag{9.122}$$

and substituting (9.120) to (9.122) into (9.119) and using the slowly varying envelope approximations

$$\left|\frac{d^2\bar{\rho}}{dr_s^2}\right|\ll|k_s^2u_s|,\left|k_s\frac{d\bar{\rho}}{dr_s}\right| \tag{9.123}$$

equation (9.119) becomes

$$\left(\frac{\partial}{\partial r_s}+\frac{\Gamma_B}{v_s}\right)\bar{\rho}=\frac{ik_s\gamma_e(\hat{x}_1\cdot\hat{x}_2)}{4v_s^2}A_1A_2^*e^{i\Delta k\,r_s} \tag{9.124}$$

where $\Delta k=|\mathbf{k}_1-\mathbf{k}_2-\mathbf{k}_s|$.

This is the material excitation equation. Then the coupled optical wave equations for $A_1$ and $A_2$ can be obtained by identifying the nonlinear polarization associated with the density change $\Delta\rho$. This is given by

$$P_{\text{NL}}=\Delta\varepsilon\,E=\left(\frac{\partial\varepsilon}{\partial\rho}\cdot\Delta\rho\right)E \tag{9.125}$$

Since $\Delta\rho$ oscillates as $e^{-i(\omega_1-\omega_2)t}$, we have

$$\frac{P_{\text{NL}}^{\omega_1}}{2}=\frac{\partial\varepsilon}{\partial\rho}\left(\frac{\bar{\rho}}{2}\right)\left(\frac{A_2}{2}\right) \tag{9.126}$$

and

$$\frac{P_{NL}^{\omega_2}}{2} = \frac{\partial \varepsilon}{\partial \rho}\left(\frac{\bar{\rho}}{2}\right)^* \left(\frac{A_1}{2}\right)$$

The nonlinear coupled-wave equations for $A_1$ and $A_2$ are, from (9.48),

$$2ik_1 \frac{\partial A_1}{\partial r_s} = \frac{-\mu_0 \omega_1^2}{2} \frac{\partial \varepsilon}{\partial \rho} \bar{\rho} A_2 e^{-i\Delta k r_s} \tag{9.127}$$

$$2ik_2 \frac{\partial A_2}{\partial r_s} = \frac{-\mu_0 \omega_2^2}{2} \frac{\partial \varepsilon}{\partial \rho} \bar{\rho} A_1 e^{i\Delta k r_s} \tag{9.128}$$

where we have replaced $z$ by $r_s$ as the reference coordinate for propagation. We have also ignored the $\nabla^2 A$ term with our plane wave approximation. With a little rewriting, (9.127) and (9.128) become

$$\frac{\partial A_1}{\partial r_s} = -\alpha_1 A_1 + ig_1 \bar{\rho} A_2 e^{-i\Delta k r_s} \tag{9.129}$$

$$\frac{\partial A_2}{\partial r_s} = -\alpha_2 A_2 + ig_2 \bar{\rho}^* A_1 e^{+i\Delta k r_s} \tag{9.130}$$

and (9.124) becomes

$$\frac{\partial \bar{\rho}}{\partial r_s} = \frac{-\Gamma_B}{v_s} \bar{\rho} + ig_s A_1 A_2^* e^{i\Delta k r_s} \tag{9.131}$$

In (9.129) and (9.130) we have included the phenomenological loss terms $-\alpha_1 A_1$ and $-\alpha_2 A_2$, respectively, to account for the losses (e.g., absorptions or random scatterings) experienced by $A_1$ and $A_2$ in traversing the medium.

The coupling constants $g_1$, $g_2$, and $g_s$ are given by

$$g_1 = \frac{\mu_0 \omega_1^2}{4k_1}\left(\frac{\partial \varepsilon}{\partial \rho}\right) \tag{9.132}$$

$$g_2 = \frac{\mu_0 \omega_2^2}{4k_2}\left(\frac{\partial \varepsilon}{\partial \rho}\right) \tag{9.133}$$

$$g_s = \frac{k_s \gamma_e}{4v_s^2} \tag{9.134}$$

Because of the nonlinear couplings among $A_1$, $A_2$, and $\bar{\rho}$, solutions to (9.129) to (9.131) are quite complicated. A good glimpse into stimulated

Brillouin scattering may be obtained if we first solve for the density wave amplitude $\bar{\rho}$ in a perturbative manner.

Ignoring the $r_s$ dependence of $A_1$ and $A_2$, (9.131) can be integrated to give

$$\bar{\rho} = \left( \frac{g_s A_1 A_2^*}{\Delta k - i\Gamma_B / v_s} \right) e^{i \Delta k r_s} \tag{9.135}$$

Substituting the complex conjugates of (9.135) into (9.130), we get

$$\frac{\partial A_2}{\partial r_s} = -\alpha_2 A_2 + \frac{g_2 g_s |A_1|^2 (\Gamma_B / v_s) A_2}{(\Delta k)^2 + (\Gamma_B / v_s)^2} + \frac{i \Delta k g_2 g_s |A_1|^2 A_2}{(\Delta k)^2 + (\Gamma_B / v_s)^2} \tag{9.136}$$

If $|A_1|^2 = \text{const}$ (i.e., the pump beam is so strong that its depletion is negligible), the solution for the generated wave $A_2$ at frequency $\omega_2 = \omega_1 - \omega_s$ is an exponential function, with a gain constant

$$\begin{aligned} G_{sB} &= \frac{g_2 g_s |A_1|^2 (\Gamma_B / v_s)}{(\Delta k)^2 + (\Gamma_B / v_s)^2} - \alpha_2 \\ &= g_{sB} - \alpha \end{aligned} \tag{9.137}$$

In other words, starting from some weak initial (spontaneous) Brillouin scattering, greatly enhanced stimulated emission at $\omega_2$ will be produced as it propagates along with the pump wave at $\omega_1$; a strong sound wave at $\omega_s$ will also be generated. As shown in Figure 9.17, the directions of propagation of these waves are related by the phase-matched condition ($\Delta \mathbf{k} = 0$, which gives $\mathbf{k}_1 = \mathbf{k}_2 + \mathbf{k}_s$ or $\mathbf{k}_1 - \mathbf{k}_2 = \mathbf{k}_s$).

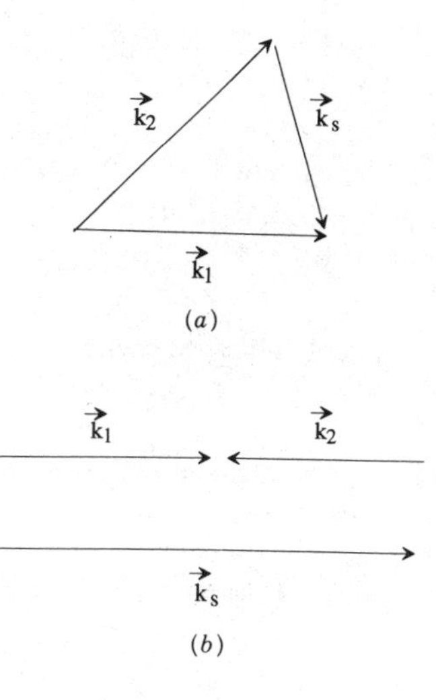


**Figure 9.17** (*a*) Wave vector phase matching of the pump wave $\mathbf{k}_1$, the generated wave $\mathbf{k}_2$, and the sound wave $\mathbf{k}_s$. (*b*) Direction of incident and stimulated Brillouin waves. Optimal gain occurs if the generated wave is counterpropagating to the incident pump wave.

In Brillouin scattering, owing to the large difference between the acoustic and optical frequencies ($\omega_s \ll \omega_1, \omega_2$), usually $\omega_2 \approx \omega_1$ and therefore $k_2 \approx k_1$. From Figure 9.16, $k_2 \approx k_1$ means that

$$k_s \approx 2k_1 \sin\left(\frac{\theta}{2}\right) \tag{9.138}$$

which is maximal at $\sin(\theta/2) = 1$ (i.e., $\theta = 180°$).

Since $g_s$ in $G_{sB}$ is proportional to $k_s$, we therefore conclude that Brillouin gain is maximal if the generated wave is backward propagating. The maximum value for $g_{sB}$ (for $\Delta k = 0$, $\theta = 180°$) is given by

$$g_{sB} = \frac{g_s g_2 |A_1|^2}{(\Gamma_B / v_s)} \tag{9.139}$$

$$= \frac{\gamma_e \mu_0 \omega_{12}^2 k_1}{16 v_s \Gamma_B k_2} \left(\frac{\partial \varepsilon}{\partial \rho}\right) |A_1|^2 \tag{9.140}$$

As one may expect, the gain is governed mainly by the electrostrictive coefficient $\gamma_e$, $(\partial\varepsilon/\partial\rho)$, and the incident laser intensity $|A_1|^2$. Note that since $\gamma_e = \rho_0(\partial\varepsilon/\partial\rho)$, the factor $\gamma_e(\partial\varepsilon/\partial\rho)$ in (9.140) can be written as $\rho_0(\partial\varepsilon/\partial\rho)^2$ or $(\gamma_e^2/\rho_0)$.

## REFERENCES

1. Y. R. Shen, "Principles of Nonlinear Optics." Wiley, New York, 1984.
2. R. W. Boyd, "Nonlinear Optics." Academic Press, San Diego, 1992.
3. C. Flytzanis, *in* "Quantum Electronics" (H. Rabin and C. L. Tang, eds.), Vol. 1, Part A, p. 30. Academic Press, New York, 1975.
4. J. D. Jackson, "Classical Electrodynamics," p. 119. Wiley, New York, 1963.
5. N. Bloembergen, "Nonlinear Optics," p. 69. Benjamin, New York, 1965.
6. I. C. Khoo and S. T. Wu, "Optics and Nonlinear Optics of Liquid Crystals." World Scientific, Singapore, 1993.
7. See, for example, I. C. Khoo and W. Wang, Effects of side diffractions and phase modulations on phase conjugations in a Kerr medium. *IEEE J. Quantum Electron.* **QE-27**, 1310 (1991), and references therein.
8. I. C. Khoo and P. Zhou, Transient multiwave mixing in a nonlinear medium. *Phys. Rev. A* **41**(3), 1544 (1990); for transient two-wave mixing effect, see V. L. Vinetskii, N. V. Kukhtarev, S. G. Odulov, and M. S. Soskin, Dynamic self-diffraction of coherent light beams. *Usp. Fiz. Nauk* **129**, p. 113 (1979); *Sov. Phys.—Usp.* (*Engl. Transl.*) **22**(9), 742 (1979).

9. I. C. Khoo, J. Y. Hou, T. H. Liu, P. Y. Yan, R. R. Michael, and G. M. Finn, Transverse self-phase modulation and bistability in the transmission of a laser beam through a nonlinear thin film. *J. Opt. Soc. Am. B* **4**, 886 (1987).
10. I. C. Khoo, P. Y. Yan, T. H. Liu, S. Shepard, and J. Y. Hou, Theory and experiment on optical transverse intensity bistability in the transmission through a nonlinear thin (nematic liquid crystal) film. *Phys. Rev. A* **29**, 2756 (1984).
11. I. C. Khoo, Sukho Lee, P. G. LoPresti, R. G. Lindquist, and H. Li, Isotropic liquid crystalline film and fiber structures for optical limiting application. *Int. J. Nonlinear Opt. Phys.* **2**, No. 4 (1993). This is a special issue, together with Vol. 2, No. 3, on "Optical limiters, switches and discriminators."
12. I. P. Bartra, R. H. Enns, and D. Pohl, Stimulated thermal scattering of light. *Phys. Status Solid.* **48**, 11 (1971).

## CHAPTER 10

# NONLINEAR OPTICAL PHENOMENA OBSERVED IN LIQUID CRYSTALS

### 10.1 A SURVEY OF NONLINEAR OPTICAL PHENOMENA OBSERVED IN LIQUID CRYSTALS

Because of the various mechanisms for optical nonlinearities present in the ordered as well as isotropic phases of liquid crystals, almost all nonlinear optical phenomena have been observed. Some of these phenomena were studied for their novelty; others have been developed into diagnostic tools or practical devices. In accordance with the basic mechanisms involved, these observed effects are grouped together under the following general headings.

1 Self-focusing, self-defocusing, and self-phase modulation
2 Optical wave mixing including harmonic generation, beam amplification, and phase conjugation
3 Optical bistability, instability, switching, and limiting
4 Stimulated scattering
5 Nonlinear waveguiding

In view of the rapid advances made by several research groups, it is likely that the following summary will be outdated shortly. Accordingly, we limit our attention here to only *exemplary* studies which are fundamentally interesting and/or practically important. Furthermore, we will focus our attention on recent studies. The reader is referred to the literature quoted in the two major review articles (1, 2) for earlier work.

### 1. Self-Focusing, Self-Defocusing, and Self-Phase Modulation

*Nematic Phase* A quantitative study of the role played by the laser wave front curvature and other geometrical/optical parameters on the transmitted far-field intensity profile (3); demonstration of the self-defocusing effect for optical limiting applications (4, 4a).

*Isotropic Phase* Study of the interplay between liquid crystal reorientation time scales and transient self-focusing effects (5).

### 2. Optical Wave Mixings

#### a. Degenerate Optical Wave Mixing

*Isotropic Phase* Use of molecular reorientational nonlinearity in optical phase conjugation (6); study of the influence of molecular structure on molecular reorientational nonlinearity (7).

*Nematic Phase* Optical phase conjugation (8); optical beam amplifications involving orientational and thermal nonlinearities (9); polarization switching and beam amplification employing the intensity-dependent birefringence of liquid crystals (10).

#### b. Nondegenerate Optical Wave Mixing

*Isotropic Phase* Theory and experimental demonstration of optical third-harmonic generation in cholesteric liquid crystal (11).

*Nematic Phase* Second-harmonic generation (12).

*Smectic-C* Phase* Second-harmonic generation (13).

### 3. Optical Bistability and Switching

*Nematic Phase* Cavityless bistability employing self-phase modulation (3); nonlinear Fabry–Perot bistability effect (14); intrinsic optical bistability (15); nonlinear optical switching near the total internal reflection state (16, 17) bistability in guided wave geometry (18).

**4. Stimulated Scattering** Nematic and cholesteric phase (19); smectic-A phase (20); isotropic phase (4a, 5, 20).

***5. Nonlinear Waveguiding*** Fiber–fiber coupling (21; see also 4a, 18).

The references as well as the phenomena quoted previously are by no means complete; the reader therefore should consult the references quoted therein and in the preceding chapters for others. New references will also be cited in the following discussion.

In the next few sections we discuss a few topics that are both fundamentally interesting and potentially useful for developing coherent optical devices.

## 10.2 OPTICAL PHASE CONJUGATION

Self-starting optical phase conjugation (SSOPC), in which a single incident laser beam generates its phase-conjugated replica via some optical wave mixing effect in a nonlinear optical material, is a fundamentally interesting and practically useful process. Usually, the signal originates as some coherently scattered noise from the pump laser beam (e.g., owing to scatterers in a crystal, spontaneous Brillouin scattering, etc.). This noise signal interacts with the pump beam and grows into a strong coherent signal. This phenomenon is commonly observed in stimulated Brillouin scattering involving high-power pulsed lasers (22) and in photorefractive materials with low-power cw lasers (23). More recently, the self-pumped phase conjugation effect has also been observed in resonant media, using frequency-shifted signal and pump field or degenerate four-wave mixing processes (24).

### 10.2.1 SSOPC by Stimulated Thermal Scattering

Observation of self-pumped phase conjugation in a nematic liquid crystal was recently reported by two groups (25, 26). Figure 10.1 depicts the experimental setup reported in Khoo et al. (25); the insert in Figure 10.1 shows the wave vector matching condition for the degenerate wave mixing process. In this process the input beam traverses the medium twice, once as an incident field $\mathbf{E}_1$ (along the $\mathbf{K}_1$ direction) and then, upon reflection from the optical system behind the sample, as a reflected field $\mathbf{E}_2$ (along the $\mathbf{K}_2$ direction). The fields $\mathbf{E}_1$ and $\mathbf{E}_2$ are not necessarily coherent with respect to each other.

The mechanism for generating a phase-conjugated signal is as follows. As shown in the insert of Figure 10.1, the incident beam generates a noise source field $\mathbf{E}_3$ (along $\mathbf{K}_3$) which is coherent with respect to $\mathbf{E}_1$. Accordingly, $\mathbf{E}_1$ and $\mathbf{E}_3$ can interfere with each other and produce an index grating. For the planar aligned nematic liquid crystal film used in this experiment, where the director axis $\hat{n}$ is normal to the plane of the figure (i.e., parallel to $\mathbf{E}_1$, $\mathbf{E}_2$, $\mathbf{E}_3$ and $\mathbf{E}_4$), the index grating is associated with the temperature-dependent extraordinary refractive index $n_e(T)$. Similarly, the reflected field $\mathbf{E}_2$ will interfere with its coherent noise $\mathbf{E}_4$ and produce an index grating.

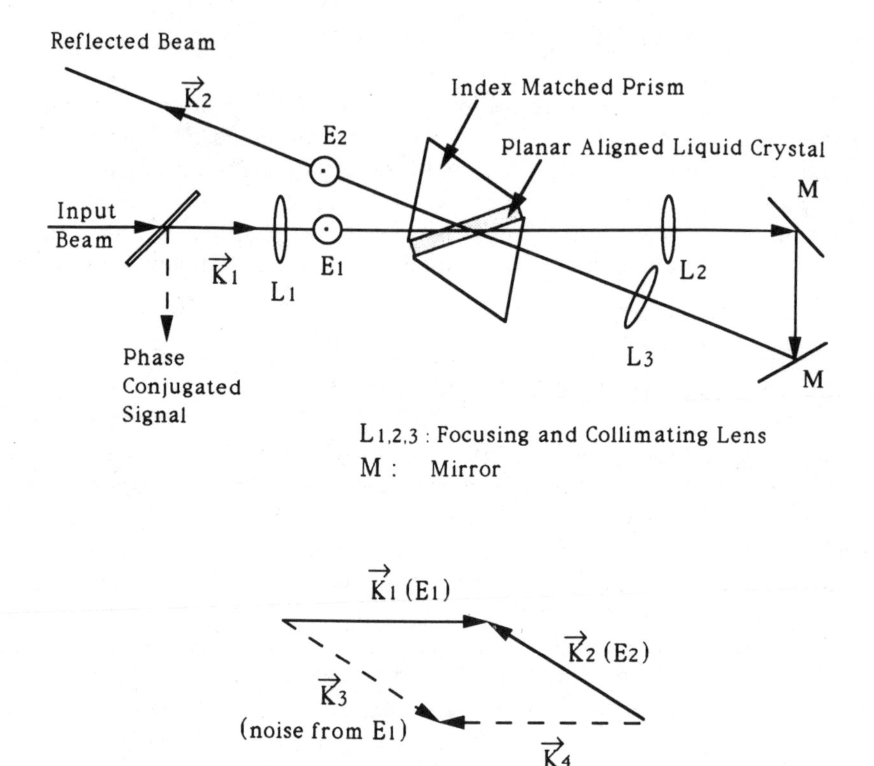


**Figure 10.1** Experimental setup for observing self-starting phase conjugation effect. Insert shows the wave vector matching diagram.

In general, the scattered noise $\mathbf{E}_3$ and $\mathbf{E}_4$ from the two "input" beams $\mathbf{E}_1$ and $\mathbf{E}_2$ contains various temporal and spatial frequency components (cf. Chapters 3 and 5). However, for a given crossing angle between the incident and reflected fields, only the scattered noise components ($\mathbf{K}_3$ and $\mathbf{K}_4$) which obey the wave vector matching condition depicted in Figure 10.1 will be able to share a common grating and experience growth via the wave mixing effects.

The detailed dynamics and mechanisms of these parametric processes between the two sets of waves ($\mathbf{E}_1, \mathbf{E}_3$) and ($\mathbf{E}_2, \mathbf{E}_4$) are similar to the two-wave stimulated scattering effects discussed in the following section. Some theoretical discussions have also been presented in Antipov and Dvoryaninov (26).

In the experiment reported in Khoo et al. (25), the liquid crystal used is a 150-$\mu$m planar aligned dye-doped 5CB cell enclosed by prisms with a refractive index of 1.73. The dye used is D16 (from EM Chemicals) at a concentration of 0.5% by weight. The laser pulse is derived from an electronically chopped argon laser (5145 Å), with a square pulse duration of 2 ms. The laser is linearly polarized, with the polarization vector parallel to the director axis of the liquid crystal (i.e., it "sees" $n_e$, the extraordinary index). For 5CB, $n_e \approx 1.68$. The laser is obliquely incident on the cell, as shown in Figure 10.1, so that the effective interaction length within the liquid crystal is about 1 mm. As a result of the reflection loss at the glass–liquid crystal interface and scattering loss in the liquid crystal, only 5% of the incident beam is transmitted. The transmitted beam is recollimated and focused back into the cell, making an angle $\theta \approx 1°$ with the incident beam. The spot diameter of the input laser at the liquid crystal is 0.5 mm, whereas the reflected beam spot size is about five times smaller; at the overlap region, the intensities of the two beams are thus roughly equal. A beam splitter placed in the path of the incident and reflected beams is used to monitor simultaneously the phase-conjugated and reflected signals.

A phase conjugation reflection signal, which propagates along the reverse of the pump beam $\mathbf{E}_1$, becomes visible (see Figure 10.2*b*) when the input laser power is about 400 mW and the cell temperature is within 5° of the phase transition temperature. Typically, the phase conjugation reflectivity is about 2% at an input laser power of 800 mW. *In spite of the aberrations imparted by the input lens* and other optics, which lead to strong aberrations and large divergence on the incident and reflected beams, the phase-conjugated signal is spatially of the same quality as the input beam and has about the same divergence.

Similar observations of the SSOPC effect are reported by Antipov and Dvoryaninov (26), where the liquid crystal used is a very thick (millimeter) cell filled with nematic liquid crystals. The cell is placed between two electrodes, and an external low-frequency ac voltage is applied to create and stabilize some alignment of the highly scattering cell.

The observed dynamics of the self-starting phase conjugation process is governed by the optical nonlinearity and scattered noise amplitude (27), as well as the thermal grating buildup time. In general, the observed signal as a function of time is of the form given in Figure 10.3; the onset time $\tau^*$ is dependent on the noise amplitude, and the buildup time is dependent on the thermal grating response time.

Figure 10.4 shows the oscilloscope traces of the SSOPC signal as a function of the input pump power. Both the onset time and buildup time are shortened as the input pump power is increased. The total time it takes for the signal to build up to the maximum can be as short as 0.5 ms, at a pump power of about 800 mW. This shortening of the buildup and onset times has also been observed as the sample temperature is increased toward $T_c$ (25).

The fundamental process responsible for the self-starting phase conjugation effect discussed previously is the transient two-beam coupling via the thermal index change as discussed in Chapter 9 (28). This explains why the SSOPC signal recorded in Figure 10.4 eventually diminishes after reaching a peak value. The signal beam amplification constant in nematic liquid crystals

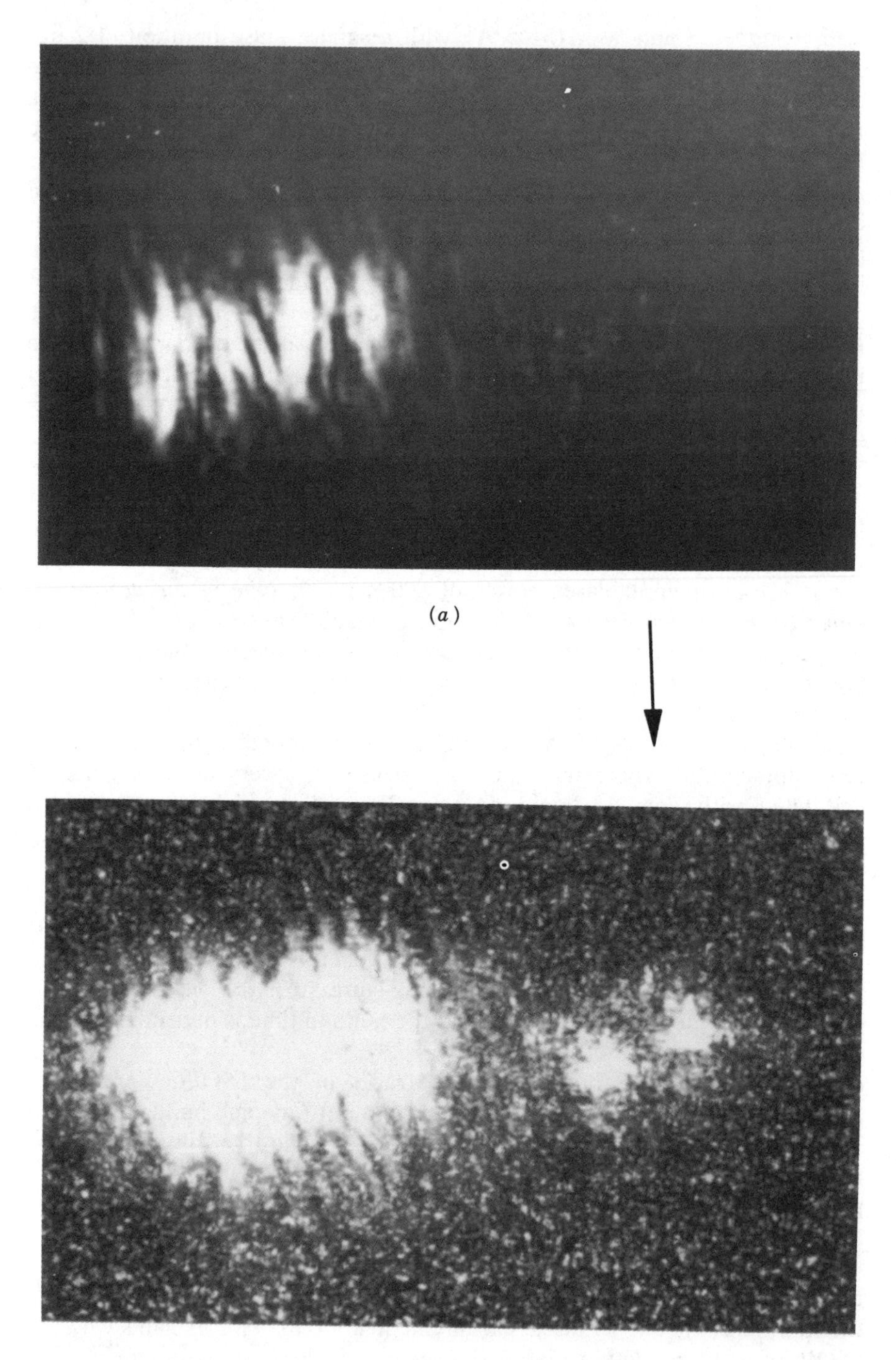

(*a*)

(*b*)

**Figure 10.2** Photograph of the reflected beam and phase conjugation signal. (*a*) At low input power, there is no phase-conjugated signal (arrow indicates location). (*b*) At higher input power, phase-conjugated signal (arrow) appears. Double reflections are due to the glass slide beam splitter used.

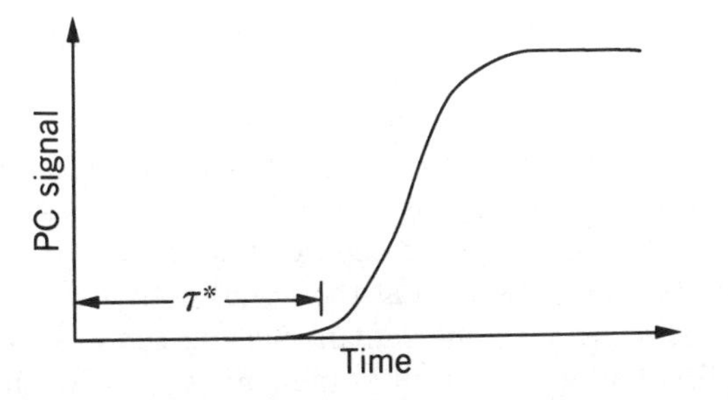


**Figure 10.3** Typical dynamics of self-starting optical phase conjugation process [27].

(9, 28) for visible–infrared laser pulses can be as high as 10 per 200 $\mu$m. Accounting for loss, this gives an absolute gain factor of 5 per 200 $\mu$m (i.e., an exponential gain factor of about 1.5). Therefore, for an interaction length greater than 1 mm in the self-starting phase conjugation experiment described previously, the signal exponential gain factor can be higher than 7, resulting in the generation of a coherent phase-conjugated signal.

There are some important details that remain to be documented. These include the dynamics of the transient beam amplification effects in liquid crystals and the quantitative roles of the (coherently) scattered noise in liquid

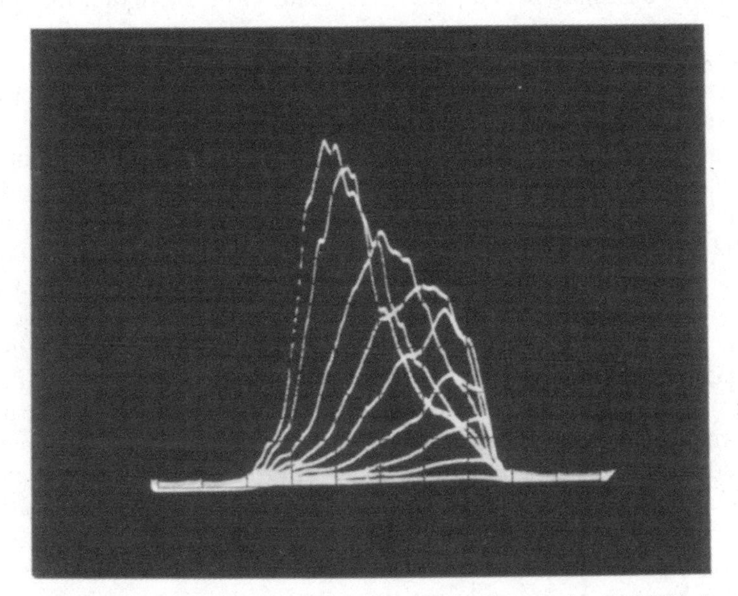

**Figure 10.4** Oscilloscope traces of the time evolution of the SSOPC signal as a function of input pump power ranging from 200 mW (lowest curve) to 800 mW (uppermost curve), showing the shortening of the onset and the build-up times. Time scale is 0.5 ms/div.

crystals in these forward beam amplification and self-starting phase conjugation effects. Some of these preliminary experimental studies, along with theoretical modeling of the dynamics of these processes, are described in Khoo et al. (28). Since the phase conjugation effect originates from the speckle noise source, the theory dealing with the effect of speckle noise in the dynamics of the self-starting phase conjugation process as reported in Khoo et al. (27) could be modified to treat the liquid crystal case. Scattering noise has also been shown to be important in placing some limitations on phase conjugation and holography, as well as mediating phase conjugation in highly disordered media (29).

### 10.2.2 SSOPC by Stimulated Orientational Scattering

The stimulated thermal wave mixing effect discussed previously can be used for applications in a broad spectral range, owing simply to the broadband birefringence of nematic liquid crystals. The required absorption constant for optimizing the processes can be tailored with appropriate dye doping. The principal drawback is that the efficiency is highly dependent on the proximity of the phase transition temperature $T_c$ and the requirement of very stable temperature control.

On the other hand, laser-induced nematic axis reorientation effects do not require such proximity to the phase temperature. Furthermore, the orientational fluctuations naturally provide efficient coupling between the ordinary and extraordinary waves. Indeed, we have recently succeeded in generating self-starting phase conjugation, using the stimulated ordinary–extraordinary $(e-o)$ wave mixing effect described next.

The process of $e-o$ wave mixing via stimulated molecular reorientations in nematic liquid crystals is analogous to the noise-initiated beam amplification and phase conjugation effects described in the preceding section. In this case the scattered noise from the incident beams originates from *orientational fluctuations* which, in conjunction with the birefringence of the nematics, give rise to orthogonally polarized scattered waves. Consider, for example, the interaction geometry depicted in Figure 10.5 in which a linearly input laser is incident normal to a planar aligned nematic film (i.e., an incident $o$ wave). Because of the orientational fluctuation $\Delta \mathbf{n}$, the scattered noise will contain an extraordinary wave component. Since the input and scattered waves are coherent, they could interfere and produce a moving grating along the direction of wave propagation. The moving grating is caused by the frequency difference $\Omega = \omega_o - \omega_e$ between the generated $e$ wave $(\omega_e)$ and the input $o$ wave $(\omega_o)$. The grating constant is given by $|q| = (\omega / c)(n_e - n_o)$. Depending on the duration of the input laser compared to the orientational grating relaxation time $\tau \sim \eta / Kq^2$ ($K$ is the elastic constant and $\eta$ is the viscosity), this process falls into the transient or steady-state regime.

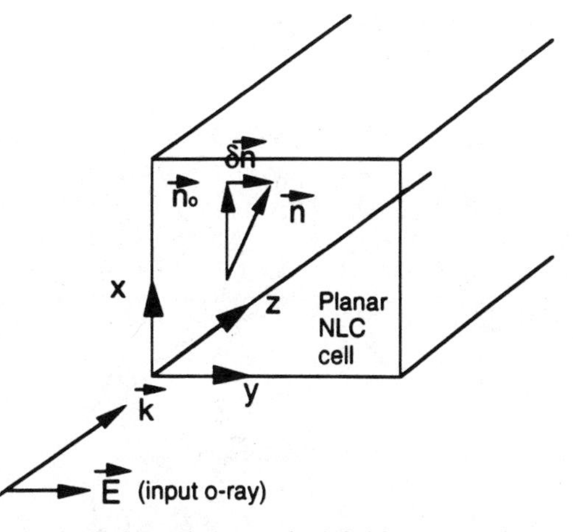


**Figure 10.5** Schematic depiction of the optical field propagating as an $o$ wave in a planar nematic liquid crystal cell. $\delta\mathbf{n}$ is the orientational fluctuation of the director axis. $\mathbf{n}$ is the perturbed director axis.

More quantitatively, consider the equation governing the laser-induced director axis reorientation $\delta n_y$:

$$\gamma \frac{\partial \delta n_y}{\partial t} - K_2 \frac{\partial^2 \delta n_y}{\partial z^2} = \frac{\Delta\varepsilon}{16\pi}\left(E_x E_y^* + E_x^* E_y\right) \tag{10.1}$$

where $\gamma$ is the viscosity coefficient, $K_2$ is the elastic constant, $\delta n_y$ is the director axis fluctuation, $\Delta\varepsilon$ is the optical dielectric anisotropy ($\Delta\varepsilon = \varepsilon_{\parallel} - \varepsilon_{\perp}$), $E_x$ is the amplitude of the incident optical electric field, and $E_y$ is the amplitude of the scattered cross-polarized field.

The coupling of the incident and scattered waves is, in the steady state, described by the coupled-wave equations (cf. Chapter 9):

$$\frac{d^2 E_x}{dZ^2} + k_e E_x = -\frac{\omega_e^2}{c^2}\Delta\varepsilon\, \delta n_y\, E_y \tag{10.2}$$

$$\frac{d^2 E_y}{dZ^2} + k_o E_y = -\frac{\omega_o^2}{c^2}\Delta\varepsilon\, \delta n_y\, E_x \tag{10.3}$$

Solving for the director axis fluctuation $\delta n_y$ in the steady state and (10.2) and (10.3) for an input $e$ wave ($E_x$) gives a steady-state solution for $E_y$ of the form

$$|E_y|^2 \sim \exp gz \tag{10.4}$$

The exponential gain coefficient $g$ is given by

$$g = \frac{\omega_o^2}{c^2} \frac{\Delta\varepsilon^2 |E_x(0)|^2}{16\pi K_2 q^2 k_e} \frac{\Omega\Gamma}{\Gamma^2 + \Omega^2} \tag{10.5}$$

where $\Omega = \omega_e - \omega_o$.

The maximum gain $g_{\max}$ occurs at $\Omega = \Gamma$ and is given by

$$g_{\max} = \frac{k\,\Delta\varepsilon^2}{4cn_o n_e K_2 q^2} I \tag{10.6}$$

Consider, for example, the following typical (experimental) parameter values: $\lambda = 0.5\ \mu\text{m}$, grating constant $\Lambda = 2\pi/q = 10\ \mu\text{m}$, sample thickness $d = 100\ \mu\text{m}$, Frank elastic constant $K_2 = 3 \times 10^{-7}$ dyne, $\Delta\varepsilon = 0.6$, $n_o = 1.5$, and $n_e = 1.7$ (for the nematic liquid crystal E7). Then we have

$$g_{\max} d = 2.5 \times 10^{-4} I\ (\text{W/cm}^2) \tag{10.7}$$

where $I$ is the intensity of the incident (pump) beam in units of watts per square centimeter.

Therefore, using a pump intensity of about 20 $\text{kW/cm}^2$ (which can be easily obtained by a lightly focused laser beam with powers on the order of 1 W), one could obtain an amplification factor of exp(5). This would be sufficient for converting the spontaneous $o$-polarized noise into a coherent signal beam.

Transient effects have been investigated by Zeldovich et al. (30), where a fairly high power pulsed ruby laser is required to generate observable signals. The more interesting and practically useful one is the steady-state case analyzed previously, corresponding to an input laser pulse duration longer than $\tau$. Typically, for the grating constant $\lambda_q = 2\pi/|q| = \lambda/\Delta n \sim 3\ \mu\text{m}$, the corresponding orientational relaxation time constant is on the order of about $\tau \approx 10$ ms, using typical liquid crystal parameters ($K \approx 3 \times 10^{-7}$ dyne, $\gamma \sim 1.2$ P). Using laser pulse durations much longer than this will allow one to observe the steady-state version of the $o-e$ ray stimulated scattering effects at relatively low laser power. This is indeed proven in some recent preliminary experimental studies conducted in the author's laboratory (31).

Figure 10.6 depicts the experimental setup. A linearly polarized low-power $Ar^+$ laser pulse of ms duration is focused onto a planar aligned 100-$\mu$m-thick nematic liquid crystal (E7) either as an $e$ or an $o$ wave. A cross polarizer at the exit monitors the generated orthogonally polarized wave. Figure 10.7*a*

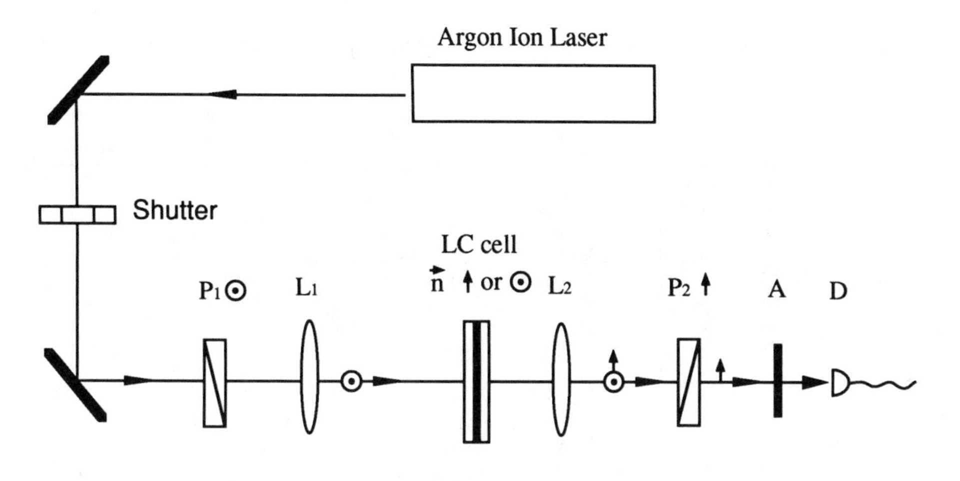


| | |
|---|---|
| Laser : | CW Ar gon Ion laser, 5145 Å |
| LC cell: | Planar, E7, 200 μm |
| $P_{1,2}$ : | Polarizers |
| $L_{1,2}$: | Focusing Lens, f1=10cm, f2=5cm |
| A: | Attenuator |
| D: | Photo-detector, connected to oscilloscope |

**Figure 10.6** Experimental setup for observing $o-e$ wave stimulated scattering. Sample is located at the focal plane of the input lens.

shows the experimental observations; indeed, above a certain threshold, there is significant stimulated scattering of the input $e$ wave into the $o$-wave component. It is important to note that, even without optimization, the threshold power for observing the effect is very small—only 400 mW, compared to a typical stimulated scattering threshold power in the kilowatt and megawatt range. Figure 10.7*b* shows the reverse of the stimulated scattering process (i.e., the incident laser is an $o$ wave, and it stimulates the scattering of an $e$ wave). The threshold for this $o-e$ scattering process is found to be similar to the $e-o$ process.

These stimulated orientation scattering effects were utilized for generating self-starting optical phase conjugation (32). The experimental setup is shown

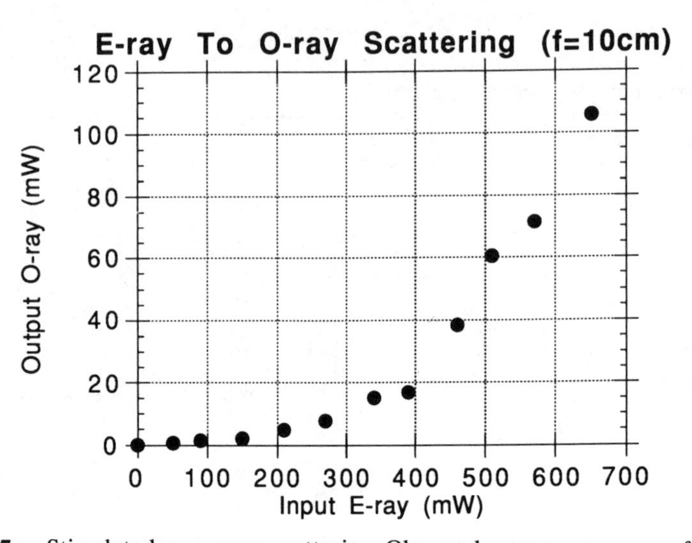


**Figure 10.7*a*** Stimulated $o-e$ wave scattering. Observed $o$-wave power as a function of the input $e$-wave power, showing a nonlinear dependence above a certain threshold input value.

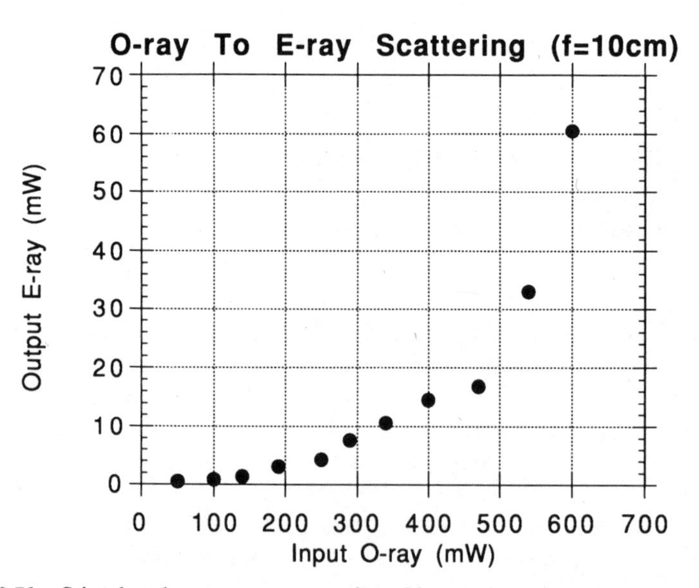


**Figure 10.7*b*** Stimulated $e-o$ wave scattering. Observed $e$-wave power as a function of the input $o$-wave power, showing a nonlinear dependence above a certain threshold input value.

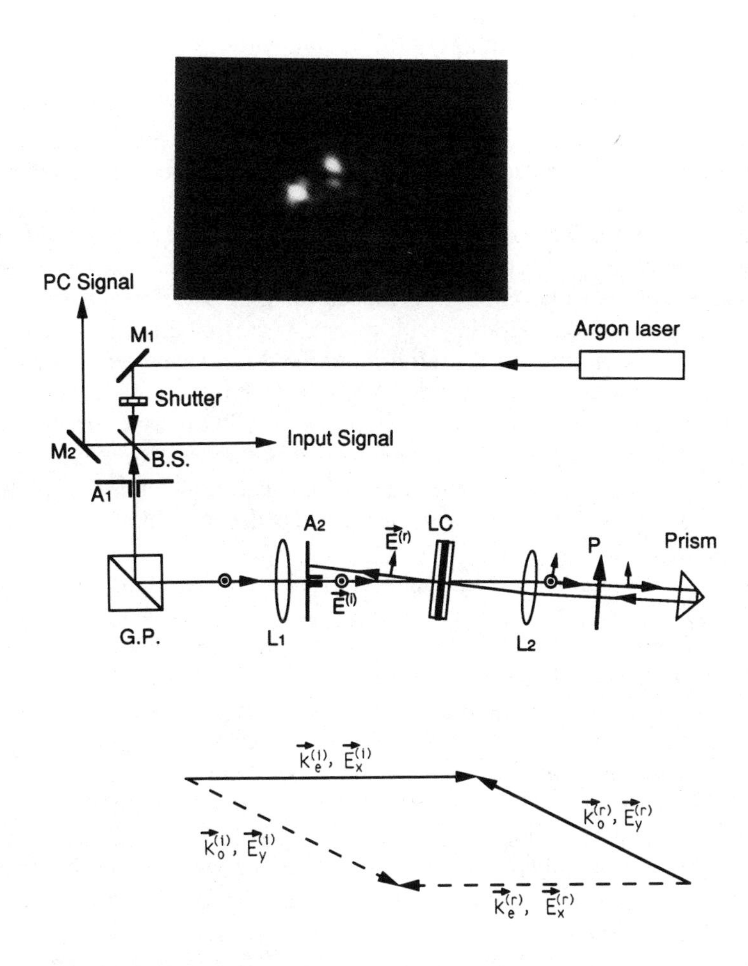


**Figure 10.8** Experimental setup for $e-o$ phase conjugation. Photograph shows the observed signal (arrow). Also shown is the $e-o$ wave vector phase-matching condition. Spot diameter of the laser at the location of the sample is about 50 $\mu$m. Crossing angle is 3°.

in Figure 10.8. The liquid crystal used is E7 (from EM Chemicals), which has a nematic-isotropic phase transition temperature of 63°C. The material has negligible absorption ($< 0.01$ cm$^{-1}$) at the argon laser wavelength (5145-Å line) used. A planar aligned nematic liquid crystal film of 100 $\mu$m thickness is made by sandwiching the E7 between two rubbed polymer-coated glass slides. The experiment is conducted at room temperature without the use of temperature cell.

A linearly polarized cw argon laser operating at the 5145-Å line is electronically chopped to yield pulses of variable millisecond duration. It is focused onto the nematic liquid crystal sample with its electric field polarization vector **E** parallel to the director axis $\mathbf{n}_o$ (i.e., an *e* wave). The transmitted beam is reflected, focused back onto the sample, where it intersects the incident beam at a crossing angle in air of 3°. The polarization of the reflected beam is rotated so that it is orthogonal to the polarization direction of the incident beam (i.e., an *o* wave).

The insert in Figure 10.8 depicts the wave vector matching condition for the case where the incident wave $E_x^{(i)}$ is an *e* wave. The scattered *o*-wave component $E_y^{(i)}$ is coherent with respect to, and interferes with, $E_x^{(i)}$ to produce an orientational grating with a wave vector **q**. Energy is transferred from $E_x^{(i)}$ to $E_y^{(i)}$ via the stimulated orientational scattering effect mentioned previously. Similarly, the reflected *o* wave $E_y^{(r)}$ interacts with its scattered *e*-wave component $E_x^{(r)}$ with an orientational grating **q** which matches that produced by the incident wave. These processes thus reinforce one another, leading to coherent signal output; that is, the scattered "noise" signals $E_y^{(i)}$ and $E_x^{(r)}$ will grow into a coherent beam when the laser power exceeds the threshold for SOS.

The SSOPC signal is photographed at an observation plane located about 5 m away. The onset dynamics of the SSOPC signal is monitored by a photodiode and recorded on a storage oscilloscope. The threshold power for SSOPC is similar to that required for the forward SOS effect in the same sample. When the power of the incident beam is small ($<300$ mW), only speckle noise appears on the observation plane. Above an input pump power of 300 mW, a well-defined beam of the phase-conjugated signal becomes clearly visible, as shown in the insert in Figure 10.8. The phase-conjugated signal spatial quality is similar to the input laser beam and has approximately the same divergence, in spite of the aberrations imposed by all the optics in front of the nematic cells. At an input power of 1 W, the maximum efficiency of the SSOPC effect is about 5%.

The dynamics of the self-starting phase conjugation and the forward stimulated *o-e* scattering process are shown in Figures 10.9*a* and 10.9*b*, respectively. Below the thresholds, only background noises, due mainly to random orientational fluctuation-induced cross-polarization scattering, are detected. The coherent signals are detected as monotonically increasing signals with build up times on the order of 45 ms (inpur power of 300 mW) and 30 ms (input power of 560 mW) for the $e-o$ self-starting phase conjugation (Figure 10.9*a*), and 7 ms for the $e-o$ stimulated scattering process (Figure 10.9*b*). The latter is in agreement with the theoretical estimate given previously.

As in most self-starting optical phase conjugation originating from speckle noise (27), the onset dynamics of the signal consists of two regimes: an onset time $\tau^*$ and a buildup time $\tau$. The length of the onset time is dependent on the noise characteristics and amplitude, whereas the buildup time is dependent on the material response time. For a grating constant of about 9 $\mu$m (wave mixing angle of 3°), the orientational grating response time is given by

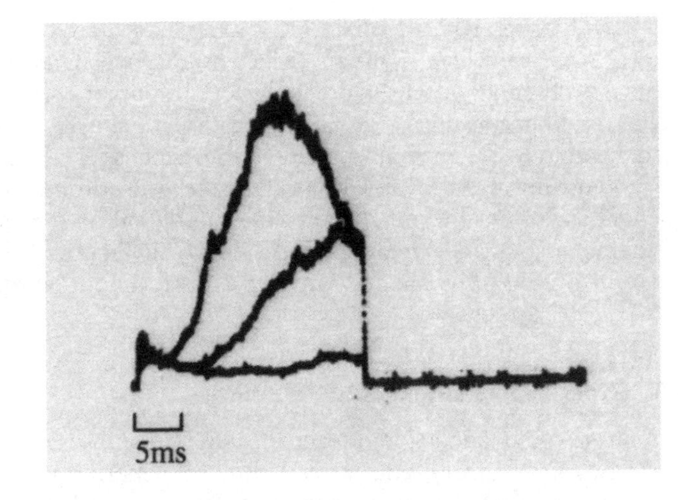


**Figure 10.9a** Observed buildup dynamics of the $e-o$ phase conjugation signal for three input powers: 300 mW (lowest trace showing noise only), 400 mW (signal appears) and 560 mW (upper trace; signal appears earlier). Time scale is 5 ms/div. Pulse duration is 25 ms.

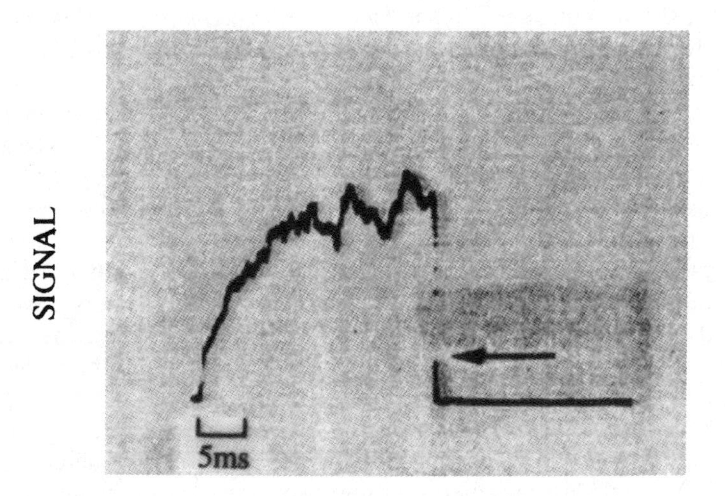


**Figure 10.9b** Oscilloscope trace of the time evolution of the stimulated $o$-wave signal for an input $e$-wave square pulse power of 500 mWatt. Input pulse duration = 25 ms. Arrow indicates the background noise level.

$\tau_r \approx \gamma / K_2 q^2$. Using typical values of $\gamma \approx 1.2$ P, $K_2 = 5 \times 10^{-7}$ dyne, and $q = 2\pi / 10$ $\mu$m, we get $\tau_r \approx 50$ ms, which is in good agreement with the experimental observation (cf. Figure 10.9*a*). For a larger crossing angle (i.e., a smaller grating constant), the SOS response time will be considerably reduced by the $q^{-2}$ dependence but at the expense of higher threshold power [cf. equation (10.6) which shows that the gain will be reduced by the $q^{-2}$ dependence]. On the other hand, the onset time is dependent on the noise characteristics and amplitudes. As shown in Figure 10.9*a*, the onset times are measured to be about 15 ms and 5 ms for input power of 300 mW and 560 mW, respectively.

### 10.2.3 Other Phase Conjugation Studies

Optical phase conjugation studies using degenerate four-wave mixing and/or transient two-wave mixing effects have also been performed with infrared lasers by Huignard et al. and spatially partially coherent light by Leith et al. (8). Using moderate-power cw or millisecond pulsed $CO_2$ lasers, Huignard and Khoo et al. (9) have observed large beam amplification and phase conjugation self-oscillation effects. One could extend these studies to practical devices such as self-pumped oscillators, phase conjugators, and other real-time holographic devices in spectral regimes not accessible by other materials. One major drawback of using liquid crystals for infrared (around 10 $\mu$m) optical wave mixing applications is the generally large absorption losses in this spectral regime. Typically, they range from 40 to 100 cm$^{-1}$, which imposes a severe limitation on the interaction length (i.e., less than 200 $\mu$m). In the mid-infrared regime (1 to 5 $\mu$m), nematics are less absorptive; typically, the absorption contrast is about 10 cm$^{-1}$, and thus longer interaction lengths are possible. Since there are few competitive materials in these spectral regimes that could provide SSOPC, we could expect nematic liquid crystals to play an important role in future studies of SSOPC involving near- and mid-infrared lasers.

In a different but equally important context, nematic liquid crystals have been employed in the study of the speckle noise effect in optical phase conjugation using coherent laser beams (8). By reducing the spatial coherence of the laser with rotating ground glass, these workers have demonstrated a high-resolution, low-noise phase conjugation imaging technique, which is complementary to the multiple-exposure method devised by Huignard et al. (34).

Figure 10.10 depicts the experimental setup used by Leith et al. and Khoo (8). A cw low-power ($\approx 200$ mW) $Ar^+$ ion laser is collimated through a rotating ground glass and then split into two beams (beam diameter $\approx 3$ mm): a strong reference beam and an object beam that illuminates an object (e.g., a slit). These two beams are combined at the image plane of the lens system ($L_3$ and $L_4$) where a nematic liquid crystal film is placed. The thermal nonlinearity of an MBBA sample is used. A mirror $M_4$ reflects the reference beam to provide a counterpropagating reading beam, which generates a

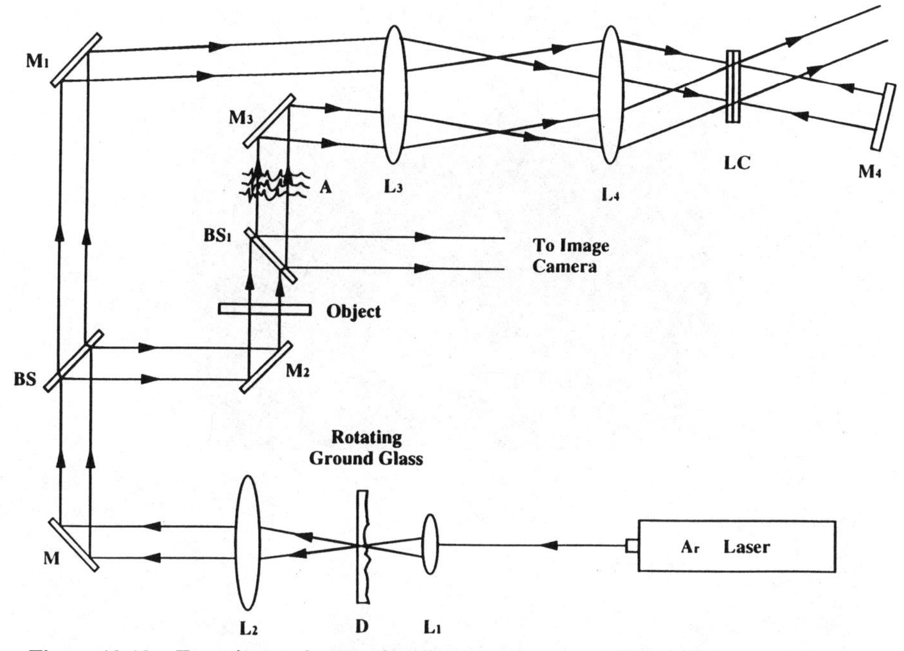


**Figure 10.10** Experimental setup for phase conjugation with spatially partially coherent laser.

phase-conjugated signal that propagates back along the object beam. After traversing the aberrator (a corrugated glass plate), the phase-conjugated signal is monitored via a beam splitter. Typically, the phase-conjugated reflection efficiency is on the order of a few percent at the level of reference beam intensity used.

Besides demonstrating the ability to correct for distortion, this study also demonstrates that the use of a spatially incoherent laser beam (resulting from the rotating ground glass) will eliminate speckle noise that plagues imaging processes with coherent laser beams. Figure 10.11*a* shows a photograph of a phase-conjugated reconstructed image of the laser beam with a spatially coherent light (i.e., no use of rotating ground glass); the coherent noise produces serious interference and degradation of the signal. On the other hand, as shown in Figure 10.11*b*, the reconstructed signal with spatially incoherent light is evidently free of such coherent noise and gives a rather well-defined image. Since liquid crystals, owing to their large birefringence and orientational fluctuations, are inherently noisy in terms of random scatterings, this particular method of reducing noise effects, as well as those devised by the Huignard group, are clearly important for phase conjugations or other real-time nonlinear optical holographic imaging processes involving liquid crystal films.

Stimulated scattering and phase conjugation effects have also been briefly observed in a smectic-A liquid crystal film [20], where the nonlinearity

(*a*)

(*b*)

**Figure 10.11** (*a*) Reconstructed image of the laser beam with a spatially coherent light. Noise effects have been greatly exaggerated for illustration purposes. (*b*) Reconstructed image of the laser beam with a spatially partially coherent laser showing little background noise.

involved is the laser-induced density change; that is, it is a stimulated Brillouin scattering process. These density effects are also responsible for the interference effects observed in the diffraction from nanosecond laser pulse induced gratings in smectic liquid crystal film. To date, however, there has not been much activity on optical wave mixing studies in the smectic-A phase.

In the isotropic phase optical phase conjugation effects have been reported by Fekete et al. (6), Madden et al. (7) (discussed in Chapter 6), and the author's group (34), which will be discussed in the next section.

## 10.3 SELF-FOCUSING, SELF-DEFOCUSING, AND OPTICAL LIMITING

### 10.3.1 Introduction

In earlier studies of self-focusing effects in liquid crystals, the main emphasis was on the understanding of fundamental phenomena in nonlinear optics or liquid crystals. In Wong and Shen (5), for example, the isotropic phase of liquid crystals is used because the molecular reorientation time is tunable over a wide range by varying the proximity of the temperature to $T_c$. By using laser pulses of pulse duration within this range, Wong and Shen (5) have studied the process of self-focusing under various temporal conditions, ranging from the steady-state to the transient regimes.

In the nematic phase, the fact that the liquid crystal film is thin but highly nonlinear allows one to employ the so-called nonlinear diffraction theory discussed in Chapter 9. Studies of this nonlinear diffraction phenomenon have provided not only detailed verification of the nonlinear diffraction process, but also practically useful techniques or processes (3–4a, 35, 36). In Khoo et al. (4, 4a), for example, the self-phase modulation effect is used for demonstrating the optical limiting (cf. Figure 10.12*a*) of a cw infrared laser. The self-focusing effect has also been employed for nanosecond laser pulse self-limiting action, as we will discuss in more detail in the next section (35, 36).

More specifically, the fact that self-phase modulation effects in thin nematic films can be quantitatively described by a nonlinear diffraction theory allows one to quantitatively characterize the optical limiting process. Using the formalism presented in Chapter 9, one can easily calculate the threshold value for the optical limiting effect to set in [i.e., the input intensity or fluence (energy/area) value above which the detected on-axis output signal deviates from a linear dependence (cf. Figure 10.12*b*)]. A straightforward numerical evaluation of the detected on-axis power using (9.99) for the setup depicted in Figure 9.14 shows that, typically, the threshold value corresponds to when the magnitude of the nonlinear on-axis phase shift $\phi_{NL}$ $(r = 0)$ is on the order of $\pi$, although it varies somewhat for different beam diameter at the sample (cf. Figure 10.12*c*). This phase shift value is also the

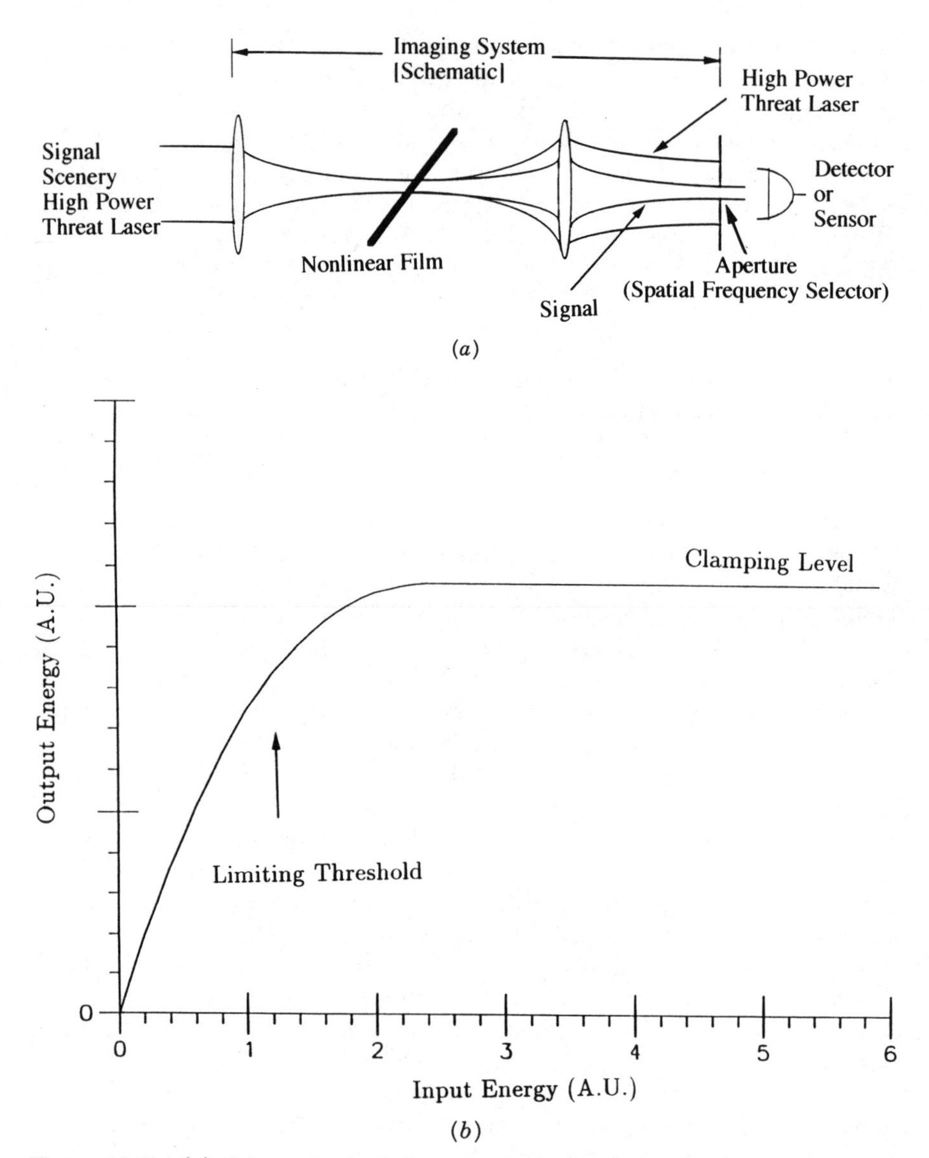


**Figure 10.12** (*a*) Schematic depiction of self-limiting application in an imaging system using self-defocusing effect. (*b*) A typical output/input dependence showing optical self-limiting action. (*c*) Magnitude of the nonlinear on-axis threshold phase shift required for self-limiting for various beam diameters. Parameters used are the same as those for Figure 9.15.

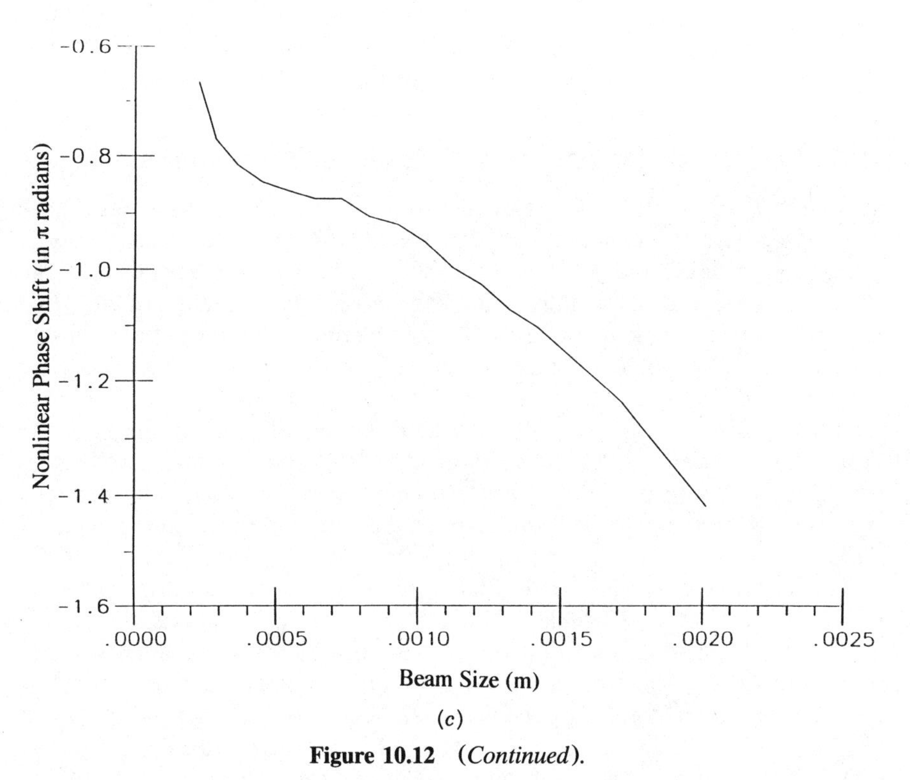


(*c*)

**Figure 10.12** (*Continued*).

point at which the transmitted self-phase modulated intensity distribution begins to assume ring or donut forms (cf. Figures 9.12 and 9.13). With this information on the value of the phase shift and other geometrical material and temporal parameters (needed for cases involving pulsed lasers as will be discussed presently), one can optimize the design of an optical self-limiter using nematic liquid crystal films or other nonlinear thin films.

It is also obvious from the nonlinear diffraction theory given in Chapter 9 that the on-axis intensity distribution at the observation plane located in the far field will vary as the nonlinear film is moved around the focal plane of the input lens. If the film is nearer to the input lens than the focal distance, the incident beam wave front curvature is negative. On the other hand, if the film is farther away from the lens than the focal length, the incident beam wave front curvature is positive. Accordingly, as the film is moved about the focal plane, the observed on-axis intensity will move through maxima and minima, depending on whether the nonlinearity is positive ($n_2 > 0$) or negative ($n_2 < 0$). This beam wave front curvature dependence has been labeled as a distance ($z$ from the input lens) dependent effect, and the technique is called $z$ scan (37). Using this so-called $z$-scan technique, Janossy (38) has measured the orientational nonlinearity of dye-doped liquid crystals.

### 10.3.2 Optical Limiting Action of Nanosecond Laser Pulses

The optical self-limiting effects of nanosecond laser pulses have previously been observed with nematic liquid crystal films (39). Nematic films, however, possess some disadvantageous features such as polarization selectivity, large scattering loss, and short interaction length. Isotropic liquid crystal (ILC) nonlinear optical responses are not polarization selective; optical losses caused by orientational fluctuations are of negligible magnitude, allowing much longer interaction lengths than in the nematic phase. Since ILCs are fluid like, thick cells can be easily fabricated. In fact, as discussed in the latter part of this section, we have also fabricated long liquid crystal cored fibers which exhibit optical limiting and switching effects with greatly reduced thresholds. The following is a summary of the results reported in Khoo et al. (36).

The laser used in the self-limiting experiment is the second harmonic ($\lambda = 0.532$ $\mu$m) of a $Q$-switched Nd:Yag laser; the pulse duration is 20 ns. The output beam size is magnified by a telescope to form a collimated beam of 1 cm diameter. The input and transmitted laser pulses are detected by fast photodiodes and a storage scope. The peak heights of the laser pulses, which provide a measure of the laser energies, are recorded and tabulated. The sample cell used is 1 mm thick and is filled with an isotropic liquid crystal mixture TM74A (from EM Chemicals) doped with different concentrations of the dye D16 to increase its absorption. The sample is maintained at room temperature (23°C) which is about 10°C above the phase transition temperature of the mixture.

Figure 10.13 shows a typically observed output versus input curve for a slightly doped sample ($\alpha \approx 2$ cm$^{-1}$), demonstrating that optical limiting sets in at an input energy of about 1 mJ. The laser beam diameter of the sample is about 2 mm. This corresponds to a threshold fluence of 0.12 J/cm$^2$. The threshold fluence varies as the laser beam diameter is varied. For example, it rises to a rather large value of about 5 J/cm$^2$ at a beam diameter of 28 $\mu$m. These variations are summarized in Figure 10.14. The results for the slightly doped sample show that the threshold fluences stay roughly in the 1.8- to 1.0-J/cm$^2$ range for beam diameters ranging from 150 to 45 $\mu$m.

In this range of beam sizes, the observed threshold fluences for the more absorbing dyed sample ($\alpha \approx 10$ cm$^{-1}$) are generally lower by a factor of about 4. The results are therefore consistent with the interpretation that absorptive thermal nonlinearity is responsible for the self-limiting action. Note that for beam diameters ranging from 40 to 8 $\mu$m, the threshold fluences increase from 2 to 8 J/cm$^2$, which are generally more than an order of magnitude higher than those observed at larger laser beam diameters.

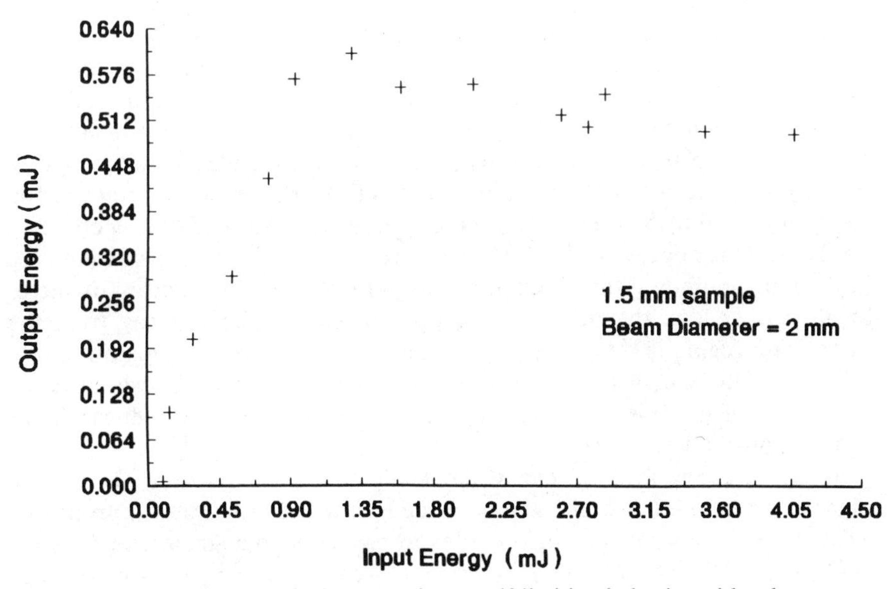


**Figure 10.13** Observed typical output/input self-limiting behavior with a beam spot diameter of 2 mm on an isotropic liquid crystal film.

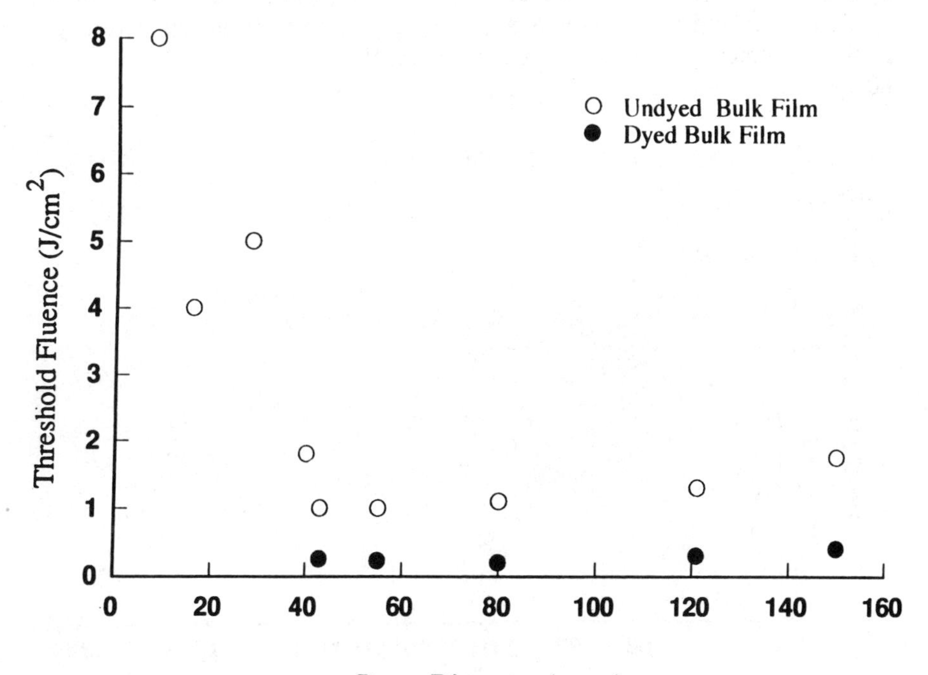


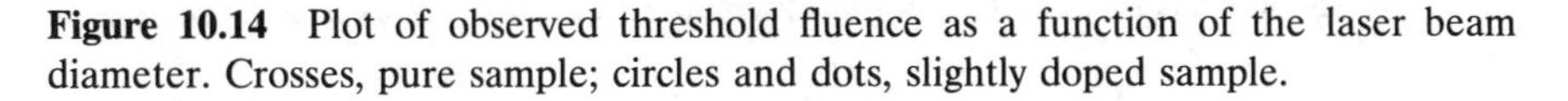


**Figure 10.14** Plot of observed threshold fluence as a function of the laser beam diameter. Crosses, pure sample; circles and dots, slightly doped sample.

As explained in Chapter 8, laser heating of the ILC gives rise to a density change $\Delta\rho$, which is responsible for the change in index. The main contribution in a time scale of 20 ns or less (i.e., the laser pulse length used is 20 ns) is the density component which is propagative (the typical sound velocity is on the order of 1500 m/s). It is thus possible that the induced index change (density fluctuation) could propagate away from the laser–ILC interaction region for a tightly focused beam, thereby diminishing its defocusing effect on the laser. This appears to be a plausible explanation for the drastic increase in the optical limiting threshold at small spot diameter. For a beam diameter of 40 $\mu$m or less, the density wave propagates completely away from the interaction region in a time of 26 ns or less; at 8 $\mu$m, the propagation time is only 5 ns. This explains the observed increase in the limiting threshold value, as a result of the less effective temporal overlap between the induced index change and the laser pulse.

It is interesting to note that while smaller laser beam size would require higher threshold fluence, the ability of the induced density wave to propagate away from the interaction region helps to reduce its damaging effect on the ILC sample. We have conducted a preliminary study of laser-induced damage on ILCs [defined as the point where a bubble (40) is created in the ILC]. Figure 10.15 depicts the dependence of the so-called "damage" threshold as a function of the beam diameter in a 1.5-mm-thick cell, showing that the damage threshold increases very substantially for small beam diameter. At a beam diameter of 28 $\mu$m, for example, the damage fluence is about 80 J/cm$^2$.

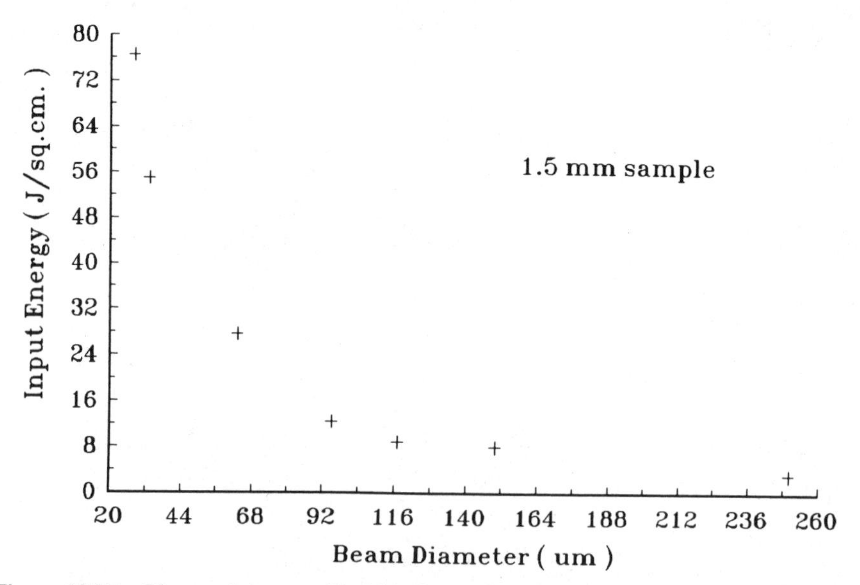


**Figure 10.15** Observed damage "bubble formation" fluence versus input laser diameter.

### 10.3.3 Self-Limiting with Isotropic Liquid Crystal Fiber

It is well known that the nonlinear optical responses of a material will be greatly enhanced, and the efficiency of the nonlinear phenomena under study could be greatly enhanced, if guided wave geometries are employed (41, 42). Recent studies have been conducted on liquid crystal cored fibers where several nonlinear optical phenomena were observed at relatively much lower optical threshold fluence (35). In particular, passive all-optical limiting actions have been found to occur at threshold fluences a few times smaller than their thin film and bulk cell counterparts. Concomitant to these observations, stimulated backscattering and phase conjugation effects also appear.

In these studies liquid crystal cored optical fibers are fabricated by filling glass capillaries with an isotropic liquid crystal (Figure 10.16*a*). Fiber arrays composed of a parallel assembly of these liquid crystal cored fibers could also be fabricated by filling commercially available capillary arrays (Figure 10.16*b*).

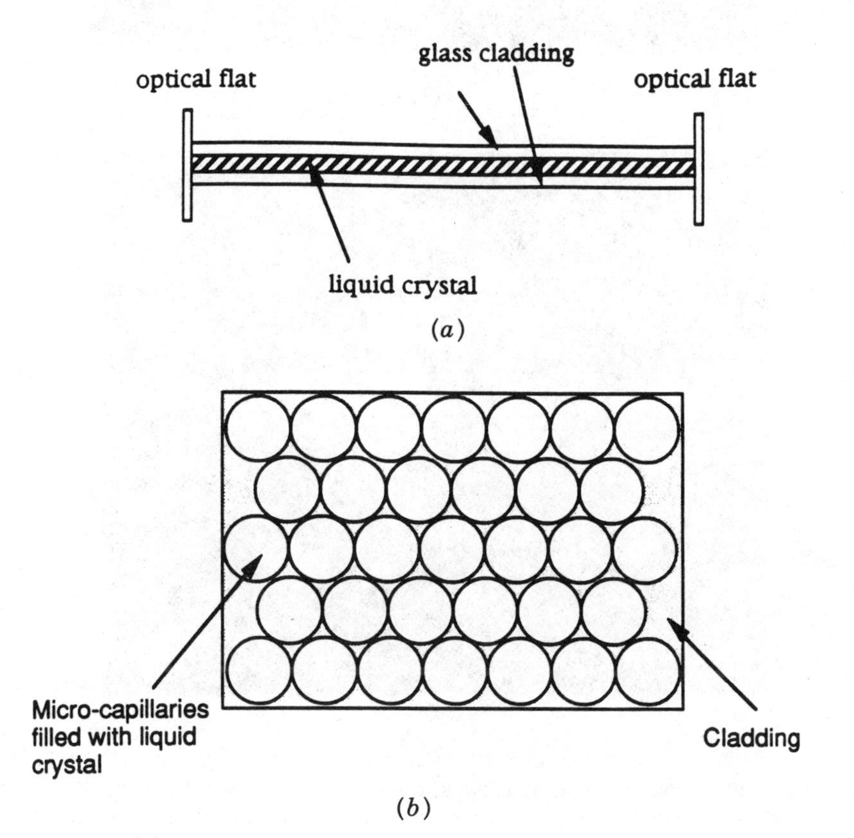


**Figure 10.16** (*a*) Schematic depiction of the construction of a liquid crystal cored optical fiber. (*b*) A liquid crystal fiber array. The core diameter can vary from 10 to 100 $\mu$m.

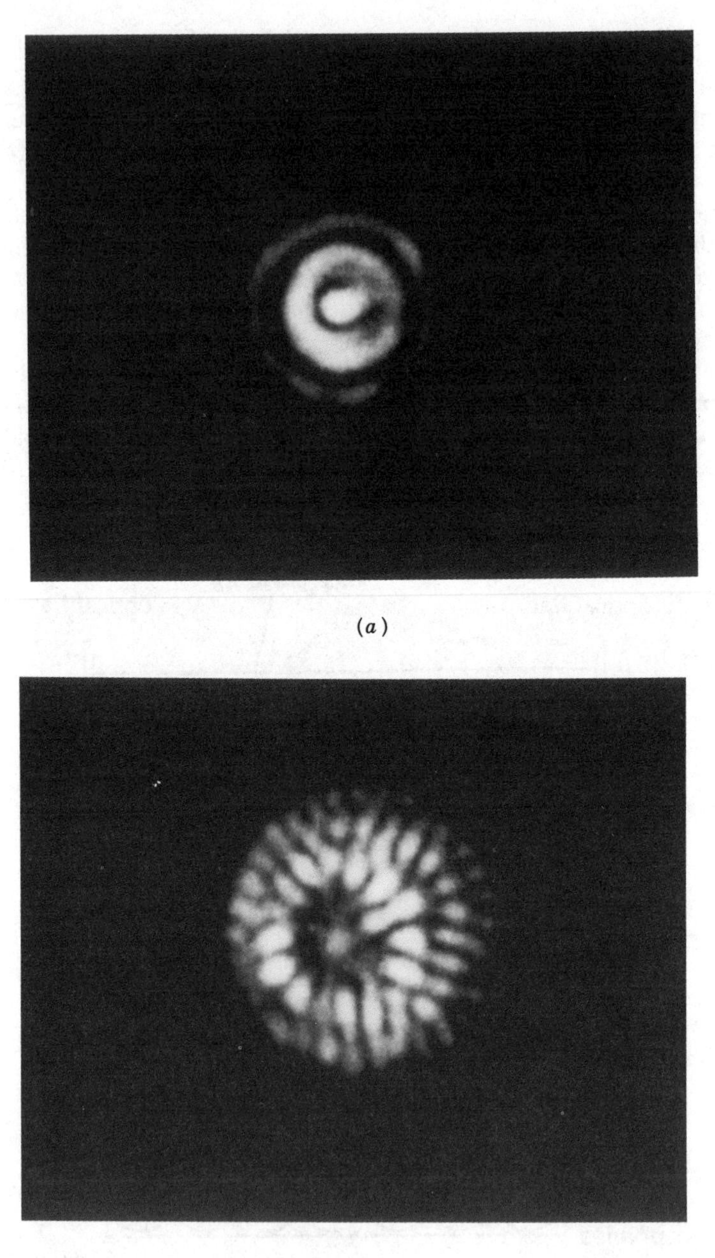

(*a*)

(*b*)

**Figure 10.17** Photograph of the far-field transmitted multimode intensity profile of a 20-ns Nd:Yag second-harmonic laser pulse through a 26-$\mu$m-diameter liquid crystal fiber. (*a*) Laser beam is centrally focused onto the fiber. (*b*) Laser beam is incident at a slightly off-axis angle. Both show multimode excitation.

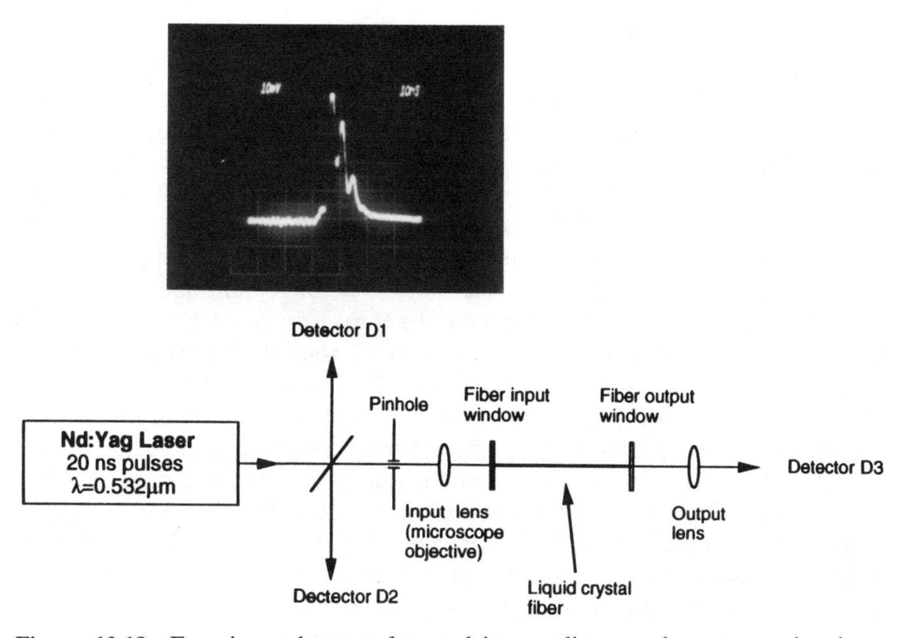


**Figure 10.18** Experimental setup for studying nonlinear pulse propagation in a liquid crystal fiber. The photograph shows the laser temporal profile.

For controlled light coupling into the filled fiber, the input and exit end surfaces of the liquid crystal cored fiber should be maintained flat. This is achieved by sealing the ends with a glass flat (cf. Figure 10.16*a*).

In the study reported in Khoo et al. (35), two types of liquid crystals have been used. One is a commercially available mixture of cholesteric liquid crystals in the isotropic phase (TM74A, from EM Chemicals). The other is 5CB (also from EM Chemicals). TM74A is isotropic at room temperature, whereas 5CB has a clearing temperature of about 35°C and has to be kept in a heated temperature cell to operate in the isotropic phase. Both liquid crystals give similar results. The typical refractive index of these glass capillaries and capillary arrays ranges from 1.45 to 1.53. On the other hand, the typical refractive index of isotropic phase liquids is about 1.54. Thus these liquid crystal cored glass capillaries function as nonlinear optical fiber waveguides. Figures 10.17*a* and *b* are photographs of the (linearly) transmitted far-field intensity distribution of a low-power 20-ns Nd:Yag second-harmonic laser pulse showing a multimode operation.

Under nanosecond laser pulse excitation, the principal nonlinear optical mechanism involved in the forward propagation is laser-induced density changes and therefore negative refractive index changes. This process, in the fiber geometry, leads to extended defocusing of the beam out of the guiding core, resulting in an optical self-limiting effect in the transmission. In a fiber array input image light rays are guided toward the output end where the image intensity distribution can be scanned by some sensor or detection system (e.g., a charge-coupled device). Optical limiting action and protection of the sensor or detector against high-power laser radiation (which traverses the system as an intense ray in one of the fibers) could utilize the self-defocusing effect mentioned previously. The defocusing effect is enhanced by the long interaction length defined by the input and the output ends of the array. The actual operation and principles of such imaging systems are understandably very complex, but obviously depend to a very large extent on the performance or working characteristics of a constituent fiber.

Figure 10.18 shows the experimental setup. The second harmonic of a 20-ns Nd:Yag laser pulse ($\lambda = 0.532\ \mu m$) of roughly Gaussian transverse intensity distribution is focused by a lens of 1 cm focal length into a liquid crystal fiber. The focal spot is located near the front entrance plane. The location of the sample entrance plane from the input lens is adjusted so that the diameter of the focused laser beam matches the fiber core diameter. The temporal profile of the laser is shown in the insert in Figure 10.18; there is some slight "ripple" owing to the presence of a second weaker longitudinal mode. The input, transmitted, and backscattered pulses are detected by fast photodiodes and recorded by a storage oscilloscope.

At low-input laser energies, the transmitted pulse shape is identical to the input. At high-input energies, the transmitted pulse shows obvious signs of a limiting effect similar to those discussed in Chapter 9. The insert in Figure 10.19 shows a typical waveform of the "limited" transmitted pulse, demonstrating greatly reduced transmission of the later portion of the pulse. As plotted in Figure 10.19, the transmitted detected energy versus the input also exhibits this deviation from linearity. A typical limiting behavior with a threshold of 2 $\mu J$ for the particular fiber used (core diameter, 26 $\mu m$; length, 3 cm) is observed. The (linear) scattering and absorptive losses of the fiber are quite small; most of the losses (about 20%) are due to interface reflection and coupling losses.

For this case we note that the threshold input fluence on the liquid crystal fiber is about 0.3 $J/cm^2$ (focal spot diameter, 26 $\mu m$; threshold energy, 2 $\mu J$). On the other hand, studies in bulk nematic or isotropic liquid crystal samples, as discussed in the preceding section, have shown that the threshold fluences for identical laser pulses at this focused spot diameter are at least 10 times larger (cf. Figure 10.14) owing to the greatly reduced interaction region between the laser and the induced density and index change in the bulk thin film samples. The present observation, corroborated by several other measurements in fibers of similar or smaller core diameter, shows that the extended region of interaction and the confined transverse dimension provided by the liquid crystal fiber waveguide are very effective in reducing the threshold fluence for tightly focused geometry.

Above an incident laser energy of about 60 $\mu J$, a stimulated backscattering signal is observed. Since the effects are observed at similar thresholds (between 60 and 80 $\mu J$) in both TM74A and 5CB, whose absorption con-

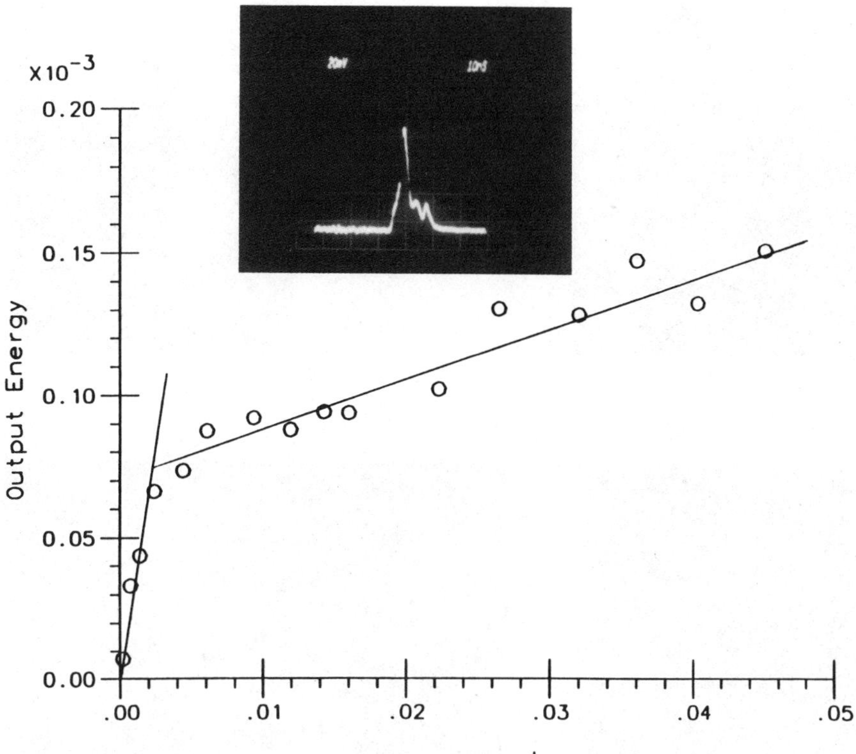


**Figure 10.19** Observed optical self-limiting effect in the transmission through an isotropic phase liquid crystal fiber. The detected output is expressed in an arbitrary unit; actual transmission (at low input power) of the fiber is about 80%. The photograph shows the temporal profile of the transmitted laser pulse through the liquid crystal fiber.

stants $\alpha$ at 0.532 $\mu$m are very different ($\alpha$ for TM74A is about 0.1 $cm^{-1}$, whereas $\alpha$ for 5CB is much smaller than 0.1 $cm^{-1}$), the effect is attributed to stimulated Brillouin scattering (SBS). An interesting feature of the observed effect is the aberration correction property associated with the phase conjugation property of SBS (22). Figure 10.20*a* shows the highly aberrated ordinary reflection from the fiber and (microscope objective) input lens system. On the other hand, the phase-conjugated SBS signal, as shown in Figure 10.20*b*, is a well-defined highly collimated beam which resembles the input beam pattern (Figure 10.20*c*). Another interesting feature of the SBS signal is that the pulse is temporally a much narrowed one, as shown in Figure 10.20*d*. This narrowing effect is attributed to the transient nature of the SBS process, owing to the rather short length (3 cm) of the fiber used (20, 43). These effects observed in liquid crystalline fibers are similar to

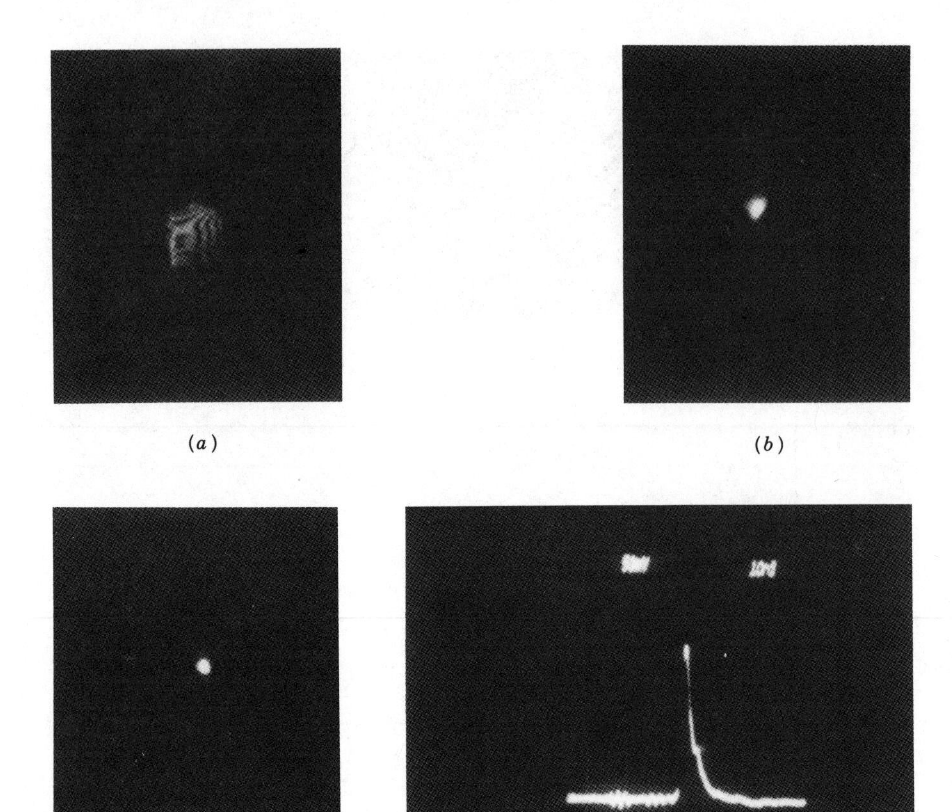


**Figure 10.20** Photograph of the laser spatial profile. (*a*) Highly aberrated plane reflection from the fiber input end. (*b*) Well-collimated phase-conjugated stimulated backscattering. (*c*) Incident laser for comparison. (*d*) Oscilloscope trace of the stimulated backscattering signal.

previous observations made in bulk liquid crystalline samples [5,20]. The main and important difference is the much lower threshold powers involved, due mainly to the extended tightly confined interaction length provided by the guided wave geometry.

## 10.4 HARMONIC GENERATIONS

Harmonic generations occupy a special place in the field of nonlinear optics; not only were they the first observed nonlinear optical effect (44), they are also the most widely researched and used in current lasers and electrooptic technology. While most commercially available materials for harmonic generations are inorganic crystals, research on organic materials is currently underway (45). This is due to the fact that organic molecules are usually large and anisotropic and may possess advantageous characteristics for optoelectronic applications.

Studies of harmonic generations in liquid crystals are still in the research (versus development) stage; both second- and third-harmonic generation effects have been observed in various liquid crystalline phases. Since the work by Shelton and Shen (11) on third-harmonic generations in liquid crystals, there has been very little work in this area, presumably because there are plenty of other materials for efficient third-harmonic generations, where the basic material physics are also interesting and challenging.

On the other hand, liquid crystal *molecules* are well known to be highly anisotropic and non-centrosymmetric and possess large second-order nonlinear molecular polarizability. For a typical liquid crystal such as 8CB (4′-*n*-octyl-4-cyanobipenyl), the polarizability is measured to be on the order of $25 \times 10^{-32}$ esu (46), which is much larger than the molecular polarizability of KDP. However, when these molecules assemble themselves in the liquid crystalline phases, they tend to assume configurations where this centroasymmetry is reduced to vanishing value [e.g., by having the polar directions of molecules (or molecular layers) lined up in opposite ways]. As a result, although the individual molecular polarizability of the liquid crystals is quite large, liquid crystals have not been shown to be efficient harmonic generators.

The centrosymmetry may, however, be broken by the application of an applied dc electric field (12); such symmetry can also be broken on a surface either as a freely suspended film or by a surface alignment modification technique that induces the flexoelectric effect (47). More recently, Sukhov and Timashev have shown that the centrosymmetry can also be broken optically (48). As explained in Chapter 9, the main obstacle in getting efficient harmonic generation then is the phase matching of the fundamental and second-harmonic wave vectors.

In the work by Saha and Wong (12), phase matching is achieved by the birefringent dispersion method discussed in Chapter 9, which is very commonly used for second-harmonic generations in nonlinear crystals. The fundamental wave propagates as the extraordinary ray and the harmonic wave as the ordinary ray. A dc electric field of 15 kV/cm is applied perpendicularly to the director axis of a planar nematic (5CB, *p*-*n*-pentyl-*p*′-cyanobiphenyl) sample. The observed harmonic signal as a function of the angle of deviation from the phase-matched direction is shown to be in good agreement with theory. From this experiment it appears that if the right geometry is chosen and if one uses the less "lossy" smectic phase, larger interaction lengths and perhaps higher harmonic generation efficiency could be attained.

Another possibility of breaking the centrosymmetry of liquid crystals is to make use of the flexoelectric effect (47, 48) (for a detailed explanation, see 48–50). Basically, if a nematic liquid crystal undergoes orientational deformations of the splay or bend type (49), a spontaneous polarization, the so-called "flexoelectric effect," will occur, resulting in a second-order nonlinear susceptibility. In the work by Sukhov and Timashev (48), a spatially periodic orientational distortion with a wave vector **q** is created by stimulated orientational scattering (cf. Section 10.2.2) of an *o* wave $\mathbf{E}_R$ into an *e*-polarized wave $\mathbf{E}_s$, where $\mathbf{E}_R$ is derived from a ruby laser. A weakly focused Nd:Yag laser fundamental beam at 1.06 $\mu$m is then incident on the sample, propagating along the $\mathbf{K}_H$ direction; its second harmonic is generated in the same direction with a wave vector $\mathbf{K}_N$. Phase matching is achieved by matching the orientational distortion grating wave vector **q** with the phase mismatch $\mathbf{K}_H - 2\mathbf{K}_N$ [i.e., $q(\alpha_0) = K_H - 2K_N$, where $\alpha$ is the (phase matched) angle made by $\mathbf{K}_R$ (the wave vector of the incident ruby laser) with the nematic director axis (51)]. Such a phase-matching method is analogous to the one employed by Shelton and Shen (11), where the pitch of the cholesteric liquid crystal plays the equivalent role of the periodic wave vector **q**.

The experimental results obtained by Sukhov and Timashev are in good agreement with their theoretical model. However, in analogy to the work by Shelton and Shen, these harmonic generation studies are more novel than useful, in terms of possible practical applications. There has also been recent research on second-harmonic generations in smectic and ferroelectric liquid crystals (13), but the observed conversion efficiency is still quite small compared to currently commercially available inorganic crystals.

Using liquid crystals with large anisotropies and some ingenious way of aligning the molecules (electrically or optically) to induce large second-order nonlinearity, it is likely that some of these novel studies could eventually result in practically competitive harmonic generators.

## 10.5 ALL-OPTICAL SWITCHING

Just as the linear electrooptical effects can be utilized in various electrically controlled switching processes, optically induced refractive index changes can be applied in optooptical switchings. These switching processes may be in the form of mixings of several waves or self-actions. In these processes the fundamental nonlinear optical parameter involved is the intensity-dependent *phase shift*, which involves a combination of the nonlinear index change and the optical path length. Such intensity-dependent shifts are also responsible for bistabilities, differential gain, transistor action, and so forth in nonlinear Fabry–Perot processes (14).

Optical switchings can also be mediated by the *refractive index change* alone. One good example is the nonlinear interface switching process (52, 53). Consider Figure 10.21. If medium 2 possesses an intensity-dependent index, then the transmission/reflection at the interface will also be intensity dependent. If medium 1 is of a higher index than medium 2, a total internally reflected (TIR) optical field in medium 1 may switch over to the

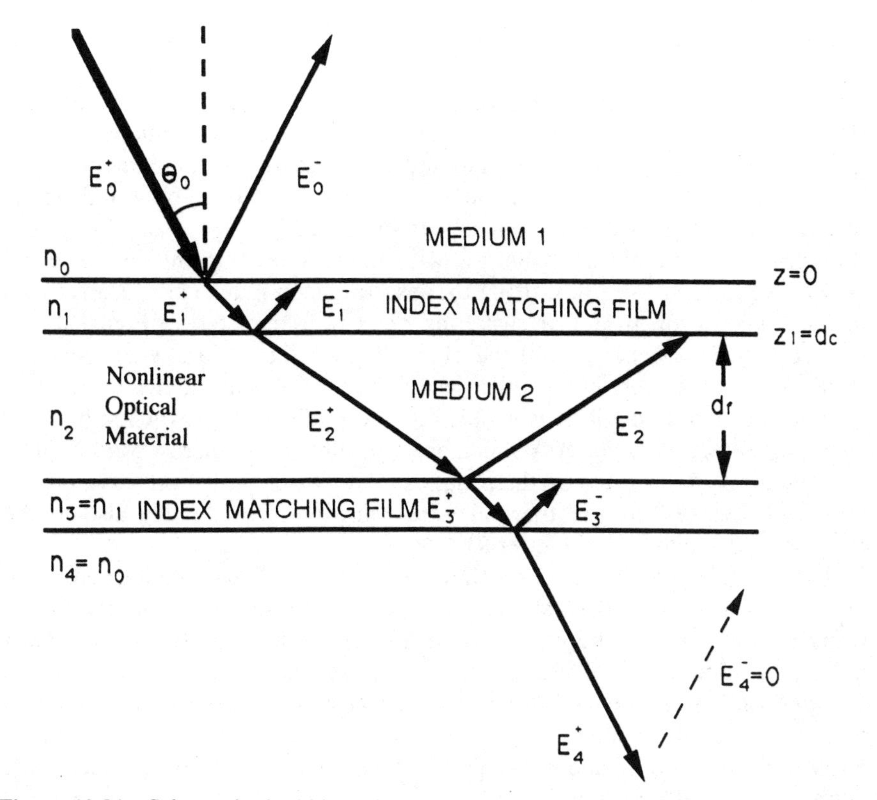


**Figure 10.21** Schematic depiction of total internal reflection ↔ transmission switching by a dielectric (medium 1) cladded nonlinear material (medium 2). Two optional thin films of index $n_1$, on both sides of the nonlinear material, are used for maximizing the transmission of the device in the transmission state. Also shown are the optical electric fields for the reflected, the transmitted, and the incident light.

transmission state if medium 2 possesses a positive optical nonlinearity (i.e., its refractive index is an increasing function of the optical intensity). On the other hand, if medium 2 possesses a negative nonlinearity, the reverse switching process is possible (i.e., from a transmissive to a total reflective state). With a proper choice of "antireflection" coating material of refractive index $n_1$, the device could be highly transmissive at low incident light intensity and could switch to the TIR state at higher incident laser power (54).

Nematic liquid crystals possess both positive and negative signs of nonlinearities associated with the ordinary and extraordinary thermal index changes, respectively. Furthermore, both index gradients are extraordinarily large near the phase transition temperature. This translates into much smaller optical power (or fluence) needed to turn on the switching processes.

A principal concern in these liquid-crystal-based switching devices is the response time. However, unlike electrooptical effects which involve the molecular reorientations of liquid crystals in their nematic phase and are slow, nonlinear liquid crystal optical switching devices can actually respond quite quickly. This is because the amount of refractive index change required to turn on the device can be created in a very short time by a sufficiently intense laser pulse. An example has been discussed in Khoo et al. (39), where the self-defocusing and the resulting switching to a low-transmission state can be achieved in nanoseconds, using the laser-induced index change. Likewise, in the transmission to a TIR switching process as depicted in Figure 10.21, the required index change and therefore the switching to a low-transmission state can be achieved in the nanosecond time scale, using the thermal or density effects of nematic or isotropic liquid crystals. For *Q*-switched infrared laser pulses (e.g., $CO_2$ lasers), which are typically in the microsecond regime, nematic films of thickness on the order of a few wavelengths will serve as very good switches (53) as the thermal time constants are typically also in the microsecond regime (cf. Chapter 7).

The experiments reported in Khoo et al. (16, 55) have confirmed these observations. Figure 10.22 shows an experimental setup for observing the transmission TIR switching process. In the experiment involving $CO_2$ laser pulses, the liquid crystal used is E7 ($n_{\parallel} = 1.75$ and $n_0 = 1.55$), sandwiched between ZnSe prisms ($n = 2.64$) which are transparent at both the visible and infrared spectral regimes. The sample is homeotropically aligned and is 83 $\mu$m thick. The incident laser is *p*-polarized, and thus it "sees" the extraordinary refractive index of the liquid crystal. Note that $dn_e/dT$ is negative, thus the system could undergo transmission- > TIR switching.

The incident angle is set 2° away from the TIR ($\theta - \theta_{TIR} = -2°$). It is estimated that, owing to the reflection loss at the air–prism and prism–liquid crystal interfaces, especially near the TIR state, only 20% of the incident laser is effectively incident on the liquid crystal. This fact emphasizes the importance of the antireflection coating (54) design proposed specifically for a ZnSe prism–liquid crystal transmission–TIR cell.

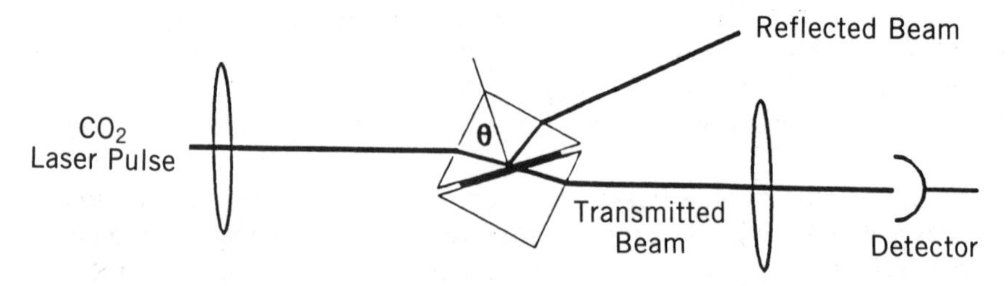


**Figure 10.22** Schematic diagram of transmission–TIR switching/limiting experimental setup.

**Figure 10.23** Oscilloscope traces of the transmitted infrared laser pulse. Incident pulse width is 130 ms. The traces correspond to increasing input power of 0.46, 1.45, 2.06, 2.72, and 3.05 W.

Figure 10.23 shows five oscilloscope traces of the transmitted light corresponding to input powers of 0.46 W (lowest curve), 1.05 W, 2.06 W, 2.72 W, and 3.05 W (uppermost curve). At high input power (2.06 W), the transmission of the later portion of the pulse is diminished. The decrease becomes more rapid as the input power is raised. At an input power of 3 W, a "switch-off" time of about 10 ms is measured. This decrease in the response time with increasing input power is in good agreement with the quantitative theory given in Khoo et al. (16).

Since the operating temperature is 22°C (room temperature), which is far from the nematic–isotropic transition temperature of E7 ($T_c = 60$°C), this long switch-off time is understandable. If the temperature of the sample is raised to near $T_c$, the required temperature rise for the required refractive index change will be smaller (because of a much larger thermal refractive index gradient near $T_c$); the switching time will be shorter. Correspondingly, the required laser energy fluence (in joules per square centimeter) will also decrease. It is estimated in Lindquist et al. (55) that the threshold fluence required for switching microsecond $CO_2$ lasers can be as small as 0.25 J/cm$^2$.

From these preliminary experiments and the analysis based on the nonlinear Fabry–Perot effect given in references (16) and (53), it is clear this type of optical switching device can be optimized to meet practical self-limiting device applications. Parameters that can be changed in the device using nematic film include the refractive indexes of the prism and the liquid crystals, incident angle, laser power, liquid crystal film thickness, absorption rate, and temperature. For a liquid crystal the switch-on time due to the thermal effect can be as short as nanoseconds, or microseconds at temperatures near $T_c$. The longer response time in the latter case is due to the critical slowing down of the order parameter $S$ near $T_c$ as discussed in Chapters 2 and 7.

An alternative material that can be used for last transmission–TIR switching is an isotropic liquid crystal, such as the TM74A mixture used in the self-defocusing limiting effect. The nonlinearity involved is the density decrease caused by laser heating (i.e., a negative refractive index change). Accordingly, a 25-$\mu$m-thick sample is made (55) that consists of a TM74A liquid crystal sandwiched between two ZnSe prisms ($n = 2.402$). $Q$-switched $CO_2$ laser pulses, of total pulse duration on the order of 10 $\mu$s as shown in Figure 10.24*a*, are focused (at a spot diameter of 250 $\mu$m) on the sample and directed at incidence angles just below the TIR condition. Figures 10.24*a–c*

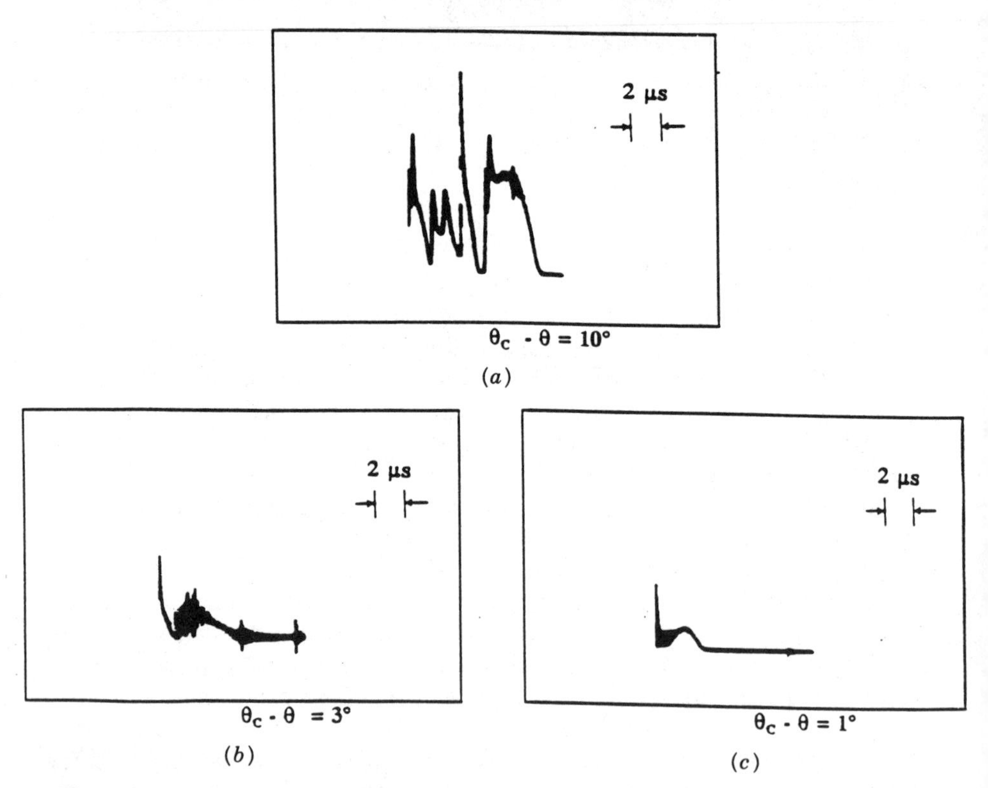


**Figure 10.24** Photographs of the transmitted $CO_2$ laser pulses through a transmission–TIR device for various incident angles near the TIR condition: (*a*) $\theta - \theta_c = 10°$; (*b*) $\theta - \theta_c = 3°$; (*c*) $\theta - \theta_c = 1°$.

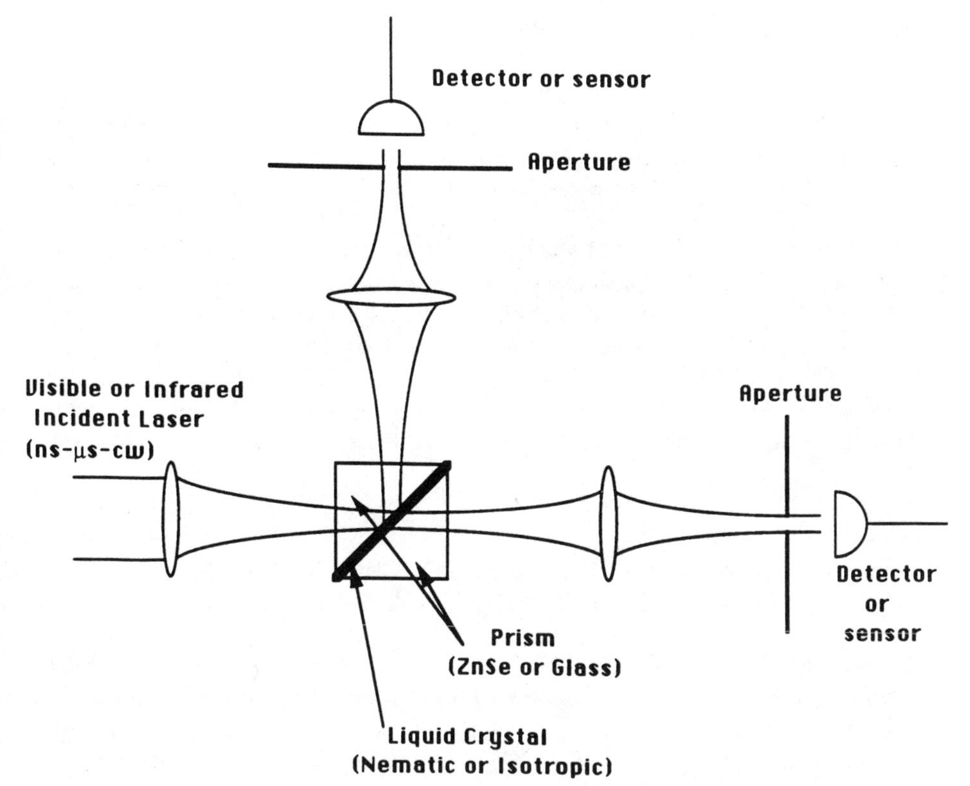


**Figure 10.25** Schematic of TIR–transmission switching/limiting device. Note that with the lens system, it can also give rise to a self-defocusing limiting effect in either the transmission or the reflection mode.

show the corresponding transmitted $CO_2$ laser pulse. At $\theta - \theta_c = 10°$, the transmitted pulse resembles the input pulse (cf. Figure 10.24*a*). As we approach $\theta_c$, as shown in Figures 10.24*b* and *c*, the later parts of the laser pulses are observed to be greatly attenuated, as a result of the (negative) index induced by the front part of the pulse. The switching threshold fluence is measured to be about 0.25 J/cm$^2$.

To reduce the transmission–TIR switching/limiting threshold, one possibility is to combine it with a self-defocusing effect (SDE) as shown in Figure 10.25. The process now has the added advantage that the defocusing effect, in conjunction with the aperture placed in front of the detector, would further lower the transmission of the high-power harmful radiation. Note that the limiting process (by SDE) will continue even as the liquid crystal turns into liquid (when the cell effectively functions as a total reflector), in contrast to self-limiting using SDE alone. Greater dynamic ranges as well as lower switching thresholds are therefore expected of such so-called hybrid cells.

Recent study (56) has also shown that prior to transmission–TIR switching, the incident laser will suffer an even more severe intensity-dependent phase shift, the so-called tunneling phase shift (TPS), associated with the increasing optical path length of the laser through the liquid crystal film. These TPS effects will further contribute to defocusing of the laser and lower the switching threshold for the self-limiting/switching configuration depicted in Figure 10.25.

From the point of view of waveguiding, the switching of the TIR state to the transmissive state as discussed previously may be viewed as an intensity-dependent optical propagation mode change; that is, the system switches from a guided (total reflection) to a lossy mode (transmission loss). In more well-defined guided wave geometries, such as fibers and planar waveguides (18), these intensity-dependent guided wave switching phenomena, with nematic liquid crystals as the nonlinear claddings, have also been observed.

## 10.6 LASER-INDUCED ENHANCED CONDUCTIVITY AND ORIENTATIONAL PHOTOREFRACTIVE WAVE-MIXING EFFECTS

In this section, we discuss a recently observed nonlinear electrooptical phenomenon in nematic liquid crystal films under the combined action of an applied dc electric field and an impinging optical field. As is well known from previous studies (57), dc field-induced current flows in nematic liquid crystals, which possess anisotropic conductivities, could lead to flows and nematic axis reorientation, and the creation of a space-charge field. The charged carriers responsible for the electrical conduction come from impurities present in the otherwise purely dielectric nematic liquid crystal. If these impurities are photoionizable, an incident optical intensity (e.g., an intensity grating created by the interference of two coherent optical beams (cf. Figures 10.26$a$–$b$)), it is possible, therefore, to create a space-charge density grating via the photoinduced spatial conductivity anisotropy. The space-charge field thus created, together with the applied dc field, will cause a refractive index change through any of the field-induced index change effects. In particular, these fields could give rise to director-axis reorientation, and effectively, a new mechanism for optical nonlinearity.

This process of photoinduced charged-carrier generation, space-charge field, and refractive index change is analogous to photorefractive (PR) effect occurring in electrooptically active materials. There is, however, an important difference. In those so-called photorefractive materials, such as $BaTiO_3$, the induced index change $\Delta n$ is linearly related to the total electric field $E$ present (58):

$$\Delta n = r_{\text{eff}} E \tag{10.8}$$

The sign and magnitude of the effective electrooptic coefficient $r_{\mathrm{eff}}$ depends on the symmetry class of the crystal and the direction of the electric field.

Nematic liquid crystals, on the other hand, possess centrosymmetry. The field-induced refractive-index change is quadratically related to the electric field, that is,

$$\Delta n = n_2 E^2 \tag{10.9}$$

A useful feature of such quadratic dependence is that it allows the mixing of the applied dc field with the space-charge field for enhanced director-axis reorientation by the latter.

Recent studies [59] conducted in the author's laboratory have provided evidence that these nonlinear electrooptical processes could result in very large optical nonlinearity. The nonlinearity is nonlocal because of the phase shift between the space-charge fields and the optical-intensity grating functions, and it gives rise to strong two-beam coupling effects at very modest applied dc field strength ($\approx 100$ V/cm) and optical power, in contrast to polymers and inorganic PR crystals, which require high field strength. This is mainly due to the large optical birefringence and dielectric anisotropy of nematics, and the easy susceptibility of the director-axis orientation to external fields. The nonlinear index coefficient $n_2(I)$ associated with this process is on the order of $10^{-3}$ cm$^2$/W.

A quantitative theory remains to be developed for the actual mechanisms responsible for the following observations. Nevertheless, in light of the novelty of these effects, and their potential applications in optical processing and holographic storage, we shall conclude this book with a summary of the principal results obtained in our preliminary studies.

Several nematic liquid crystals have been studied, including pure single-constituent nematics such as 5CB (Pentyl-Cyano-Biphenyl), E7, E63 and E46 (which are mixtures of several nematogens), all purchased from EM Chemicals, Hawthorne, NY. They all yield similar results. In the following discussions, we report results obtained with 5CB. Figures 10.26*a*–*b* show two nematic alignments and electrode configurations used. In Figure 10.26*a* the liquid crystal is homeotropically aligned, with the director axis $\hat{n}$ perpendicular to the transparent (ITO) coated electrode windows, which are treated by the surfactant HTAB (Hexadecyl-triMethyl-Ammonium Bromide). On the other hand, Figure 10.26*b* shows a planar-aligned sample, with director axis parallel to the ordinary glass windows. The glass windows are coated with PVA (Poly-Vinyl-Alcohol) that is rubbed in a unidirectional manner.

The laser used is the 5145 Å line of a linearly polarized Argon laser. It is divided into two coherent beams of almost equal power and then combined on the liquid crystal sample at a small wave-mixing angle. For the equivalent directions of polarizations and propagation and applied field shown in Figure 10.26*a*–*b*, we have observed the same effects and parametric depen-

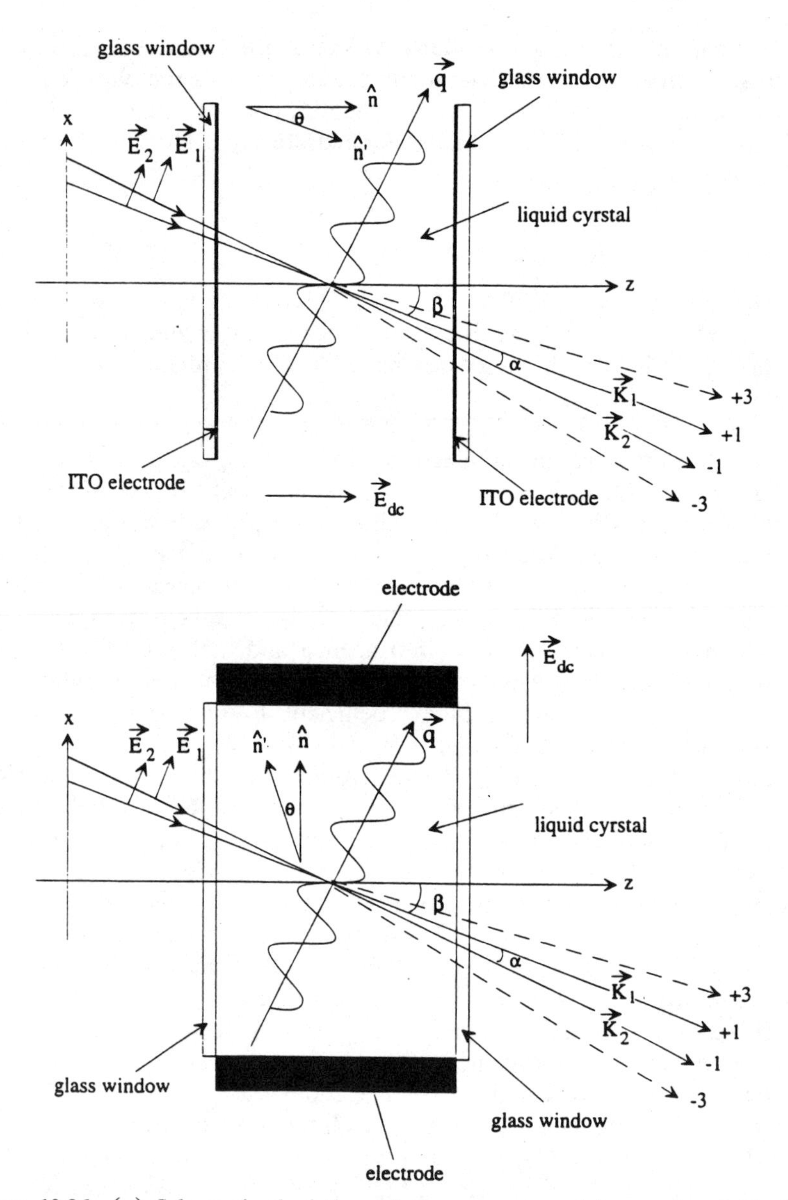


**Figure 10.26** (*a*) Schematic depiction of the geometry of interacting fields in a homeotropically aligned dye-doped nematic liquid crystal film. The wave mixing angle $\alpha$, the reorientation angle $\theta$, and the beam propagation $\beta$ are measured within the nematic film. $\pm 1$, $\pm 3$ are the transmitted input beams and their self-diffractions, respectively. (*b*) Geometry of interacting fields in a planar aligned sample.

dences. This eliminates the possibility of contribution from the ITO electrode or the surface treatment materials.

Using a linearly polarized He-Ne laser as a probe, we have determined that only extraordinary-wave lasers could produce the following effects (i.e., if $\beta = 0$, or if the He-Ne probe is polarized in the $\hat{y}$-direction, we observe null effect). Polarization dependence and the requirement that a dc-applied field be present eliminate thermal effects and other phase-change or order-disorder transitions as possible contributing mechanisms (60). If an ac field (frequency $\gtrsim 10^2$ Hz) is used, we also observe null effect. The electrooptically induced refractive index change is associated with director-axis reorientation in the $x-z$ plane.

Figure 10.27 shows the dependence of the current flowing through the nematic as a function of the applied dc field. When the nematic is illuminated by the (10 mW) Argon ion laser (5145 Å), a detectable change in conductivity from the dark state (conductivity is on the order of $10^{-8}$ Ω/cm) is observed at an applied voltage of 1 V. This conductivity change becomes

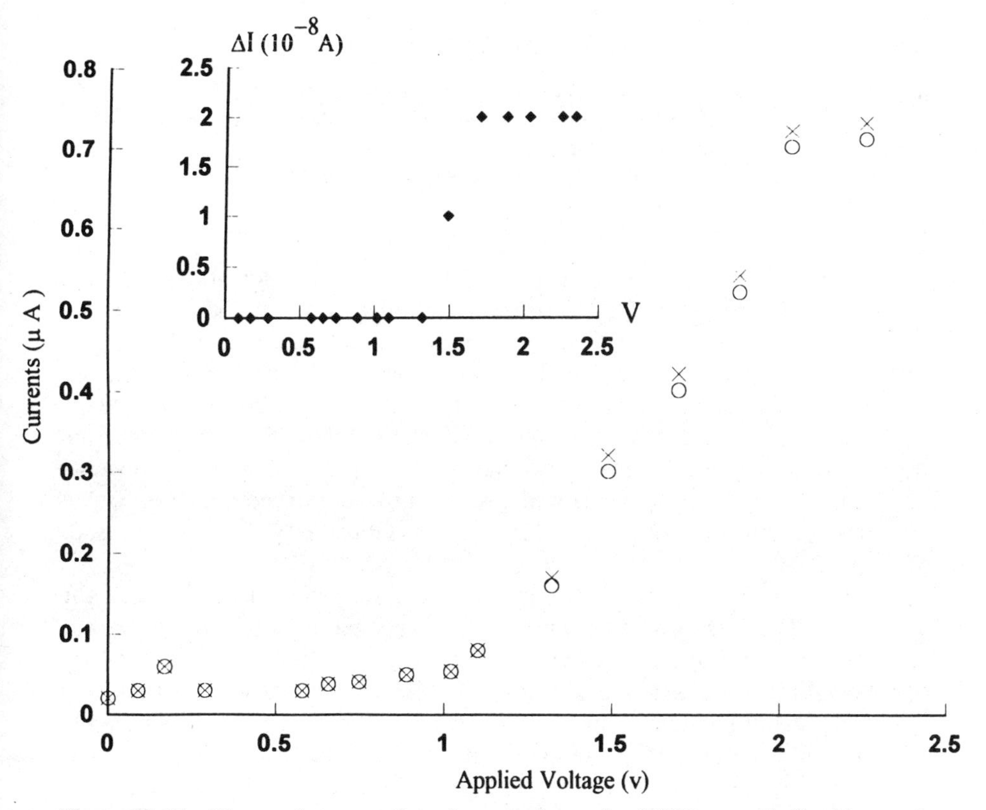


**Figure 10.27** Observed current flow through the undyed 5CB nematic liquid crystal sample. Circles, dark state; Crosses, under laser illumination.

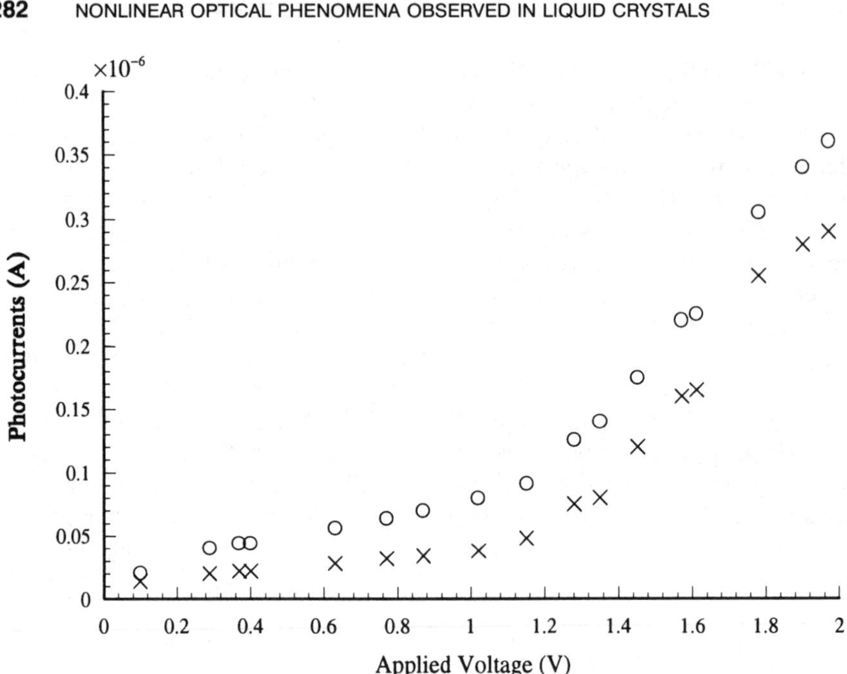


**Figure 10.28** Observed dependence of the current across the dyed (Rhodamine 6G) nematic liquid crystal on the applied voltage. Circles, dark state; crosses, values obtained when the sample is illuminated by a 10 mW laser. Note the increase in conductivity for both cases above an applied voltage of 1.15 V.

very pronounced if traces of dye are dissolved in the nematics. We have found that dichroic dyes D2, the laser dye Methyl Red, and Rhodamine 6G (R6G) are quite effective in enhancing the photoinduced conductivity change. In such dyed systems, the charged carriers arise from heterolytic dissociation (61) of the photoexcited dye molecules upon their relaxation to the ground states in a polar solvent such as 5CB. Figure 10.28 shows how the dc current varies as the applied voltage across the dyed sample is increased. Above an applied voltage of 1.5 V, the rate of change shows a marked increase. This is due to field ionization of the dye dopants and impurities in the liquid crystals (57).

Figures 10.29$a$–$b$ show typical diffraction patterns on the exit side of the undoped film. For this case, the optical propagation angle $\beta$ within the nematic medium is 0.5 radian. The wave mixing angle $\alpha$ is $1.8\times10^{-3}$ radian, corresponding to a grating constant of 278 $\mu$m. Visible diffractions are observed for an incident laser power as weak as $\approx 10$ mW for V = 1.5 Volt. The diffractions decay in a time of about 6 s when one beam is blocked. The time constant is consistent with the orientation decay constant $\tau_r = \gamma/k(\pi^2 d^{-2} + q^2) \approx 4$ s for $\gamma = 0.1$ poise, $K = 0.3\times10^{-6}$ dynes, $d = 100$ $\mu$m,

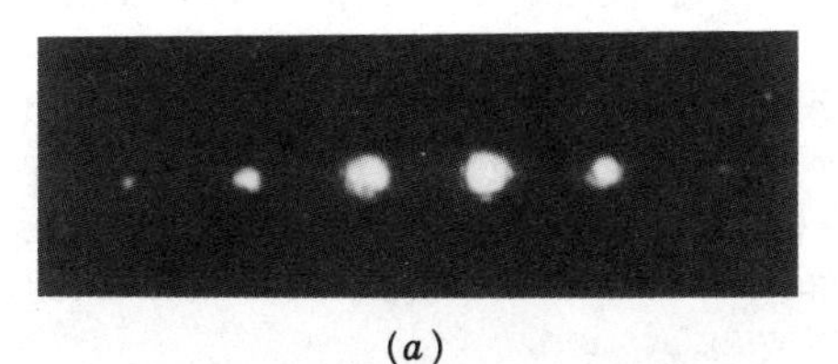

(*a*)

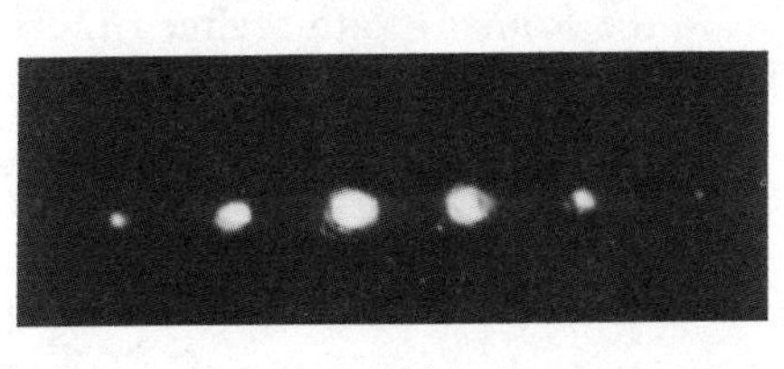

(*b*)

**Figure 10.29** Photographs of the transmitted inputs ($\pm 1$ beams) and their first-order self-diffractions ($\pm 3$ beams). The second-order diffractions are also visible by eye. Experimental conditions are: V = 1.5 Volt, input beam power about 10 mW in each beam; $\beta = 0.5$ radian, $\alpha = 1.8 \times 10^{-3}$ radian. (*a*) $\mathbf{E}_z$ along $+\hat{z}$ direction; (*b*) $\mathbf{E}_z$ along $-\hat{z}$ direction.

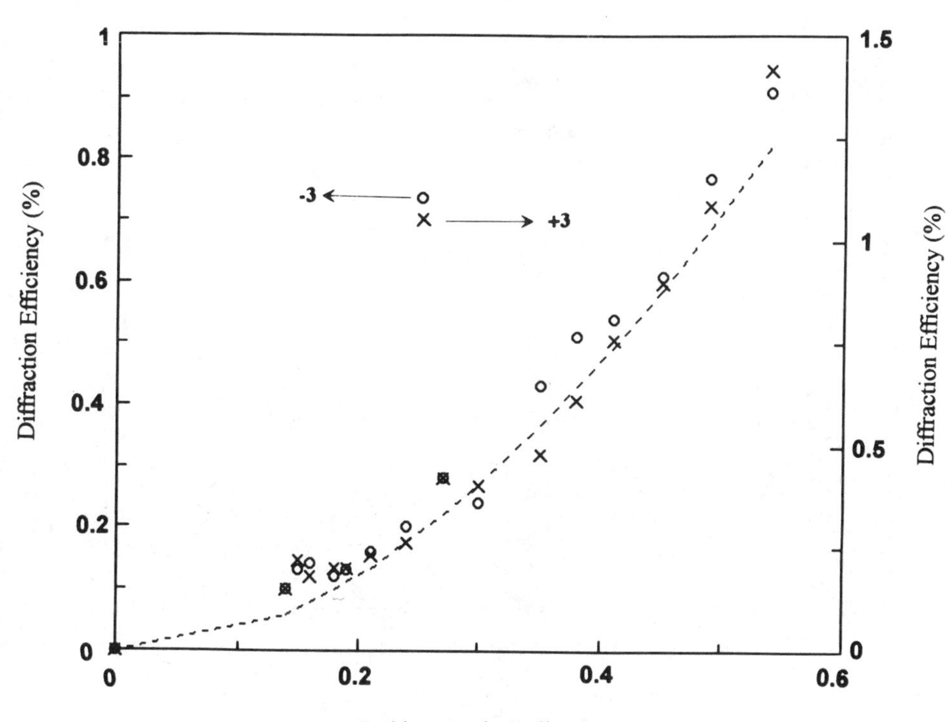


**Figure 10.30** Dependence of the $\pm 3$ beam diffraction efficiency on the angle $\beta$. Experimental conditions: Input power in $-1$ beam: 11.2 mW; power in $+1$ beam: 8.8 mW; V = 1.5 Volt, $\mathbf{E}_z$ along $+\hat{z}$ direction. Undyed sample.

$q = 2\pi/(278\ \mu m)$. Roughly the same relaxation times are observed if the $(\pm 1)$ beams are present, and the applied dc field is turned off. The effects vanish if the applied voltage is turned off or if $\beta = 0$. The $\beta$- and V-dependences of the side diffractions ($\pm 3$ beams) are shown in Figure 10.30 and 10.31, respectively. Notice that the voltage dependence of the diffraction is closely parallel to the photo-induced conductivity change (cf. Figure 10.27). On the other hand, the $\beta$-dependence is consistent with the reorientational mechanism to be described in the following section.

For $\beta = 0.5$ radian, the diffraction efficiency is measured to be about 1% (cf. Figure 10.30). Since the diffraction is in the Raman–Nath regime ($d\lambda \gg 2\pi q^{-1})^2$), the diffraction efficiency $\eta$ is proportional to $(\Delta n_e \pi d/\lambda)^2$, where $\Delta n_e$ is the index-grating amplitude. Using $d = 100\ \mu m$, $\lambda = 0.5145\ \mu m$, $\eta = 0.01$, one obtains $\Delta n_e = 0.16 \times 10^{-3}$. Since the optical intensity $I$ 0.2

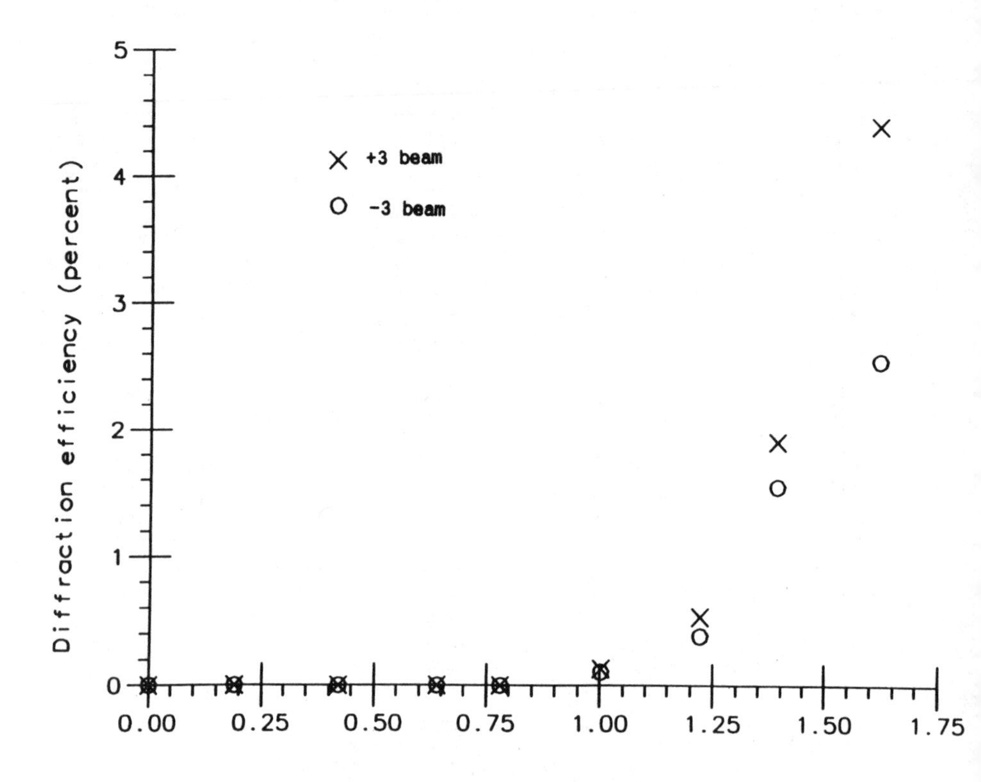


Figure 10.31 Dependence of the $\pm 3$ beam-diffraction efficiency on the applied dc voltage at $\beta = 30.8°$; $P_{-1} = 11.2$ mW; $P_{+1} = 8.8$ mW, $\mathbf{E}_z$ in $+\hat{z}$ direction. Undyed sample.

$W/cm^2$ (10 mW in a laser spot diameter of 2.5 mm), the nonlinear index coefficient $n_2 = \Delta n_e / I$ is thus $0.8 \times 10^{-3}$ $cm^2/W$.

In general, dyed samples give, as expected, higher diffraction efficiencies than the undoped ones. An important feature of these electrooptically written gratings in dyed sample is their rather long lifetimes. The grating persists for ~1 hour after all the incident optical and applied fields are turned off. To restore and optically read the grating, one needs simply to turn on the dc field. Such long lifetime is probably due to the large amount of charged carriers generated in the dyed sample, and the high resistivity of nematics.

The nonlocality (i.e., the phase shift $\Delta\phi$ between the optical index and the intensity function) gives rise to beam-coupling effect (cf. Section 9.5). For the conditions used in obtaining Figures 10.29$a$–$b$ (i.e., V = 1.5 ($E_z$ in the $+\hat{z}$ direction), the +1 beam is observed to experience gain (from 6 mW to 7.7 mW), while the −1 beam experiences a loss (from 7.6 mW to 4 mW). If the applied electric-field direction is reversed (which effectively introduce a further phase shift of $\pi$ to $\Delta\phi$), the reverse occurs, namely, beam +1 experiences a loss (from 6 mW to 3.8 mW), while beam −1 is amplified (from 7.6 mW to 9.6 mW). The exponential gain per unit length is estimated to be $0.25/100\ \mu m = 25\ cm^{-1}$.

A detailed theoretical discussion on these nonlinear electrooptical effects is clearly outside the scope of this section. We give here a qualitative phenomenological analysis. The conductivity change $\Delta\sigma$ associated with an optical intensity function $I_{op} = I_0(1 + \cos q\chi')$ (unity modulation) is of the form $\Delta\sigma = \Delta\sigma_0(1 + \cos q\chi')$ where $\Delta\sigma_0$ is the conductivity difference between the illuminated and the dark state (cf. Figure 10.2). In the presence of an applied dc field $\mathbf{E}_a = (0, 0, E_a)$, a space-charge field $\mathbf{E}_{sc}$ is set up via the Carr–Helfrich effect (57, 60). Since the space field $\mathbf{E}_{sc}$ is related to the space-charge density $\rho(\rho \propto \Delta\sigma)$ by $\nabla \cdot \mathbf{E}_{sc} = \rho/\varepsilon$, it is phase-shifted with respect to the $\Delta\sigma$ grating function (i.e., we have

$$\mathbf{E}_{sc} = (E_{sc} \cos\beta, 0, E_{sc} \sin\beta) \cos(q\chi' + \Delta\phi) \tag{10.10}$$

where $\Delta\phi$ is the phase-shift between the space-charge field, and the $\Delta\sigma$ or optical intensity function $\chi'$ is the spatial coordinate along the $\mathbf{q}$ direction).

$\mathbf{E}_{sc}$ and $\mathbf{E}_a$ together creates a torque on the nematic (6, 7):

$$\mathbf{f}_E = \frac{\Delta\varepsilon}{4\pi}(\hat{n} \cdot \mathbf{E})(\hat{n} \times \mathbf{E}) \tag{10.11}$$

where $\mathbf{E} = (\mathbf{E}_{sc} + \mathbf{E}_a)$ where $\Delta\varepsilon$ is the dc dielectric anisotropy. This field-induced torque is counteracted by elastic torques due to splay, twist, or bend distortions in the nematic. Following the usual nematogen theory, the resulting director-axis reorientational profile ($\theta = \theta_0 \cos(q\eta + \Delta\phi)$ + dc and other terms) and the index profile ($\Delta n = \Delta n_e \cos(q\eta + \Delta\phi) + \text{dc}$ and other terms) could be calculated in a straightforward manner. Under the one-elastic constant and small $\theta_0$ approximation, this yields

$$\Delta n_e \approx (n_\parallel - n_\perp)\frac{n_\parallel}{n_\perp}(\sin 2\beta)\theta_0 \tag{10.12}$$

and

$$\theta_0 \sim \frac{\Delta\varepsilon E_a E_{sc}}{16\pi k}\cos\beta[dz - z^2] \tag{10.13}$$

Notice that the quadratic ($E^2$) dependence of the field-induced torque on the molecule (cf. equation (10.11)) is manifested in the appearance of the field product $E_a E_{sc}$ in equation (10.13).

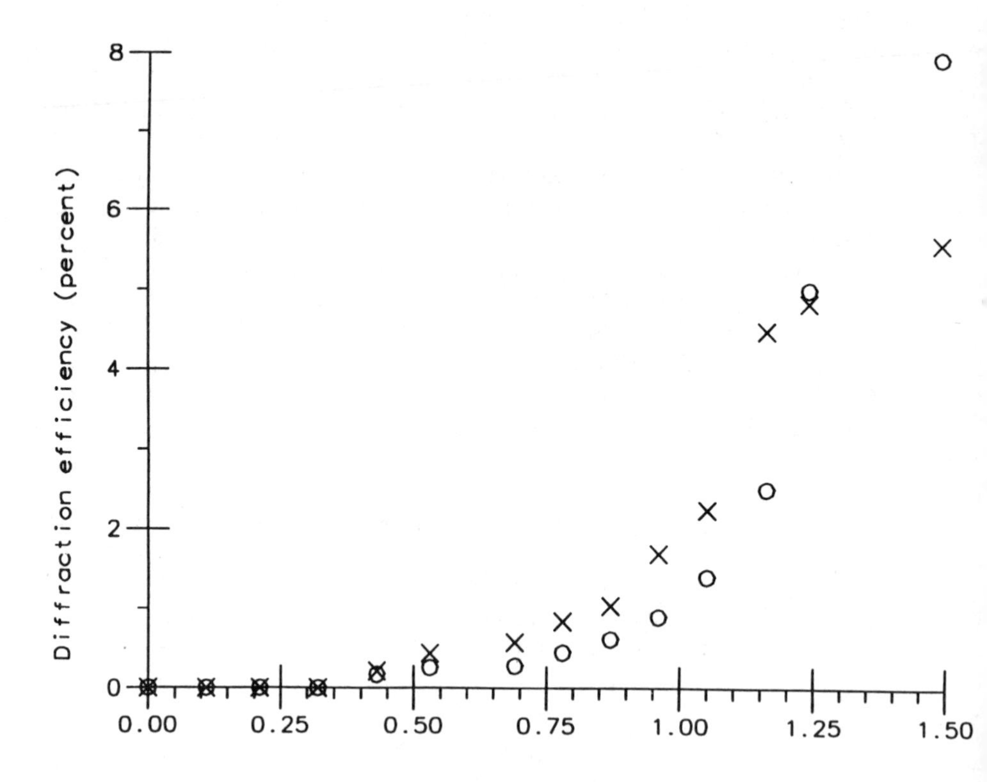


**Figure 10.32** Typical observed dependence of the first-order diffraction efficiencies of the dyed sample as a function of the applied dc voltage. $\mathbf{E}_a$ is along the $+z$ direction. (Crosses for $+3$ diffraction; circles for $-3$ diffraction.) Incident $+1$ beam power is 11.2 mW; $-1$ beam power is 8.8 mW. $\beta = 30.8°$.

This qualitative theory predicts that the $\beta$-dependence of the diffraction efficiency, which is proportional to $(\Delta n_e)^2$, is of the form $(\sin^2\beta\cos^4\beta)$. This is in good agreement with the experimental results (cf. Figure 10.30). The dependence of $\eta$ on the applied voltage V is, from equations (10.12) and (10.13), of the form $V^2(E_{sc})^2$. Since $E_{sc}$ depends on the photoinduced conductivity change, which in turn varies sharply with the applied voltage (cf. Figure 10.27), we expect the diffraction efficiency to vary just as sharply. This is indeed observed experimentally, as shown in Figure 10.31 for the pure sample, and Figure 10.32 for the dyed sample.

Clearly, a more quantitative theory similar to the conductivity-anisotropy–induced space-charge field effects given in (57) is needed in order to assess these space-charges and fields quantitatively. One could, nevertheless, employ the approximate analysis above to compare the reorientational effect of these optically induced dc fields with purely optical-field effect. For the same interaction geometry as in Figures 10.26$a$–$b$, and setting $E_a = 0$, one gets (6)

$$\theta_0(\text{opt}) \sim \frac{\Delta\varepsilon_{\text{op}}}{8\pi K} E_{\text{op}}^2 \sin 2\beta (dz - z^2) \tag{10.14}$$

where $\Delta\varepsilon_{\text{op}}$ is the optical dielectric anisotropy; $\Delta\varepsilon_{\text{op}} \approx 0.6$ for 5CB. Comparing (10.13) and (10.14), we get

$$\frac{\theta_0}{\theta_0(\text{opt})} = \frac{\Delta\varepsilon}{\Delta\varepsilon_{\text{op}}} \frac{[E_a E_{sc}]}{[E_{\text{op}}^2] \sin\beta} \tag{10.15}$$

For 5CB, $\Delta\varepsilon = 17$, the ratio $\Delta\varepsilon / \Delta\varepsilon_{\text{op}}$ is therefore 28. For $\beta = 0.5$, we thus have

$$\frac{\theta_0}{\theta_0(\text{opt})} \sim 40 \left[ \frac{E_a E_{sc}}{E_{\text{op}}^2} \right] \tag{10.16}$$

In our experiments, we found that in order to observe the same diffraction with purely optical fields (e.g., 1% as shown in Figure 10.29), an optical power of 200 mW (intensity of 3.2 $\text{W/cm}^2$) is needed. This corresponds to $\Delta\varepsilon_{\text{op}} E_{\text{op}}^2$ of $1400(\text{V/cm})^2$. In other words, the photoinduced and applied dc field combination, $\Delta\varepsilon E_a E_{sc}$ is $\approx 1400\ (\text{V/cm})^2$. Since $E_a = 150$ V/cm, $\Delta\varepsilon E_{sc}$ is a meager 9 V/cm. The key factor responsible is simply the much larger magnitude of $\Delta\varepsilon$ compared to $\Delta\varepsilon_{\text{op}}$. The preceding analysis also shows that such orientational photorefractive effects, arising from the photoinduced space-charge fields, could be greatly improved by employing more efficient photoionizable dopants, and by choosing liquid crystalline materials and field-interaction geometry.

## 10.7 CONCLUSION

In conclusion, liquid crystals are complex and wonderful nonlinear optical materials. On the other hand, nonlinear optics is also complex and interesting. Both fields have now reached the mature stage. Many interesting and useful materials, processes, phenomena, applications, and devices are now awaiting our exploration.

## REFERENCES


1. I. C. Khoo, Nonlinear optics of liquid crystals. In "Progress in Optics" (E. Wolf, ed.), Vol. 26. North-Holland Publ., Amsterdam, 1988.
2. N. V. Tabiryan, A. V. Sukhov, and B. Ya. Zeldovich, The orientational optical nonlinearity of liquid crystals. *Mol. Cryst. Liq. Cryst.* **136**, 1 (1986); V. F. Kitaeva and A. S. Zolot'ko, Optically induced Freedericksz effect. *Laser Res. USSR* **10**(4), 275 (1989).
3. I. C. Khoo, J. Y. Hou, T. H. Liu, P. Y. Yan, R. R. Michael, and G. M. Finn, Transverse self-phase modulation and bistability in the transmission of a laser beam through a nonlinear thin film. *J. Opt. Soc. Am. B* **4**, 886 (1987); P.-Ye Wang, H.-J. Zhang, and J.-H. Dai, Laser-heating-induced self-phase modulation, phase transition, and bistability in nematic liquid crystals. *Opt. Lett.* **13**, 479 (1988); I. C. Khoo, P. Y. Yan, T. H. Liu, S. Shepard, and J. Y. Hou, Theory and experiment on optical transverse intensity bistability in the transmission through a nonlinear thin (nematic liquid crystal) film. *Phys. Rev. A* **29**, 2756 (1984).
4. I. C. Khoo, R. R. Michael, and G. M. Finn, Self-phase modulation and optical limiting of a low power $CO_2$ laser with a nematic liquid-crystal film. *Appl. Phys. Lett.* **52**, 2108 (1988).

4a. I. C. Khoo, Sukho Lee, P. G. LoPresti, R. G. Lindquist, and H. Li, Isotropic liquid crystal film and fiber structure for optical limiting application. *Int. J. Nonlinear Opt. Phys.* **2**, No. 4 (1993).

5. See, for example, G. K. L. Wong and Y. R. Shen, Transient self-focusing in a nematic liquid crystal in the isotropic phase. *Phys. Rev. Lett.* **32**, 527 (1974); see also D. N. Ghosh Roy and D. V. G. L. N. Rao, Optical pulse narrowing by backward stimulated Brillouin scattering. *J. Appl. Phys.* **59**, 332 (1986).
6. D. Fekete, J. Au Yeung, and A. Yariv, Phase conjugate reflection by degenerate four wave mixing in a nematic crystal in the isotropic phase. *Opt. Lett.* **5**, 51 (1980).
7. P. A. Madden, F. C. Saunders, and A. M. Scott, Degenerate four-wave mixing in the isotropic phase of liquid crystals: The influence of molecular structure. *IEEE J. Quantum Electron.* **QE-22**, 1287 (1986).
8. I. C. Khoo and S. L. Zhuang, Wave front conjugation in nematic liquid crystal film. *IEEE J. Quantum Electron.* **QE-18**, 246 (1981); F. Sanchez, P. H. Kayoun, and J. P. Huignard, Two-wave mixing with gain in liquid crystal at 10.6 m

wavelength. *J. Appl. Phys.* **64**, p. 26 (1988); see also L. Richard, J. Maurin, and J. P. Huignard, Phase conjugation with gain at $CO_2$ laser line from thermally induced grating in nematic liquid crystals. *Opt. Commun.* **57**, 365 (1986); E. N. Leith, Hsuen Chen, Y. S. Cheng, G. J. Swanson, and I. C. Khoo, Coherence reduction in phase conjugation imaging. *In* "Proceedings of the 5th Rochester Conference on Coherence and Quantum Optics" (E. Wolf and L. Mandel, eds.), p. 1155. Plenum, London, 1984.

9. I. C. Khoo and T. H. Liu, Theory and experiments on multiwave-mixing-mediated probe beam amplification. *Phys. Rev. A* **39**, 4036 (1989); T. H. Liu and I. C. Khoo, Probe beam amplification via degenerate optical wave mixing in a Kerr medium. *IEEE J. Quantum Electron.* **QE-23**, 171 (1987); I. C. Khoo, P. Y. Yan, G. M. Finn, T. H. Liu, and R. R. Michael, Low power (10.6 $\mu$m) laser beam amplification via thermal grating mediated degenerate four wave mixings in a nematic liquid crystal film. *J. Opt. Soc. Am. B* **5**, 202 (1988).
10. I. C. Khoo, Optical amplification and polarization switching in a birefringent nonlinear optical medium: An analysis. *Phys. Rev. Lett.* **64**, 2273 (1990).
11. J. W. Shelton and Y. R. Shen, Phase-matched third-harmonic generation in cholesteric liquid crystals. *Phys. Rev. Lett.* **25**(1), p. 23 (1970); see also *Phys. Rev. A* **5**, p. 1867 (1972).
12. S. K. Saha and G. K. Wong, Phase-matched electric-field-induced second-harmonic generation in a nematic liquid crystal. *Opt. Commun.* **30**, 119 (1979); see also Z.-C. Ou-Yang and Y.-Z. Xie, Theory of second-harmonic generation in liquid crystals. *Phys. Rev. A* **32**, 1189 (1986); A. V. Sukhov and R. V. Timashev, Optically induced deviation from central symmetry; lattices of quadratic nonlinear susceptibility in a nematic liquid crystal. *JETP Lett.* (*Engl. Transl.*) **51**(7), 415 (1990); see also N. B. Baranova and B. Ya. Zeldovich, *Dokl. Akad. Nauk SSSR* **263**, 325 (1982); *Sov. Phys.—Dokl.* (*Engl. Transl.*) **27**, 222 (1982).
13. J. Y. Liu, M. G. Robinson, K. M. Johnson, D. M. Wabba, M. B. Ros, N. A. Clark, R. Shao, and D. Doroski, Second harmonic generation in ferroelectric liquid crystals. *Opt. Lett.* **15**, 267 (1990); see also A. Taguchi, Y. Oucji, H. Takezoe, and A. Fukuda, *Jpn. J. Appl. Phys.* **28**, 997 (1989); M. I. Barnik, L. M. Blinov, A. M. Dorozhkin, and N. M. Shtykov, Generation of the second optical harmonic induced by an electric field in nematic and smectic liquid crystals. *Sov. Phys.—JETP* (*Engl. Transl.*) **54**, 935 (1981).
14. I. C. Khoo, J. Y. Hou, R. Normandin, and V. C. Y. So, Theory and experiment on optical bistability in a Fabry-Perot interferometer with an intracavity nematic liquid-crystal film. *Phys. Rev. A* **27**, 3251 (1983).
15. A. J. Karn, S. M. Arakelian, Y. R. Shen, and H. L. Ong, Observation of magnetic-field-induced first-order optical Freedericksz transition in a nematic film. *Phys. Rev. Lett.* **57**, 448 (1986); see also S.-H. Chen and J. J. Wu, Observation of first order Freedericksz transition in a nematic film induced by electric and optical fields. *Appl. Phys. Lett.* **52**, 1998 (1988).
16. I. C. Khoo, Ping Zhou, R. R. Michael, R. G. Lindquist, and R. J. Mansfield, Optical switching by a dielectric cladded nematic film. *IEEE J. Quantum Electron.* **QE-25**, 1755 (1989).
17. I. C. Khoo, W. Wang, F. Simoni, G. Cipparrone, and D. Duca, Experimental studies of the dynamics and parametric dependences of total-internal-reflection

to transmission switching and limiting effects. *J. Opt. Soc. Am. B* **8**, 1464 (1991); R. A. Innes and J. R. Sambles, Optical nonlinearity in a nematic liquid crystal using surface plasmon-polaritons. *Opt. Commun.* **64**, 288 (1987).

18. H. Vach, C. T. Seaton, G. I. Stegeman, and I. C. Khoo, Observation of intensity-dependent guided waves. *Opt. Lett.* **9**, 238 (1984); see also E. S. Goldburt and P. St. J. Russell, Nonlinear single-mode fiber coupler using liquid crystals. *Appl. Phys. Lett.* **46**, 338 (1985).
19. A. V. Sukhov, Stimulated orientational backscattering and attendant phenomena in cholesteric liquid crystal. *Mol. Cryst. Liq. Cryst.* **185**, 227 (1990).
20. I. C. Khoo, R. R. Michael, and P. Y. Yan, Simultaneous occurrence of phase conjugation and pulse compression in stimulated scatterings in liquid crystal mesophases. *IEEE J. Quantum Electron.* **QE-23**, 1344 (1987).
21. I. C. Khoo and R. Normandin, Nonlinear liquid crystal fiber-fiber coupler for switching and gating operation. *J. Appl. Phys.* **65**, 2566 (1989); see also Z. K. Ioannidis, I. P. Giles, and C. Bowry, All fiber optical intensity modulators with liquid crystals. *Appl. Opt.* **30**, 328 (1991).
22. B. Ya. Zeldovich, N. F. Pilipetsky, and V. V. Shkunov, "Principles of phase conjugation," Springer Ser. Opt. Sci., Vol. 42. Springer-Verlag, Berlin, 1985; K. D. Ridley and A. M. Scott, "Self-pumped Brillouin enhanced four-wave mixing in a medium with a small phase mismatch" *Opt. Commun.* **76**, 406 (1990); "Frequency detuning in Brillouin induced four-wave mixing" *Int. J. Nonlinear Opt. Phys.* **1**, 563 (1992); see also R. W. Boyd, "Nonlinear Optics." Academic Press, San Diego, 1992.
23. J. Feinberg, "Self-pumped continuous-wave phase conjugator using internal reflection" *Opt. Lett.* **7**, 486 (1982); J. O. White, M. Cronin-Golomb, B. Fischer, and A. Yariv, "Coherent oscillation by self-induced gratings in the photorefractive crystal $BaTiO_3$" *Appl. Phys. Lett.* **40**, 450 (1982); M. Cronin-Golomb, B. Fisher, J. O. White, and A. Yariv, "Theory and application of four-wave mixing in photorefractive media" *IEEE J. Quantum Electron.* **QE-20**, 12 (1984); "Passive phase conjugator mirror based on self-induced oscillation in an optical ring cavity" *Appl. Phys. Lett.* **42**, 919 (1983).
24. C. J. Gaeta, J. F. Lam, and R. C. Lind, "Continuous-wave self-pumped optical phase conjugation in atomic sodium vapor" *Opt. Lett.* **14**, 245 (1989). M. Vallet, M. Pinard, and G. Grynberg, "Nonfrequency-shifted self-pumped phase-conjugate mirror using sodium vapor" *ibid.* **16**, 1071 (1991).
25. I. C. Khoo, H. Li, and Yu Liang, Self-starting optical phase conjugation in dyed nematic liquid crystal with stimulated thermal scattering effect. *Opt. Lett.* **18**, 3 (1993).
26. O. L. Antipov and D. A. Dvoryaninov, Parametric generation phase conjugation of intersecting laser beams in a layer of nematic liquid crystal containing a dye. *JETP Lett.* (*Engl. Transl.*) **53**, 611 (1991).
27. See, for example, I. C. Khoo, N. Beldyugina, H. Li, A. V. Mamaev, and V. V. Shkunov, Onset dynamics of self-pumped phase conjugation from speckled noise. *Opt. Lett.* **18**, 473 (1993); see also A. V. Mamaev and A. A. Zozulya, "Dynamics and stationary states of a photorefractive phase conjugate semilinear mirror" *Opt. Commun.* **79**, 373 (1990); V. T. Tikhonchuk, M. G. Zhanuzakov, and A. A. Zozulya, "Stationary states of two coupled double phase conjugate mirrors" *Opt. Lett.* **16**, 288 (1991).

28. See also I. C. Khoo, H. Li, and Y. Liang, Dynamics of transient probe beam amplifications via coherent multiwave mixing in a local nonlinear medium nematic liquid crystal. *IEEE J. Quantum Electron.* **29**, 2972 (1993).
29. R. Mcgraw, D. Rogovin, and A. Gavrielides, "Light-scattering limitations for phase conjugation in optical Kerr media" *Appl. Phys. Lett.* **54**, 199 (1989); V. E. Kravtsov, V. I. Yudson, and V. M. Agranovich, "Theory of phase conjugation of light in disordered nonlinear mirror" *Phys. Rev. B* **41**, 2794 (1990); E. N. Leith, C. Chen, H. Chen, D. Dilworth, J. Lopez, J. Rudd, P.-C. Sun, J. Valdmanis, and G. Vossler, "Imaging through scattering media with holography" *J. Opt. Soc. Am. A* **9**, 1148 (1991).
30. B. Ya. Zeldovich, S. K. Merzlikin, N. F. Pilipetskii, and A. V. Sukhov, Observation of stimulated forward orientational light scattering in a planar nematic liquid crystal. *JETP Lett.* (*Engl. Transl.*) **41**, 515 (1985).
31. I. C. Khoo, Y. Liang, and H. Li, Polarization switching by stimulated orientational scattering in birefringent nonlinear nematic liquid crystal film. *IEEE-Lasers Electro-Opt. Soc. Tech. Meet.*, San Jose, CA, 1993. p. 321.
32. Y. Liang, H. Li and I. C. Khoo, "Broadband low-power self-starting phase conjugation with stimulated orientational scattering in nematic liquid-crystal film," Technical Digest-Conference on Lasers and Electro-Optics, Anaheim, CA (1994).
33. G. Cipparrone, V. Carbone, C. Versace, C. Umeton, R. Bartolino, and F. Simoni, Optically induced chaotic behavior in nematic liquid crystal film. *Phys. Rev. E* **47**, 3741 (1993).
34. J. P. Huignard, J. P. Herriau, P. Aubourg, and E. Spitz, Phase-conjugate wavefront generation via real-time holography in $Bi_{12}SiO_2$ crystals. *Opt. Lett.* **4**, 21 (1979).
35. I. C. Khoo, H. Li, P. G. LoPresti, and Yu Liang, Observation of optical limiting and back scattering of nanosecond laser pulses in liquid crystal fibers. *Opt. Lett.* **19**, 530 (1994).
36. I. C. Khoo, Sukho Lee, P. G. LoPresti, R. G. Lindquist, and H. Li, Isotropic liquid crystalline film and fiber structures for optical limiting application. *Int. J. Nonlinear Opt. Phys.* **2**, No. 4 (1993). This is a special issue, together with Vol. 2, No. 3, on "Optical limiters, switches and discriminators."
37. M. Sheik-Bahae, A. A. Said, and E. W. Van Stryland, "High-sensitivity single-beam $n_2$ measurement" *Opt. Lett.* **14**, 955 (1989).
38. I. Janossy and T. Kosa, The influence of anthraquinone dyes on optical reorientation of nematic liquid crystals. *Opt. Lett.* **17**, 1183–1185 (1992).
39. I. C. Khoo, R. R. Michael, and P. Y. Yan, Theory and experiments on optically induced nematic axis reorientation and nonlinear effects in the nanosecond regime. *IEEE J. Quantum Electron.* **QE-23**, 267 (1987).
40. These bubbles will float away from the interaction region in a few seconds; in this sense, the "damage" to the isotropic liquid crystal sample is thus a non-catastrophic one. For the mechanisms of these bubble formation in ordinary liquids, see, for example, Y.-X. Yan and K. A. Nelson, "Impulsive stimulated light scattering. I. General Theory" *J. Chem. Phys.* **87**, 6240 (1987), and references therein; D. F. Gaitan, L. A. Crum, and C. C. Church, "A study of the timing of sonoluminescence flashes from stable cavitation" *J. Acoust. Soc. Am.* **84**, S36 (1988).

41. G. I. Stegeman and S. T. Seaton, Nonlinear integrated optics. *J. Appl. Phys.* **58**, R57–R78 (1985); see also H. G. Winful, Nonlinear optical. phenomena in single-mode fibers. *In* "Optical-Fiber Transmission" (E. E. Basch, ed.), pp. 179–228, and references therein. Howard W. Sams and Co., 1987; G. P. Agarwal, "Nonlinear Fiber Optics." Academic Press, San Diego 1989.
42. See, for example, R. H. Stolen and H. W. K. Tom, Self-organized phase-matched harmonic generation in optical fiber. *Opt. Lett.* **12**, 57 (1987); see also M. C. Farris, P. St. J. Russell, M. E. Fermann, and D. N. Payne, Second-harmonic generation in an optical fiber by self-written $\chi^{(2)}$ grating. *Electron. Lett.* **23**, 322 (1987); Y. Aoki and K. Tajima, Stimulated Brillouin scattering in a long single-mode fiber excited with a multi-mode pump laser. *J. Opt. Soc. Am. B* **5**, 358 (1988).
43. D. N. G. Roy and D. V. G. L. N. Rao, Optical pulse narrowing by backward stimulated Brillouin scattering. *J. Appl. Phys.* **59**, 232 (1986); see also G. K. L. Wong and Y. R. Shen, Transient self-focusing in a nematic liquid crystal in the isotropic phase. *Phys. Rev. Lett.* **32**, 527 (1974).
44. P. A. Franken, A. E. Hill, C. W. Peters, and G. Weinreich, Generation of optical harmonics. *Phys. Rev. Lett.* **8**, 18 (1962).
45. D. S. Chemla and J. Zyss, eds., "Nonlinear Optical Properties of Organic Molecules and Crystals." Academic Press, Orlando, FL, 1987.
46. Y. R. Shen, Studies of liquid crystal monolayers and films by optical second harmonic generation. *Liq. Cryst.* **5**, 635 (1989); see also Y. R. Shen, Surface properties probed by second-harmonic and sum-frequency generation. *Nature (London)* **337**, 519 (1989).
47. S.-J. Gu, S. K. Saha, and G. K. Wong, Flexoelectric induced second-harmonic generation in a nematic liquid crystal. *Mol. Cryst. Liq. Cryst.* **69**, 287 (1981).
48. A. V. Sukhov and R. V. Timashev, Optically induced deviation from central symmetry; lattices of quadratic nonlinear susceptibility in a nematic liquid crystal. *Pisma Zh. Eksp. Teor. Fiz.* **51**(7), 364 (1990).
49. P. G. deGennes, "The Physics of Liquid Crystals." Clarendon Press, Oxford, 1974.
50. S. M. Arkelyan, "Optically induced absence of a center of symmetry in bulk nematic liquid crystals: the piezoelectric mechanism of second harmonic generation" *Sov. Phys. Phys. Solid State* **26**, 806 (1984).
51. B. Ya Zel'dovich et al., "Observations of orientational stimulated light scattering in the mesophase of a nematic liquid crystal" *Sov. Phys.—Dokl. (Engl. Transl.)* **28**, 1038 (1983).
52. A. E. Kaplan, Theory of hysteresis reflection and refraction of light by a boundary of nonlinear medium. *Sov. Phys.—JETP (Engl. Transl.)* **45**, 896 (1977).
53. See Khoo et al. (16, 17); see also I. C. Khoo and P. Zhou, Dynamics of switching total internal reflection to transmission in a dielectric cladded nonlinear film. *J. Opt. Soc. Am. B* **6**, 884 (1989).
54. P. Zhou and I. C. Khoo, Anti-reflection coating for a nonlinear transmission to total reflection switch. *Int. J. Nonlinear Opt. Phys.* **2**(3), 437 (1993). This is a special issue, together with Vol. 2, No. 3, on "Optical switches, limiters and discriminators."

55. R. G. Lindquist, P. G. LoPresti, and I. C. Khoo, Infrared and visible laser induce thermal and density nonlinearity in nematic and isotropic liquid crystals. *SPIE—Int. Soc. Opt. Eng.* **1692**, 148–158 (1992).

56. I. C. Khoo and P. Zhou, Nonlinear interface tunneling phase shift. *Opt. Lett.* **17**, 1325 (1992).

57. W. Helfrich, "Conduction-induced alignment of nematic liquid crystals: Basic model and stability consideration," *J. Chem. Phys.* **51**, 4092 (1969). See also reference 49.

58. See, for example, P. Gunter and J. P. Huignard, ed., "Photorefractive Materials and Their Applications," Vol. I and II, Springer-Verlag, Berlin, 1989.

59. I. C. Khoo, H. Li and Y. Liang, Observation of orientational photorefractive effects in nematic liquid crystals, *Opt. Lett.* (In press, 1994).

60. A. G.-S. Chen and D. J. Brady, "Surface-stabilized holography in an azo-dye-doped liquid crystal," *Opt. Letts.* **17**, 1231 (1992); W. M. Gibbons, P. J. Shannon, S. T. Sun and B. J. Swetlin, "Surface-mediated alignment of nematic liquid crystals with polarized laser light," *Nature* **351**, 49 (1991); K. Ichimura, Y. Suzuki, T. Seki, Y. Kawanishi, T. Tamaki and K. Aoki, "Reversible alignment change of liquid crystals induced by photochromic molecular film," *Jap. J. Appl. Phys. Suppl.* **28**, 289 (1989); I. C. Khoo, H. Li and Y. Liang, "Optically induced extraordinarily large negative orientational nonlinearity in dye-doped liquid crystal," *IEEE J. Quant. Electron.* **QE-29**, 1444 (1993). See also reference 39.

61. F. P. Shaefer, ed., "Dye-Lasers," Springer-Verlag, New York, 1974.

# INDEX